DISEASES OF FIELD CROPS: DIAGNOSIS AND MANAGEMENT

VOLUME 2

Pulses, Oil Seeds, Narcotics, and Sugar Crops

DISEASES OF FIELD CROPS: DIAGNOSIS AND MANAGEMENT

VOLUME 2

Pulses, Oil Seeds, Narcotics, and Sugar Crops

Edited by

J. N. Srivastava

Department of Plant Pathology

Bihar Agricultural University

Sabour–813210, Bhagalpur, Bihar, India

A. K. Singh

Division of Plant Pathology

Sher-e-Kashmir University of Agricultural Sciences and Technology, Jammu (J & K), India

Apple Academic Press Inc.
4164 Lakeshore Road
Burlington ON L7L 1A4
Canada

Apple Academic Press, Inc.
1265 Goldenrod Circle NE
Palm Bay, Florida 32905
USA

Exclusive worldwide distribution by CRC Press, a member of Taylor & Francis Group

Diseases of Field Crops: Diagnosis and Management, Volume 2: Pulses, Oil Seeds, Narcotics, and Sugar Crops
International Standard Book Number-13: 978-1-77188-840-0 (Hardcover)
International Standard Book Number-13: 978-0-42932-196-2 (eBook)

Diseases of Field Crops: Diagnosis and Management, Two Volumes set
International Standard Book Number-13: 978-1-77188-841-7 (Hardcover)
International Standard Book Number-13: 978-0-42932-177-1 (eBook)

Library and Archives Canada Cataloguing in Publication

Title: Diseases of field crops : diagnosis and management/edited by J. N. Srivastava (Department of Plant Pathology, Bihar Agricultural University, Sabour-813210, Bhagalpur, Bihar, India), A. K. Singh (Division of Plant Pathology, Sher-e-Kashmir University of Agricultural Sciences and Technology, Jammu, (J & K), India).
Other titles: Diseases of field crops (Burlington, Ont.)
Names: Srivastava, J. N. (Plant pathologist), editor. | Singh, A. K. (Plant pathologist), editor.
Description: Includes bibliographical references and indexes. | Content: Volume 2. Pulses, oil seeds, narcotics, and sugar crops.
Identifiers: Canadiana (print) 2020020274X | Canadiana (ebook) 20200202855 | ISBN 9781771888417 (set ; hardcover) | ISBN 9781771888400 (v.2 ; hardcover) | ISBN 9780429321771 (set ; eBook) | ISBN 9780429321962 (v.2 ; eBook)
Subjects: LCSH: Plant diseases—Molecular aspects. | LCSH: Phytopathogenic microorganisms—Control.
Classification: LCC SB732.65. D57 2020 | DDC 632/.3—dc23

CIP data on file with US Library of Congress

Apple Academic Press also publishes its books in a variety of electronic formats. Some content that appears in print may not be available in electronic format. For information about Apple Academic Press products, visit our website at **www. appleacademicpress. com** and the CRC Press website at **www. crcpress. com**

About the Editors

J. N. Srivastava, PhD
Associate Professor cum Sr. Scientist (Plant Pathology), Bihar Agricultural University, Sabour, Bhagalpur, Bihar, India

J. N. Srivastava, PhD, is currently an Associate Professor cum Senior Scientist (Plant pathology) at Bihar Agricultural University, Sabour, Bhagalpur, Bihar, India. Formerly, he was Assistant Professor Junior Scientist (Plant pathology) at Sher-e-Kashmir University of Agriculture Sciences and Technology, Jammu and Kashmir, India. He has acted as Principal Investigator and Co-Principal Investigator for several projects on biological control/ integrated disease management. In addition to teaching and research, he is also engaged in various agriculture extension activities, such as farm advisory services, including video conferencing, radio and TV talks, vocational trainings for the rural youth skill, entrepreneurship development programs, farmer training, in-service training, etc. Dr. Srivastava has published many research papers in both Indian and international journals, book chapters, extension articles, practical and technical bulletins, and leaflets. He was the recipient of many awards for achievement and teaching, including an Excellence in Teaching Award (2017), Excellence in Science Communication Award (2017), Distinguished Faculty Award (2017), Eminent Scientist Award (2018), etc. He is a member of many academic and scientific organizations and is associated with many international, national, and provincial scientific, cultural, and academic/educational bodies, and also serving as an editorial board member of the International Journal of Plant Protection. He is also a reviewer for various scientific journals.

A. K. Singh, PhD
Assistant Professor, Division of Plant Pathology, Sher-e-Kashmir University of Agricultural Sciences and Technology, Jammu (J & K), India

A. K. Singh, PhD, is Assistant Professor in the Division of Plant Pathology at Sher-e-Kashmir University of Agricultural Sciences and Technology, Jammu and

Kashmir, India. He has been engaged for more than 12 years in teaching (both undergraduate and postgraduate levels) and research and also involved in the transfer of technology through different extension activities. He has published more than 30 research papers in national and international journals of repute, one practical manual, and several book chapters and popular articles and has presented many research papers at international, national and regional symposiums and seminars. He is a recipient of several prestigious awards for his research contributions. Also, he served as a member of the editorial board of *Krishi Vikas Patrika*, published by the Directorate of Extension, SKUAST of Jammu. He is a member of several societies of plant pathologists in India.

Contents

Contributors

Kotramma C. Addangadi
Regional Research Station, S. D. Agricultural University, Bhachau –370140, Gujarat, India

Durga Prasad Awasthi
Department of Plant Pathology, College of Agriculture, Tripura–799210, India

M. S. Bhale
Department of Plant Pathology, College of Agriculture, Jawaharlal Nehru Krishi Vishwa Vidyalaya-Jabalpur, Madhya Pradesh, India

Usha Bhale
Department of Plant Pathology, College of Agriculture, Jawaharlal Nehru Krishi Vishwa Vidyalaya-Jabalpur, Madhya Pradesh, India

Jayant Bhatt
Department of Plant Pathology, College of Agriculture, Jawaharlal Nehru Krishi Vishwa Vidyalaya-Jabalpur, Madhya Pradesh, India

A. K. Chaubey
JNKVV Krishi Vigyan Kendra, Sidhi–48661, Madhya Pradesh, India

Upma Dutta
Division of Microbiolgy SKUAST-J, Jammu & Kashmir, India

N. M. Gohel
Department of Plant Pathology, B. A. College of Agriculture, Anand Agricultural University, Anand–388 110, Gujarat, India

C. Gopalakrishnan
Department of Plant Pathology, TNAU, Coimbatore-3, Tamil Nadu, India

Sachin Gupta
Division of Plant Pathology, SKUAST-J, Jammu & Kashmir, India

Anamika Jamwal
Krishi Vigyan Kendra, Kathua, SKUAST-J, Jammu & Kashmir, India

Sonika Jamwal
ACRA, SKUAST-J, Jammu & Kashmir, India

K. Jayalakshmi
ICAR-NAARM, Hyderabad, Telangana–500030, India

Manoj Kumar Kalita
Biswanath College of Agriculture, Assam Agricultural University, Biswanath Chariali, Assam, India

C. P. Khare
Division of Plant Pathology Indira Gandhi Agricultural University, Raipur, Chhattisgarh, India

U. K. Khare
Department of Plant Pathology, College of Agriculture, Jawaharlal Nehru Krishi Vishwa Vidyalaya-Jabalpur, Madhya Pradesh, India

A. Khulbe
Krishi Bhawan, ICAR, New Delhi, India

Ashish Kumar
JNKVV, College of Agriculture, Rewa, Madhya Pradesh, India

Sanjeev Kumar
Department of Plant Pathology, College of Agriculture, Jawaharlal Nehru Krishi Vishwa Vidyalaya-Jabalpur, Madhya Pradesh, India

Ranganathswamy Math
Assistant Professor, Department of Plant Pathology, College of Agriculture, Jabugam, Anand Agricultural University, Gujarat– 391155, India

P. D. Meena
ICAR-Directorate of Rapeseed-Mustard Research, Bharatpur–321 303, Rajasthan, India

K. K. Mishra
JNKVV, Zonal Agricultural Research Station, Powarkheda, Hoshangabad, Madhya Pradesh, India

Harshil V. Parmar
Department of Plant Pathology, B. A. College of Agriculture, Anand Agricultural University, Anand–388110, Gujarat, India

A. K. Patibanda
Department of Plant Pathology, Agriculture College, ANGRAU, Bapatla, Andhra Pradesh, India

B. K. Prajapati
Agricultural Research Station, S. D. Agricultural University, Aseda, Ta. Deesa, Dist. Banaskantha–385535, Gujarat, India

Lingan Rajendran
Department of Plant Pathology, TNAU, Coimbatore-3, Tamil Nadu, India

J. Raju
Plant Quarantine station, Ministry of Agriculture and Farmers Welfare, Government of India, Mangalore, Karnataka–575011, India

H. Ravindra
AINRP (Tobacco), ZAHRS, University of Agricultural and Horticultural Sciences, Shivmaogga, Karnataka–577201, India

Geeta Sharma
Junior Research Officer, G. B. P. U. A. T., Pantnagar–263145, Uttrakhand, India

K. K. Sharma
Dr. D. R. Bhumbla Regional Research Station for Kandi Area (PAU), Ballowal Saunkhri, P. O. Takarla, Tehsil-Balachaur, SBS Nagar–144521, Punjab, India

Ashish Sheera
Sam Higginbottom University of Agriculture, Technology, and Sciences, Allahabad–211007, Uttar Pradesh, India

A. K. Singh
Division of Plant Pathology, Sher-E-Kashmir University of Agricultural Sciences and Technology (SKUAST-J), Chatha, Jammu–180009, Jammu & Kashmir, India

Jai Singh
JNKVV Krishi Vigyan Kendra, Sidhi–48661, Madhya Pradesh, India

S. K. Singh
Rainfed Research Sub-Station for Sub-Tropical Fruits, Raya, Technology (SKUAST-J), Jammu–180009, Jammu & Kashmir, India

S. N. Singh
Department of Plant Pathology, College of Agriculture, Jawaharlal Nehru Krishi Vishwa Vidyalaya-Jabalpur, Madhya Pradesh, India

V. B. Singh
Rainfed Research Sub-Station for Sub-Tropical Fruits, Raya, Technology (SKUAST-J), Jammu–180009, Jammu & Kashmir, India

J. N. Srivastava
Department of Plant Pathology, Bihar Agricultural University, Sabour–813210, Bhagalpur, Bihar, India

P. K. Tiwari
Division of Plant Pathology Indira Gandhi Agricultural University, Raipur, Chhattisgarh, India

Kothakota Venkataramanamma
Regional Agricultural Research Station, ANGRAU, Nandyal–518502, Andhra Pradesh, India

V. K. Yadav
JNKVV, College of Agriculture, Ganj Basoda, Vidisha, Madhya Pradesh, India

Abbreviations

AGs	anastomosis groups
AMV	alfalfa mosaic virus
BGM	botrytis grey mold
BND	bud necrosis disease
BWYV	beet western yellows virus
BYMV	bean yellow mosaic virus
CLS	cercospora leaf spot
CMV	cucumber mosaic virus
CR	common region
DAP	di-ammonium phosphate
DAS	days after sowing
EFSA	European Food Safety Authority
ELS	early leaf spot
FLS	frogeye leaf spot
FYM	farm yard manure
GAP	good agriculture practices
GRAV	groundnut rosette assistor virus
GRV	groundnut rosette virus
HCN	hydrogen cyanide
IAA	indole acetic acid
IDM	integrated disease management
LLS	late leaf spots
MYMIV	mungbean yellow mosaic India virus
MYMV	mungbean yellow mosaic virus
NAA	naphthalene acetic acid
NSKC	neem seed kernel cake
PBNV	peanut bud necrosis virus
PEMV	pea enation mosaic virus
Pf1	pseudomonas fluorescens
PO	peroxidase
PPSMV	pigeon pea sterility mosaic virus
PSbMV	pea seed-borne mosaic virus
RCR	rolling circle replication
RFLP	restriction fragment length polymorphism

RH	relative humidity
RKNs	root-knot nematode
Sat-RNA	satellite-RNA
SND	sunflower necrosis disease
SR	sclerotinia rot
ss	single-stranded
TNV	top necrosis virus
TSV	tobacco streak virus
Tv	trichoderma viride
ULCV	Urdbean leaf crinkle virus
VAM	vesicular-arbuscular mycorrhizae
WR	white rust

Preface

Globally, enormous losses of crops are caused by plant diseases. The loss can occur from the time of seed sowing in the field to harvesting and storage. The problem of plant diseases has assumed alarming dimensions with changing patterns of intensification of growing field and horticultural crops that involve maximum exploitation of resources through multiple cropping and the extensive use of fertilizers, fungicides/pesticides, and irrigation. The overuse of fungicides/pesticides has led to the emergence of fungicide/pesticide-resistant pathogens and also causes environmental pollution. These environment changes have further increased problems for plant pathologists by the introduction of new plant diseases and situations of diverse natures covering seed-borne, air-borne, soil-borne, and vector-borne plant diseases as well as the conversion of some minor diseases into major diseases caused by several infectious agents in India and elsewhere in the world. Alteration requires a thorough knowledge to know the status of appropriate plant disease diagnosis and the exact causes of the diseases, and to find the ways and means to develop management strategies accordingly to protect our crops in ecofriendly manner.

Keeping in view the above facts and keeping in mind the new curricula recommended by the Indian Council of Agricultural Research, New Delhi, we designed this book particularly for undergraduate and postgraduate students, researchers, and course instructors.

This comprehensive book, *Diseases of Field Crops: Diagnosis and Management, Volume 2,* deals with pulses, oilseed crops, narcotics, and sugar crops. It includes 18 chapters, and each chapter deals with only one crop and includes detailed account of topics covered, introduction, symptoms, causal organisms, disease cycles, epidemiology, and management of the diseases caused by fungi, bacteria, and viruses.

The book chapters are contributed by authors who are engaged in teaching, research, and extension services and who are well-known national scientists in their respective fields. They have incorporated their experience and knowledge along with the latest information about the plant diseases with suitable photographs for easily identification of the disease.

We trust that, this book will serve as the useful resource book and guide for students, researchers, and teaching faculty and others who are engaged in agriculture and in resolving plant disease diagnosis, disease etiology, and management.

We express our sincere thanks and gratitude to the various scientists who have spared time and contributed valuable materials for this book. We also thankful to CRC Press for publishing this book to place in hand in esteemed reader.

—J. N. Srivastava, PhD
A. K. Singh, PhD

Introduction

Among the biotic stress plant diseases are causing considerable yield loss in economically important crops. The disease may attack at juvenile stage to crop maturity or harvesting of the crops. They may affect different parts of the plants, such as foliage, stem, root, flowers and seed thatinduce various types of symptoms. However, vascular system infecting pathogens causing wilt and affect the entire plant. Many pathogens survive on the stored grains which may be onward transmitted or causes spoilage. Several pathogens causing complex symptom, e.g., root rot, wilting leaf spot fruit rot which is difficult to distinguish each other.

The crop losses can be reduced by suitable control measuresbytargeting specific pathogens if theyaccurately diagnosed. Need-based application of fungicides will improve environment and economic gains. In plant disease, visual observation of infected plant continues to be the dominant methods. Several sophisticated tools are being used for disease diagnosis, which include microscopy, isolation, immunological, biochemical assay, and genome analysis.

Adaptation of single practice like fungicide application will leads development of resistance in fungal pathogens. Integrated approach will helpful in order to sustainable management of crop disease. Conventional cultural practices are effective to control various soil borne plant pathogens e.g., *Pythiumspp, Fusariumspp, Rhizoctoniaspp, Sclerotium* spp, etc. Soil solarization and mulching provides the effective efforts to manage the nematode and root infecting pathogens. Antagonistic microbial bioagents like Trichoderma, and Pseudomonas fluorescence are found effective to control the soilborne plant pathogen. Application of different group fungicides (systemic and non-systemic) provide effective control of diseases in crops. Integration of crop varieties with different genetic makeup is always pronouncing the safest way to manage the disease with high yield output.

The compiled book *"Diseases of Field Crops: Diagnosis and Management,"* volume 2, *"Pulses, Oil Seeds, Narcotics, and Sugar Crops"* deal on several aspects in different crops like pulses, oil seeds, narcotics, and sugar crops. The chapters of the book are focused on economic importance, symptom, causal organism, disease cycle, epidemiology, and suitable disease management options.

The book contains different diseases of Bengal gram crops and including common and minor disease which having greater significance. Field pea are commonly grown in sub tropical and are also temperate part of India which are affected by different disease, e.g., rust, bunts, smut, and powdery mildew spot blotch covered under the chapter. Maize is commonly grown throughout the year in Indian which affected by several diseases and causing considerable yield loss. The horse gram crop is hampered by anthracnose, powdery mildew, dry root rot, rust, aerial blight, cercospora leaf spot, and yellow mosaic disease. Lentil is commonly grown in all over in India. The crop is severely challenged by various pathogens inducing ascochyta blight, anthracnose, wilt, botrytis grey mold (BGM), powdery mildew, rust, and stemphylium blight.

Wilt, stem rot/stem blight, stem canker, powdery mildew, rust, cercospora leaf spot, halo blight, pigeon pea, sterility mosaic, and pigeon pea mosaic which discus in pigeon pea chapter.

Soyabean crops affected by several diseases, e.g., charcoal rot, rust, anthracnose, and pod blight, aerial blight, frog eye leaf spot, and yellow mosaic.

The emphasis also given on castor diseases, e.g., seedling blight, alternaria blight, cercospora leaf spot, powdery mildew, wilt, root rot/charcoal rot, gray mold/gray rot/blossom blight and bacterial leaf spot. The ground nut crop is hampered by early and late leaf spot of groundnut, rust, black hull, crown rot/ collar rot groundnut rosette virus (GRV) disease and kalahasti malady. This chapter also discuss about the Aflatoxin problem. Linseed chapter discusses about the Rust, Fusarium Wilt, Alternaria Blight, Powdery Mildew, Seedling Blight, and Root Rot, Pasmo, and Sclerotinia Stem Rot. Rapeseed mustard, Sesamum, and Sunflower crop diseases are also discussed in this book. A oil seed crops namely Taramira diseases is explained in detail. Smut Disease of Sugarcane is a very important disease in India and is also detailed in this book. In the stimulant crops, Tobacco crops are very important crops in India and all diseases of tobacco are explained in the book.

The compiled chapters are substantiated with novel figures, which are of excellent quality labelled in the book. Each chapter also provides suitable and recent references. Information provided on different aspect of disease will helpful for academician and researchers. Moreover, the book will also provide sufficient information to undergraduate and postgraduate studies in colleges and universities.

—Editors

CHAPTER 1

Current Status of Bengal Gram Diseases and Their Management Strategies

K. K. SHARMA[1] and ASHISH KUMAR[2]

[1]Dr. D. R. Bhumbla Regional Research Station for Kandi Area (PAU), Ballowal Saunkhri, Tehsil-Balachaur, SBS Nagar–144521, Punjab, India, E-mail: kksharma@pau. edu

[2]JNKVV, College of Agriculture, Rewa, Madhya Pradesh, India

Chickpea (*Cicer arietinum* L.) is the second most important pulse crop after beans (*Phaseolus vulgaris*), which is widely cultivated in India. It fixes atmospheric nitrogen and belongs to family leguminaceae (Ferguson et al., 2010). This crop is cultivated under rainfed conditions in Asian, African, American countries, etc. (Sharma et al., 2015). The global area under cultivation was estimated about 12.14 million hectare with an annual production of 11.30 million tonnes (FAOSTAT, 2012). In India, area under this crop was maximum and estimated about 8.32 million hectares giving 7.70 million tonnes of production annually (FAOSTAT, 2012). It can be grown as a main sole crop or mixed with cereals in heavy and other types of soils, respectively. The chickpea grown with limited irrigation gives good returns although, it is preferably cultivated in the area with low-rainfall during Rabi season. Though, significant yield losses have been estimated in chickpea caused by several diseases like wilt, ascochyta blight, dry root rot, collar rot, botrytis grey mold, wet root rot and chickpea stunt, but few of them cause severe losses in chickpea growing areas while others remain as minor in occurrence. Considering the economic importance, major diseases are discussed along with their management practices.

1.1 MAJOR DISEASES

1.1.1 ASCHOCHYTA BLIGHT

1.1.1.1 ECONOMIC IMPORTANCE

In India, this disease appears frequently in states *viz.,* Madhya Pradesh, Haryana, Himachal Pradesh, Bihar, Northwestern part of U. P., and Punjab. The credit goes to Butler for its description first time in 1911 in country when it appeared in Punjab, Bihar, and Jammu in epidemic form.

1.1.1.2 SYMPTOMS

The pathogen attacks on all above ground plant parts which are diagnosed generally at the late vegetative to near blooming. On leaf, the lesions are circular to elongated, sunken brown spots appear with darker margin and grayish center. On stem and pods, black and dotted pycnidial bodies can be seen in the form of concentric rings. Shoot terminals are especially liable to attack. When the lesions girdle the stem, speedy death of the stem portion above the point of girdling can be seen.

The patches of plants exhibiting wilting symptoms clearly visible in the field with the advancement of disease, which cover the entire, field subsequently (Bashir and Malik, 1998). When infection occurs at flowering or pod formation stage, it may form unhealthy pods with distorted seeds without showing any observable symptom. Moreover, infected pods may not able to produce seed at all. Under condition of heavy rains couple with favorable temperature, whole crop in the field may be killed quickly (Figure 1.1).

1.1.1.3 CASUAL ORGANISM (ASCHOCHYTA RABIEI)

Fungus produces hyaline to brown, septate mycelium. Pycnidia are spherical to globular with a clear ostiolar opening. Single celled or bicelled pycnidiospores (conidia) are hyaline, oval to oblong, straight or slightly Curved with assize of 9–20 x 3–6 micron which are produced on small conidiophores. The perethecia are globose, dark colored, 100 to 140 micron in diameter and contain asci which are typically 8 spored. The ascospores hyaline, thin walled, ellipsoid, and two celled.

FIGURE 1.1 Symptoms of aschochyta blight in chickpea.

1.1.1.4 FAVORABLE CONDITIONS

Hot and moist weather, temperature between 20 and 25°C with 60% relative humidity (RH) and dense crop cover favor the speedy spread and development of the disease (Pande et al., 2009). Rapid development of the disease also co-related with high rainfall during flowering.

1.1.1.5 DISEASE CYCLE

The fungus is externally and internally seed-borne in nature and perpetuates as pycnidial bodies on diseased crop residue which is left in the fields after harvesting. The fungal spores and hyphae both remain on soil and seeds may also provide primary inoculum. The air-borne pycnidiopores (conidia) causes secondary infection. About 6% infection from seed is quite enough to create epidemics under favorable conditions. The primary inoculum multiplies on hypocotyle, epicotyle, and near the stem base followed by the spreading on aerial parts. Dissemination of pathogen occurs through conodia and wind borne rain splashes, by contact between leaves through insects and animals fields.

1.1.1.6 MANAGEMENT

For the effective management of the ascochyta blight, integrated strategy, i.e., field sanitation, cultural, monitoring of crop, use of resistant varieties, application of chemical fungicides and biological approaches is successful. Farmers always use disease free healthy seeds and sow the seeds up to 15 cm or deeper in soil. Remove or destroy blight-infested crop residues after harvest. Bury the infected crop debris by plowing deep in the soil. Intercropping with the cereals such as chickpea-barley is the most suitable combination for reducing the diseases.

Since this fungus is externally and internally seed born, fungicidal treatment with carbendazim 50 WP + thirum 75 DS @ 2 g/kg seed of the seeds is very significant for the management of the disease. Treating the seeds in hot water at 52°C for 10 min was also found very effective to reduce the level seed borne inoculum. Growers should apply the mancozeb @ 3 g/l on the crop immediately after appearance of initial symptoms of the disease. Under the conditions like repeated rain with heavy dew, a preventative spray of chlorothalonil at the rate of 1 kg/ha should be applied 7 to 10 days before flowering whereas the same fungicide should also be sprayed at flowering under dry conditions. Spray the crop with carbendazim at the rate of 500 g/ha is advised at 7 to 10 days after spray of chlorothalonil. Under favorable weather conditions for disease, growers should continue successive sprays of aforesaid fungicides at 10 to 14 days during the major bloom and pod fill period. Farmers should grow resistant/tolerant varieties like Gaurav (H75-35), PBGI, GNG146, BG261, GNG469, GL 23094, GL 260584, GL 26069.

1.1.2 BOTRYTIS GREY MOLD (BGM)

1.1.2.1 ECONOMIC IMPORTANCE

Botrytis grey mold (BGM) is reported from worldwide and but it is quite serious disease in countries namely India, Nepal, Pakistan, Australia, Argentina including some parts of Bangladesh (Nene et al., 2012). This disease appear regularly every year in moderate to severe form in North-Indian states. It develops rapidly and spread widely under conducive weather which may cause up to 100% yield losses

1.1.2.2 SYMPTOMS

The all the above ground parts, leaves, flowers, pods, branches, and stems are attacked by the pathogen causing disease during the entire crop growth at any time but it appears in most serious form in February-March. Growing tips of the plant parts and flower buds are affected first. The initial symptom of the disease appears as chlorotic, watery spots on leaflets, petioles, growing tips, flowers, and young stems which enlarge rapidly and become grey or dark. Patches of died plants are frequently visible under highly favorable weather condition. Diseased plants show lack of pod setting as initial symptoms at maturity and such pods generally have either distorted seeds or no seeds. Under wet conditions, grey to dark brown lesions appear on stems, leaves, flowers, and pods (Figure 1.2).

FIGURE 1.2 Symptoms of botrytis grey mold blight in chickpea.

1.1.2.3 CASUAL ORGANISM

Botrytis cinerea

The fungus mycelium is hyaline, septate, and brown. Macroconidia are hyaline to pale brown, oval or globose and single celled while microconidia are smaller may readily germinate in water or even on the host surface through germ tubes under conditions of high humidity. The sporodochia of 5–5.0 μ in diameter consisting of densely interwoven brown septate hyphae and produce round to oval unicellular conidia measuring 4–8 μ diameters. It is reported that the grey mold pathogen has great diversity and can adapt under wide range of climatic conditions. Pande et al. (2006) *have* found the 4–5 pathotypes of *B. cinerea* from Northern India

1.1.2.4 FAVORABLE CONDITIONS

Pathogen may infect the host at 15–25°C (optimum 20–25°C). High moisture and RH are highly favorable for development of the disease and fungus produces profuse sporulation on dead plant parts, mainly on flowers and pods under these conditions (Pande et al., 2006).

1.1.2.5 DISEASE CYCLE

The fungus perpetuates exteriorly on the surface and internally inside the infected seed or as a saprophyte on decaying and dead crop residue or as soil-borne sclerotia. *B. cinerea* survives at least 5–6 months in infected crop debris with conidial stage. Sowing of infected seeds produce numerous spore mass on affected crop which usually establish the pathogen in new areas. The fungus may survive in infected seeds for 6–8 months, i.e., till next planting season. Seed-borne inoculum alone was unable to initiate the disease until and unless environmental conditions become favorable. Spores of this pathogen are discharged from conidiophores and disseminated by wind or splatter of rain. The pathogen produces abundant spores on Infected host in December which start secondary infection during February and March when conditions become favorable. The conditions like dense crop canopy, high humidity, and rain provide conducive environment for the development and dispersal of the disease.

1.1.2.6 MANAGEMENT

The cultivated genotypes are lacking to have a suitable degree of resistance to this disease and chemical fungicides are found unsuccessful under high disease pressure. Hence, integrated strategy to combine all the available practices including cultural, physical, mechanical, chemical, and biological may effectively reduce yield losses. The grey mold disease can be managed by using minimum dosage of fungicides and other cultural practices coupled with moderate level of host resistance (Pande et al., 2009). Cultural practices like wider row spacing or inter cropping of chickpea with mustard/toria/wheat and application of higher amount of K_2O reduces the severity of disease.

Treatment with carbendazim + thiram at the rate of 2 g/kg of seeds was found effective to eliminate seed-bone infection. Grow grey mold tolerant/resistant varieties if available, like ICCV 6853, ICCV 96859. In Bangladesh, a package, consist of growing tolerant genotype 'Avarodhi,' soil application of DAP, broader spacing between rows, use of pre-treated seeds with carbendazim + thiram at the rate of 2 g/kg of seeds and need based foliar spray of carbendazim has been developed and adopted (Bakr et al., 2005; Pande et al., 2005, 2006; Pande et al., 2009).

1.1.3 FUSARIUM WILT

1.1.3.1 ECONOMIC IMPORTANCE

Chickpea wilt caused by *F. oxysporum* f. sp. *ciceris* was reported first time by Butler in 1918. It was estimated to cause 10–15% annual yield losses worldwide (Jalali and Chand, 1992). Early wilting causes 77–94% losses while 24–65% loss was reported in case of late wilting (Haware and Nene, 1980).

1.1.3.2 SYMPTOMS

Symptoms of the disease may appear from seedling to pod formation stage as yellowing and drooping of foliage. Symptoms at seedling stage can be seen after 3–5 weeks of sowing as lighter green to yellow foliage with withered stem and affected seedlings fails to grow. All the leaves turn yellow and become straw colored subsequently. The wilted dead seedlings or adult plants in the field may be seen usually in patches. Sometime, partial wilting where only one-side branches wilt is seen. Dark brown discoloration of vascular tissues is the most typical symptoms of this disease (Figure 1.3).

1.1.3.3 CASUAL ORGANISM (FUSARIUM OXYSPORUM F. SP. CICERIS)

The fungus mycelium is profusely found inter and intracellular in the vascular tissues infected plant. It produces white mycelium on cultured on medium and a wine red pigment on starch medium. Fungus has asexual fruiting body, sporodochium which produces ample number of spores (conidia) on conidiophores. The pathogen produces hyaline, septate, sickle shaped macro conidia with tapering ends measuring 25–40 × 3–4 micron in size whereas hyaline, elliptical single or bi-celled and thin walled micro conidia measure 4–6 × 2–4 micron in size. Moreover, it also process resting thick walled chlamydospores.

FIGURE 1.3 Symptoms of fusarium wilt in chickpea.

1.1.3.4 FAVORABLE CONDITIONS

Hot and dry weather with temperature beyond 25°C and alkaline soils favor the incidence of wilt and growing same crop continuously in the infested field (monoculture) aggravate disease severity (Chauhan, 1963).

1.1.3.5 DISEASE CYCLE

Wilt fungus is primarily soil borne (Singh et al., 1974) but seed borne nature is also (Haware et al., 1978) documented. It may survive in soil in the form of thick walled chlamydospores for more than 6 years without chickpea crop (Haware et al., 1986) or on the infected residue. Although, other plants like lentil, pigeon pea and pea are reported their role as a carrier of disease without visible symptoms (Haware and Nene, 1982). The primary infection occurs through soil borne chlamydospores or mycelium. *Fusarium oxysporum* f. sp.

ciceri is more severe in light soils than heavy soils where chickpea is grown as a rainfed crop (Kotasthane et al., 1976).

1.1.3.6 MANAGEMENT

Growers must use disease free healthy seeds and avoid sowing of the crop during condition of high temperatures. Expose the infected crop residue during hot sunny days by plowing the field deep in the month of May-June reduces initial inoculum of the pathogen. Practice soil solarization for 6–8 weeks during summer months is very effective to reduce soil inoculum in field (Chauhan et al., 1988). Farmers should not grow chickpea in heavily infested field. Follow the crop rotation continuously with non-host crop e.g., rotation with sorghum for six years, to reduce soil borne inoculum and diminish the effects of wilt disease.

Treating the seeds with mixture of thiram + carbandizm (1 part+2 part) or Carboxin + thiram (1 part + 2 part) at the rate of 3 g/kg of seeds protect them from initial attack. The non-pathogenic bacteria or fungi, e.g., non-pathogenic isolates of *F. oxysporum* (Fuchs et al., 1997; Hervás et al., 1997) are reported to have their biocontrol activity for managing Fusarium wilt in chickpea. Soil application of *Trichoderma viride* @ 2.5 kg/ha (colonized on FYM) and seed treatment @ 4–5 g/kg of seeds has been found very effective to reduce disease incidence. Soil application of *T. harzianum* and *T. viride* in combination was also found effective in reducing natural incidence of chickpea wilt (Prasad et al., 2002). Farmers should preferably grow wilt tolerant/resistant cultivars like BG 3004, GL 26054, JG 2000-07, GAG 0419, GJG 0714, GJG 0814, IPC 2005-59, IPC 2005-74, IPC 2008-103, BCP 60, BCP 136, Phule G 9807, JSC 35, PBG 1 of Desi and HK 05-169, BG 3002, IPCK 113 of Kabuli.

1.1.4 DRY ROOT ROT

1.1.4.1 ECONOMIC IMPORTANCE

Mitra (1931) reported first time this disease in chickpea from India. Later, it was also reported from other areas in India where chickpea is grown. In central and southern zone of the country, Rhizoctonia dry root rot is considered as one of the most destructive disease of chickpea where it is generally cultivated under rainfed conditions (Ghosh et al., 2013).

1.1.4.2 SYMPTOMS

Dry root rot disease is generally appeared at blooming or pod formation stage which can be seen as dried plants in the field scattered irregularly. The initial sign appears as yellowing of leaves, branches, and quick drying of the plants. The affected plant exhibits drooping of from the top including petioles. Occasionally the dried plant shows chlorosis of top most leaves. In dry soils, the tap root turns dark brown, fragile, and starts rotting which results in damage of lateral and finer roots. When affected plant is pulled out from the soil, some part of tap root, i.e., its lower portion is often left there which is distinguishing character of this disease (Figure 1.4).

FIGURE 1.4 Symptoms of dry root rot in chickpea.

1.1.4.3 CASUAL ORGANISM (RHIZOCTONIA BATATICOLA)

The pathogen, *R. bataticola* survives in its asexual form as sclerotia in soil and crop debris. On artificial medium or lab conditions, it produces brown to grey colored mycelium. The young hypha is hyaline without septa which has branching at right angle with constriction near the point of origin. It produces characteristic black sclerotia which are spherical to irregular smooth and black in color (Sharma et al., 2012a). The higher rate of phenomenon like mutation, hyphal fusion and recombination in *R. bataticola* are supposed to be the reasons for its higher pathogenic and genetic variability.

1.1.4.4 FAVORABLE CONDITIONS

Higher temperature and moisture stress generally aggravate the intensity of dry root rot in areas of humid tropics. It can be appeared abruptly at optimal temperature from 25 to 30°C. Speedy increase in severity of dry root rot on chickpea has been reported at flowering and pod formation stages when day temperature goes beyond 30°C under dry soil conditions (Gurha et al., 2003).

1.1.4.5 DISEASE CYCLE

The pathogen may survive in soil up to 3–6 years on diseased plant residue in the form of mycelium or without crop residue as hard resting sclerotia. This soil borne inoculum provides the main source of infection (Baird et al., 2003; Francl et al., 1988). The symptomatic condition of disease development may occur due to chocking of xylem vessels, mechanical pressure during infection process, microsclerotia, toxins, and lytic enzymes (Sharma et al., 2004). Pathogen may also attack during emergence of cotyledons where it enters in rootlets through slight wound openings and becomes intra and intercellular. The attacked cortical cells of roots of young or adult plants disintegrate leads to severe root decay (Singh and Mehrotra, 1982). Hyphae colonize the vascular system and sclerotial bodies of *R. bataticola* plug the xylem vessels. Below ground, the necrosis of roots frequently increases without showing any above ground symptoms till flowering and pod formation stage.

1.1.4.6 MANAGEMENT

Weather and residual soil moisture in semi-arid tropical areas of the country which is not possible to control in fact; the dry root rot pathogen has also reported to attack on about half of thousand cultivable and wild species of plants therefore, Integrated approach for the management of this disease has become an effective strategy against it (Dhingani and Solanky, 2016).

The examples are available where sowing of early varieties with well-timed irrigation can escape the crop from dry root rot or lesser disease due to aforesaid reason during maturity. Adjustment in date of sowing has been found very effective which avoid the occurrence of hot weather conditions during crop maturity helpful to escape the crop from disease (Gurha et al., 2003). Rotation with non-host crops reduces the soil bore perpetual inoculum and sclerotia of the pathogen (Singh et al., 1990b). The primary inoculum can be reduced by adopting tillage practices like elimination of diseased plant residue from the field or bury it in the soil by plowing deep which helps to minimize the severity of disease. Moreover, plowing the field with a furrow turning plow reduces soil moisture stress which may reduce disease intensity.

Beneficial effect of *Trichoderma viride* has shown after seed treatment against the disease (Gurha et al., 2003; Singh et al., 1998). The disease may reduce up to some level when seeds are treated with captan or thiram at 3 g/kg of seeds. Seed treatment with bavistin and thiram has also reported to reduce incidence of dry root rot in central and southern parts of the country at farmers' field (Ghosh et al., 2013). Best way to grow resistant or tolerant varieties like PG 00110, GL 21107, CSJ 592, H 04-49, IPC 2005-28, BG 3002, IC 269792, NDG 7-702, CSJ 559, IPC 2005-66, NDG 7-702 to manage the disease is always appreciated.

1.1.5 COLLAR ROT

1.1.5.1 ECONOMIC IMPORTANCE

Collar rot is worldwide in occurrence where chickpea is cultivated including India. The yield loss in chickpea due to this disease caused by *S. rolfsii* has been estimated to cause 30 to 60 per under field conditions (Prasad, 2005).

1.1.5.2 SYMPTOMS

Wet soil with warm temperature is usually conducive for occurrence of the disease at the seedling stage and it is mostly appeared up to 6 weeks after emergence in wet soil. Diseased seedlings exhibit downward rotting from the collar region and white mycelial growth is apparently visible covering the tap root of affected and dried seedlings. The disease can be diagnosed as slight yellowing of foliage of drying plants before their death which are distributed throughout the field. In the area where rotation of chickpea is followed by paddy usually associated with this disease. Uprooted seedlings show mustard size minute sclerotia at initial stages of infection which are depressed in mycilial strands around the collar.

1.1.5.3 CASUAL ORGANISM

Sclerotium rolfsii

Typical character Mycelium of *S. rolfsii* Sacc. is floccose, not ropy, producing numerous sclerotia which are globose, pinkish dull or light to dark brown in color and 0.8–2.5 mm in diameter (Subramanian, 1971). Sharma et al. (2002) studied variability of 26 isolates of *Scleratium rolfsii* from the country in respect to colony morphology, mycelial growth, teleomorph production, size, and color of sclerotia.

1.1.5.4 FAVORABLE CONDITIONS

High soil temperature (25–30°C), moisture, and low soil organic matter at the time of sowing and seedling stage favor the disease (Mathur and Sinha, 1968). Sowing of gram followed by paddy leads to higher disease incidence.

1.1.5.5 DISEASE CYCLE

The pathogen survives as mycelium on diseased crop debris or in soil as sclerotia which function as main perpetual structure and primary inoculum. The deep-seated dormant sclerotia in soil may last up to a year while viable sclerotia on soil surface may quickly germinate after getting chemical stimuli of volatile compounds released from decaying plants (Punja, 1985). Pathogenic Sclerotia of the pathogen are dispersed through irrigation and rainwater, wind; infested soil adhered with various farm equipments and contaminated tools, infected seeds and seedling material for transplanting (Mahen et al., 1995).

1.1.5.6 MANAGEMENT

The wide host range of collar rot of pathogen of chickpea and its ability to produce huge amount of hard, resting sclerotial bodies which last in soil for many years; it is not possible to manage the disease through chemical application (Punja, 1985). Hence, the use of botanicals and biocontrol agents may be the substitute for managing the pathogen, *Sclerotium rolfsii*.

Since, disease is directly correlated with soil moisture hence, growers should maintain adequate soil moisture at sowing and seedling stages. Remove all the undecomposed organic matter while preparing seedbed to inhibit the development of pathogen. The best approach to grow resistant varieties if available. Treating the seeds with carboxin at 3 g/kg of seeds may reduce the chances of infection at seedling stage. The crop can be saved up to some level from *Sclerotium rolfsii* by treating the seeds with *Trichoderma harzianum* and *T. virens* which also improves plant growth (Mukhopadhyay, 1995). Kumar et al. (2008) reported maximum protection to the crop against collar rot pathogen when seeds were treated with combination of *T. harzianum* at 4 g/kg of seeds + carboxin at 0.5 g/kg of seeds which also improves final plant stand.

1.2 MINOR DISEASES

1.2.1 CHICKPEA STUNT

1.2.1.1 ECONOMIC IMPORTANCE

It is the disease of economic significance in crops like faba bean, lentil, pea, and chickpea which mostly affect its quality of seed. Although, it is expected to be insignificant economically however, it exhibits mild symptoms causing slight yield loss frequently. Chickpea stunt disease has highest significance for this crop due to its worldwide presence in chickpea growing countries. Yield losses have been estimated about 15–16% varies according to varieties.

1.2.1.2 SYMPTOMS

The plants infected with stunt are easily identified their arrested growth as compare to healthy plants. Affected plants show stunted growth with small leaflets, mild mottling of the foliage and yellowish or orange-brown discoloration in the field. The stems and leaves of the infected or diseased plants become stiffer and thicker than healthy plant. Brown discoloration of phloem is most characteristic symptom. Infected plants bear malformed pods if survive up to this stage. Diseased plants show premature drying (Nene et al., 2012).

Seeds obtained from infected crop rarely express dark discoloration or plants raised from infected seeds may not exhibit any visible symptom. Although, infected plants may show yellowing of newly emerged shoots, leaf mottling, tip necrosis of the shoots and overall stunted plant growth.

1.2.1.3 CASUAL ORGANISM

Bean (pea) leaf roll virus.

1.2.1.4 FAVORABLE CONDITIONS

In India, early sowing of the crop in September with wider spacing and aphid vector activity favor the incidence of stunt disease.

1.2.1.5 MANAGEMENT

Farmers should preferably grow tolerant/resistant varieties like BGM 568, HK 06-159, JGK 2006-301, IPC 2000-6, CSJ 313 to avoid quantitative and qualitative loss.

1.2.2 WET ROOT ROT (RHIZOCTONIA SOLANI)

The disease appears at seedling stage under warm temperature and high moisture in soil. The symptom of the disease appears as a distinctive blackish-brown lesion over the collar region of the stem which can spread to lower branches of the older plants. Seed treatment with thiram or benlate at 3 g/kg of seeds.

1.2.3 ANTHRACNOSE (COLLETOTRICHUM DEMATIUM)

This disease is reported only from India as a minor disease and normally problematic in early sown September crop when temperature remains high near about 25–30°C. It is seed and soil borne in nature. The disease may attack and destroy the plants during any stage of crop growth. Infected plants dried branches are distributed entire in the field (Nene et al., 2012).

1.2.4 STEMPHYLIUM BLIGHT (STEMPHYLIUM SARCINIFORME)

It is reported from few countries including India as a disease of minor significance. The symptom of this disease is generally appears from the flowering stage onwards. The noticeable symptom of the disease as defoliation of lower branches is apparent. Circular to oval necrotic spots with darker brown center and wide ash color border appear on the leaflets. Such type of elongated spots also develops on the main stems (Nene et al., 2012).

1.2.5 RUST (UROMYCES CICERIS-ARIETINI)

It is considered as a disease of minor importance which appears during crop maturity. The badly affected crop appears rusty due to presence of rust pustules containing urediospores. Initially, small, circular to ovate, brown, powdered rust pustules appear on the leaves. Severely affected plants start to dry prematurely.

1.2.6 POWDERY MILDEW (LEVEILLULA TAURICA)

It is also a minor disease of this crop reported from many countries including India. Similar to rust, powdery mildew also appears near crop maturity. It can be easily characterized with the presence of white floury growth of the fungus on aerial parts of the plant.

KEYWORDS

- **anthracnose**
- **chickpea stunt**
- ***Leveillula taurica***
- ***Stemphylium* blight**
- **urediospores**
- **wet root rot**

REFERENCES

Baird, R. E., Watson, C. E., & Scruggs, M., (2003). Relative longevity of *Macrophomina phaseolina* and associated mycobiota on residual soybean roots in soil. *Plant Dis.*, *87*, 563–566.

Bakr, M. A., Afzal, M. A., Johansen, C., MacLeod, W. J., & Siddique, K. H. M., (2005). Integrated management of botrytis grey mold of chickpea: On-farm evaluation in Bangladesh. In: '*Paper Presented in the 15th Australasian Plant Pathology Conference on Innovations for Sustainable Plant Health'* (p. 193). Deakin University Waterfront Campus, Geelong, Vic. Australian Plant Pathology Society: Canberra.

Bashir, M., & Malik, B. A., (1988). Diseases of major pulse crops in Pakistan: A review. *Tropical Pest Management*, *34*(3), 309–314.

Chauhan, Y. S., Nene, Y. L., Johansen, C., Haware, M. P., Saxena, N. P., Sardar, S., et al., (1988). Effects of soil solarization on pigeon pea and chickpea. *Research Bulletin No.11*. Patancherum A. P. 502 324, India. International Crop Research Institute for the Semi-Arid Tropics.

Chouhan, S. K., (1963). Influence of different soil temperature on the incidence of *Fusarium* wilt of gram (*Cicer arietinum* L.). *Proceeding Indian Academy of Science, 33*, 552–554.

Dhingani, J. C., & Solanky, K. U., (2016). Integrated management of root rots disease (*Macrophomina phaseolina* (Tassi.) Goid) of chickpea through bioagents, oil cakes and chemicals under field conditions in south Gujarat conditions. *Plant Archives*, *16* (1), 183–186.

FAOSTAT, (2012). *Agriculture*. Available from: http://faostat. fao. org (accessed on 13 January 2020).

Ferguson, B. J., Indrasumunar, A., Hayashi, S., Lin, M. H., Lin, Y. H., Reid, D. E., & Gresshoff, P. M., (2010). Molecular analysis of legume nodule development and autoregulation. *J. Integr. Plant Biol.*, *52*, 61–76.

Francl, L. J., Willie, T. D., & Rosembrok, S. M., (1988). Influence of crop rotation on population density of *Macrophomina phaseolina* in soil infested with *Heterodera glycines*. *Plant Dis.*, *72*, 760–764.

Fuchs, J. G., Moënne-Loccoz, Y., & Défago, G., (1997). Nonpathogenic *Fusarium oxysporum* strain Fo47 induces resistance to *Fusarium* wilt in tomato. *Plant Disease*, *81,* 492–496.

Ghosh, R., Sharma, M., Telangre, R., & Pande, S., (2013). Occurrence and distribution of chickpea diseases in central and southern parts of India. *Am. J. Plant Sci.*, *4*, 940–944.

Gurha, S. N., Singh, G., & Sharma, Y. R., (2003). Diseases of chickpea and their management. In: Ali, M., Kumar, S., & Singh, N. B., (eds.), *Chickpea Research in India* (pp. 195–227). Lucknow: Army Printing Press.

Haware, M. P., & Nene, Y. L., (1980). Influence of wilt at different growth stages on yield loss in chickpea. *Tropical Grain Legume Bull.*, *19*, 38–40.

Haware, M. P., Nene, Y. L., & Rajeshwari, R., (1978). Eradication of *Fusarium oxysporum* f. sp. *Cicero* transmitted in chickpea seed. *Phytopath.*, *68*, 1364–1367.

Hervás, A., Landa, B., & Jiménez-Díaz, R. M., (1997). Influence of chickpea genotype and *Bacillus* sp. on protection from *Fusarium* wilt by seed treatment with nonpathogenic *Fusarium oxysporum. European Journal of Plant Pathology, 10*, 631–642.

Jalali, B. L., & Chand, H., (1992). Chickpea wilt. *In:* Singh, U. S., Mukhopadhayay, A. N., Kumar, J., & Chambe, H. S., *(eds.), Plant Diseases of Cereals and Pulses (pp. 429–444) Englewood Cliffs, NJ: Prentice Hall.*

Kotasthane, S. R., Agrawal, P. S., Joshi, L. K., & Singh, L., (1976). Studies on wilt complex in Bengal gram (*Cicer arietinum* L.). *JNKVV Research Journal*, *10*(3), 25258.

Kumar, R., Mishra, P., Singh, G., & Yadav, R. S., (2008). Integration of bioagents and fungicides for management of collar rot of chickpea. *Journal of Biological Control*, *22*(2), 487–489.

Mahen, V. K., Mayee, C. D., Brenneman, T. B., & McDonald, D., (1995). Stem and pod rot of groundnut. *Information Bulletin No.44* (p. 2). ICRISAT, Patancheru 502 324 AP, India.

Mathur, S. B., & Sinha, S., (1968). Disease development in guar (*Cyamopsis psoraloides* DC) and gram (*Cicer arietinum* L.) attacked with *Sclerotium rolfsii* under different soil conditions. *Phytopath.*, *62*, 319–322.

Mitra, M., (1931). Report of the imperial mycologist. *Sci. Rep. of the Agric. Res. Inst.*, *1929–1930*, 58–71.

Mukhopadhyay, A. N., (1995). Exploitation of *Gliocladium virens* and *Trichoderma harzianum* for biocontrol seed treatment against soil borne disease. *Indian J. Mycol. Pl. Pathol.*, *25*(182), 124.

Nene, Y. L., Reddy, M. V., Haware, M. P., Ghanekar, A. M., Amin, K. S., Pande, S., & Sharma, M., (2012). *Field Diagnosis of Chickpea Diseases and Their Control: Information Bulletin No.28* (p. 60). Patancheru, A. P. 502 324, India: International Crops Research Institute for the Semi-Arid Tropics. ISBN 92-9066-199-2. Order code: IBE: 028.

Pande, S., Galloway, J., Gaur, P. M., Siddique, K. H. M., Tripathi, H. S., Taylor, P., et al., (2006). Botrytis grey mold of chickpea: A review of biology, epidemiology, and disease management. *Aus. J. Agri. Res.*, *57*, 1137–1150.

Pande, S., Sharma, M., Kumari, S., Gaur, P. M., Chen, W., Kaur, L., et al., (2009). Integrated foliar diseases management of legumes. *International Conference on Grain Legumes: Quality Improvement, Value Addition and Trade*. Indian Society of Pulses Research and Development, Indian Institute of Pulses Research, Kanpur, India.

Pande, S., Siddique, K. H. M., Kishore, G. K., Baya, B., Gaur, P. M., Gowda, C. L. L., Bretag, T., & Crouch, J. H., (2005). Ascochyta blight of chickpea: Biology, pathogenicity, and disease management. *Aust. J. Agric. Res.*, *56*, 317–332.

Prasad, R. D., (2005). Status of biological control of soil born plant pathogens. *Indian J. Agri. Sci.*, *77*(9), 583–588.

Prasad, R. D., Rangeshwaran, R., Anuroop, C. P., & Rashmi, H. J., (2002). Biological control of wilt and root rot of chickpea under field conditions. *Ann. Pl. Prot. Sci.*, *10*(1), 72–75.

Punja, Z. K., (1985). The biology, ecology, and control of *Sclerotium rolfsii*. *Annual Review of Phytopathology*, *23*, 97–127.

Sharma, B. K., Singh, U. P., & Singh, K. P., (2002). Variability in Indian isolates of *Sclerotium rolfsii*. *Mycologia.*, *97*(6), 1051–1058.

Sharma, M., Ghosh, R., & Pande, S., (2015). Dry root rot (*Rhizoctonia bataticola* (Taub.) Butler): An emerging disease of chickpea—where do we stand? *Archives of Phytopathology and Plant Protection*, *48*(13/16), 797–812.

Sharma, M., Ghosh, R., Ramesh, R. K., Upala, N. M., Chamarthi, S., Varshney, R., & Pande, S., (2012a). Molecular and morphological diversity in *Rhizoctonia bataticola* isolates causing dry root rot in chickpea (*Cicer arietinum* L.) in India. *Afr. J. Biotechnol.*, *11*, 8948–8959.

Sharma, Y. K., Gaur, R. B., & Bisnoi, H. R., (2004). Cultural, morphological and physiological variability in *Macrophomina phaseolina*. *J. Mycol. Plant Pathol.*, *34*, 532–534.

Singh, D. V., Lal, S., & Singh, S. N., (1974). Breeding gram (*Cicer arientinum* L.) to resistance to wilt. *Indian J. Genet. and Pl. Breeding*, *34*, 267–270.

Singh, P. J., & Mehrotra, R. S., (1982). Penetration and invasion of gram roots by *Rhizoctonia bataticola*. *Indian Phytopathol.*, *35*, 336–338.
Singh, R., Sindhan, G. S., Parashar, R. D., & Hooda, I., (1998). Application of antagonists in relation to dry root and biochemical status of chickpea plants. *Plant Dis. Res.*, *13*, 35–37.
Singh, S. K., Nene, Y. L., & Reddy, M. V., (1990b). Influence of cropping systems on *Macrophomina phaseolina* populations in soil. *Plant Dis.*, *74*, 812–814.
Subramanian, C. V., (1971). *Hyphomycetes: An Account of Indian Species, Except Cercosporae*. ICAR, New Delhi.

CHAPTER 2

Current Status of Chickpea Diseases and Their Integrated Management

DURGA PRASAD AWASTHI[1] and RANGANATHSWAMY MATH[2]

[1]*Department of Plant Pathology, College of Agriculture, Tripura-799210, India, E-mail: pathodurga@gmail. com*

[2]*Department of Plant Pathology, College of Agriculture, Jabugam, Anand Agricultural University, Gujarat-391155, India*

Chickpea scientifically known as *Cicer arietinum* L. is one of the most important crop grown under the family Leguminosae. There are basically two types of Chickpea grown in India one is Desi types which account for 85% of area and the second one is Kabuli types. India is one of the leading states in Chickpea Production. It is grown in different states of India like Madhya Pradesh, Rajasthan, Maharashtra, Uttar Pradesh, Karnataka, and Andhra Pradesh. Chickpea grains are eaten as raw as Dal and are grounded as flour. Tender leaves are also eaten as vegetables. Diseases are becoming a major concern for chickpea production. Major diseases in chickpea cultivation are wilt, *Ascochyta* blight, grey mold, dry root rot, wet root rot, collar rot, chickpea stunt, and rust.

2.1 WILT

2.1.1 INTRODUCTION/ECONOMIC IMPORTANCE

Chickpea wilt is reported from different regions of world like Bangladesh, Ethiopia, India, Iran, Pakistan, Nepal, Spain, etc. In India, chickpea wilt was first reported by E. J. Butler in the year 1918 later on its etiology was determined by Padwick in the year 1940. In India, wilting is one of the significant diseases of chickpea. Loss due to this disease may vary from 20 to 100%.

2.1.2 SYMPTOMS

Wilting is observed in different cultivars of chickpea at different growth stages causing significant yield loss with different degree of severity. Wilting at early stage of growth causes more loss than at later stage. In seedling stage, disease can be observed within 3 weeks of sowing. These seedlings retain their dull green color. Leaves of plant develop a greyish-green chlorosis, typically affecting lower leaves first. At adult stage infected plants shows drooping of the petioles, rachis, and leaflets which later on collapses and dies. Internally, the xylem tissues show dark-brown to almost black. Sometimes the symptom of wilting appears only one side of the plant. Seeds harvested from diseased plant are light, rough, and dull (Figure 2.1).

FIGURE 2.1 Yellowing and wilting of plants

2.1.3 CAUSAL ORGANISM

Fusarium oxysporum f. sp. *ciceri*
Kingdom: *Fungi*
Phylum: *Ascomycota*
Subphylum: *Pezizomycotina*

Class: *sordariomycetes*
Order: *Hypocreales*
Family: *Nectriaceae*

The pathogen produces macroconidia microconidia and chlamydospores. The microconidia (2.5–4.5 μm × 5–11 μm) are oval or cylindrical, straight or curved. Macroconidia (3.5–4.5 μm × 25–65 μm) are produced more sparsely than microconidia. Hyphae are septate and profusely branched. The optimum temperature for growth of fungus is 7–35°C with pH range of 4 to 9.4.

2.1.4 DISEASE CYCLE

The pathogen remains in infected plant debris in the soil or in seeds. The primary infection initiates through chlamydospores which can remain viable up to the next crop season. The pathogen perpetuates in roots as well as stems. Invasion of the roots is followed by the penetration of the epidermal cells of the host or the non-host and the development of a systemic vascular disease in host plants. Penetration occurs either through wounds or directly. Penetration is influenced by different factors like plant surface structures, activators, or inhibitors of fungal spore germination, and germ tube formation etc. Pathogen survives as mycelium and chlamydospores in seed and soil and also on infected crop residues, roots, and stem tissue buried in the soil.

2.1.5 EPIDEMIOLOGY

Desi type of chickpea is less susceptible than the Kabuli type. Root exudates influence wilt incidence. Exudates of some resistant cultivars contain antifungal compounds that suppressed spore germination of the pathogen. Landa et al. (2001) had reported that wilt development was greater at 25°C than at 20°C or 30°C. Root knot nematode may also have role in the Fusarium wilt of chickpea.

2.1.6 INTEGRATED DISEASE MANAGEMENT (IDM)

- Seed Treatment with Carbendazim @ 2.5 g/kg of seed.
- Resistant Variety: Grow resistant varieties like C-214, Avrodhi, Uday, BG-244; Pusa-362, JG-315, Phule G-5, etc.
- Soil solarization along with soil amendments with neem cake or mustard cake help in reduction of inoculums.

- Soil application of bio-control agent like *Trichoderma viride* @ 2.5 kg/ha mixed with well rotten farmyard manure (FYM) or neem seed kernel cake (NSKC) of about 100 kg. After mixing the bio agent with manure incubate it for 21 days in shed before application.
- In fields having incidence of wilt, cultivation of chickpea is to be avoided.
- The following entries were found resistant to moderately resistant under All India Coordinated Research Project on Chickpea.

Desi:	BG 3004, Tungabhadra, GL 26054, GNG 1958, GJG 0714, PG 97030, NDG 9-21, CSJ 592, IPC 2005-62, Phule G 105-10-1, GJG 0504, IC 251907, IPC 2008-103, GJG 0814, BCP 60, BCP 136, Phule G 9807, IPC 2005-59, JG 2000-07, IPC 2005-74, JSC 35, IPC 2004-68, GAG 0419
Kabuli:	HK 05169, IPCK 2005-58, HK 06-152, IPCK 2005-23, IPCK 02 (AVT2), BG 3002, IPCK 113

2.2 ASCOCHYTA BLIGHT

2.2.1 INTRODUCTION/ECONOMIC IMPORTANCE

Ascochyta blight is a harmful foliar disease which is a major constraint for chickpea production worldwide (Figure 2.2). In India, it was first reported by Butler in 1911. It is also reported from Europe, North Africa, Mediterranean Sea, Iran, Iraq, Pakistan, Portugal, Spain, USA Former USSR, Tanzania, Mexico, Bangladesh, and India. Disease was severe in states like Punjab, Bihar, Jammu, Haryana, Himachal Pradesh, Uttar Pradesh, and Rajasthan. In 1930s due to chickpea, blight there was total loss of the crop in Spain. In the undivided Punjab (now part of Pakistan) in the year 1922 to 1933, loss due to *Ascochyta* blight was estimated to be around 25% to 50% almost every year.5% to 75% loss was reported from Rajasthan in 1982. Under encouraging environmental condition, loss due to disease may goes up to 100%.

2.2.2 SYMPTOMS

The disease affects all aerial parts and can attack at any growth stage of the crop. Initial symptoms appear near the tip of young shoots and in top leave as circular spots. Lesions are circular on leaves, pods whereas these are elongated on stems. The apical twigs, branches, and stems may show girdling and plant parts above girdled portion are killed and break off. Lesions on pods are prominent and usually circular with dark margins. Fully developed lesions on pods show concentric rings of pycnidia. In early infections, the pod is blighted and fails to develop any seed while in late infections seeds become shriveled and infected.

Symptoms on the seeds appear as a brown discoloration which develop into deep, round or irregular cankers, sometimes having pycnidia. The fungus perpetuates on plant debris or trash left in soil. The diseased plant finally dries up.

FIGURE 2.2 Symptoms *Ascochyta* spot disease infected plants.

2.2.3 CAUSAL ORGANISM

Asexual Stage: *Ascochyta Rabiae*
Sexual Stage: *Didymella Rabiei*
Kingdom: *Fungi*
Phylum: *Ascomycota*
Subphylum: *Pezizomycotina*
Class: *Dothidiomycetes*
Order: *Pleorosporales*
Family: *Didymellaceae*

Mycelium appears hyaline to brownish and septate. The asexual stage of *Ascochyta rabiae* forms pear shaped pycnidia on stems, leaves as well as in pods. Pycnidium may be flat, round or globose dark brown, measuring about 100–200 μm in diameter. Each pycnidium has prominent ostiole of 30–50 μm wide. Conidia bear on conidiophores. Conidia are hyaline in color and oval to oblong in shape and measure 9–20 to 3–6 μ. Mostly conidia are one septate or in some cases non septate. At perfect stage dark colored globose pseudothecia are formed, each measuring 100–140 μm in diameter. Each contains eight ascospores which are bicelled with one cell larger than the other.

Disease is a seed borne in nature. The pathogen perennates on/in seed as pycnidia on plant debries left in field. Ascospores plays role in the initiation of disease epiphytotics. Seed is considered to be main source of primary infection in Indian. Primary inoculum multiplies on hypocotyls, epicotyl, and near the stem base. Conidia spread through raindrop splashes, insects, and cause secondary infection.

2.2.4 EPIDEMIOLOGY

Disease is favored by humidity, cool, and cloudy condition. The disease is favored at 10°C night and 20°C day temperatures respectively. Dense canopy also favors blight development. Wet weather with strong winds having 22–26°C temperature are most favorable the spread of the disease. High rainfall during crop growth also favors development of disease in epiphytotic form.

2.2.5 INTEGRATED DISEASE MANAGEMENT (IDM)

- Select healthy seeds and adopt low seed rate.
- Follow seed treatment with Thiram or Carbendazim @ 2.5 g/kg of seed.
- Burying infected crop residue reduces primary source of inoculums.
- Follow crop rotation preferably with crops other than *Fabaceae*.
- Cultivars or varieties namely G-543, Pusa-256, Gaurav, GNG-146, PBG-1, GL 260584, GL 26069, GL 23094, IPC 2005-66 are resistant to blight.
- Farm hygiene, use of resistant varieties and crop monitoring should be integrated together for management of the disease.
- Always disinfect machinery, vehicles, and boots.
- In mild attack spray Zineb, Maneb or Captan @ 2.5 g/l of water. Post infection application of Tebuconazole or Difenconazole to moderately resistant varieties gives good control in mild epidemics.

2.3 BOTRYTIS GREY MOLD

2.3.1 *INTRODUCTION/ECONOMIC IMPORTANCE*

This disease appear regularly every year in moderate to severe form in northern states of India. It causes significant damage in Uttar Pradesh.

2.3.2 *SYMPTOMS*

The disease attacks leaves, flowers, pods, branches as well as on stems. Initial symptoms include water soaking, softening of affected plant parts. Grey to dark brown lesions is formed on affected parts which are readily covered with dense fungal growth like sporophores and mycelium. The branches as well as stems are getting affected leading to breaking and killing of plants. Affected parts like leaves or flowers rot. Infected plants bear no or shivelled seeds having grey white mycelium under conducive condition.

2.3.3 *CAUSAL ORGANISM*

Botrytis cinerea
Kingdom: *Fungi*
Phylum: *Ascomycota*
Subphylum: *Pezizomycotina*
Class: *Leotiomycetes*
Order: *Helotiales*
Family: *Sclerotiniaceae*

It belongs to the phylum Ascomycota. Initial hyphae are thin, hyaline; conidiophores are light brown, erect, and septate, ramified pseudodichotomically with somewhat enlarged tips. Around 1 to 3 germ tubes are being produced by germinating conidia.

2.3.4 *DISEASE CYCLE*

Botrytis cinerea perpetuates on seed for minimum 5 years and is internally or externally seed borne in nature. The pathogen also survives on plant remains in soil surface for few months in the form of mycelia as well as sclerotia and thus it is regarded as primary source of inoculum. Secondary spread of disease is through air borne conidia.

2.3.5 EPIDEMIOLOGY

Disease appears at flowering time. Dense vegetative growth high irrigation or rain, close spacing, and spread habitat varieties favor disease development. Temperatures from 20°C to 25°C and high humidity at reproductive stage favor disease development. Relative humidity (RH), leaf wetness, and temperature are also important factors (Tripathi and Rathi, 1992; Butler, 1993; Pande et al., 2002).

2.3.6 INTEGRATED DISEASE MANAGEMENT (IDM)

- Use disease-free, healthy seeds for sowing.
- Follow seed treatment. Fungicides like carbendazim @ 2.5 g/kg of seed can be used.
- Use of *Trichoderma harzianum* and *T. viride* are reported to be effectual against *Botrytis cinerea.*
- Optimum seed rate and spacing are recommended as high plant density will increase disease severity.
- Fungicides namely captan, chlorothalonil, carbendazim, mancozeb, triadimefon, triadimenol thiabendazole, thiophanate-methyl, and thiram, are found to be effective when used as a foliar spray.
- Foliar spray with carbendazim @ 1 g/l immediately after noticing the symptoms may be given.
- Cultural operations like removal and destruction of infected crop debries, deep plowing, and late sowing helps to minimize the disease.

2.4 DRY ROOT ROT

2.4.1 INTRODUCTION/ECONOMIC IMPORTANCE

Among the several soil borne fungal diseases of Chickpea, dry root rot (*Rhizoctonia bataticola)* is extremely significant disease. The disease has being Australia, Iran, Kenya, Lebanon, Bangladesh, Myanmar, Nepal, Mexico,

Pakistan, Sudan, Turkey, Spain, and the United States. In central and southern India the diseases has become a matter of concern.

2.4.2 SYMPTOMS

Disease may appear at flowering and podding. The characteristic symptom is yellowing and rapid drying of infected plants. The tap root appears dark brown, brittle, and shows wide rotting ensuing loss of lateral roots. If uprooted some portion of tap root is often left at soil. At later stages of disease dark, minute sclerotial bodies appears exposed or inside the affected areas.

2.4.3 CAUSAL ORGANISM

Rhizoctonia bataticola
Pycnidial stage: *Macrophomina phaseolina.*

2.4.4 DISEASE CYCLE AND EPIDEMIOLOGY

Pathogen is facultative sporophyte and disease is both seed and soil borne in nature. When crop is given high temperature, moisture stress incidence of dry root rot is more. Pathogen survives as sclerotia which serve as primary source of inoculum.

2.4.5 INTEGRATED DISEASE MANAGEMENT (IDM)

- Avoid drought.
- Sow on optimum time for escaping hot weather.
- Entries/cultivars/accession numbers under AICRP on chickpea like PG 00110, GL 21107, CSJ 592, H 04-49, IPC 2005-28, BG 3002, IC 269792, NDG 7-702, CSJ 559, IPC 2005-66, NDG 7-702 are resistant.
- Carbendazim + Thiram (1:2) @ 3 g/kg of seed can be used as seed treatment.
- Soil application of bio-control agent like *Trichoderma viride* @ 2.5 kg/ha mixed with well rotten FYM or NSKC of about 100 kg. After mixing the bio agent with manure, incubate it for 21 days in shed before application.

2.5 COLLAR ROT

2.5.1 INTRODUCTION/ECONOMIC IMPORTANCE

Collar rots disease (*Sclerotium rolfsii* Sacc. [teleomorph *Athelia rolfsii*]) has the potentiality to become a serious threat for chickpea cultivation. Under conducive conditions, at seedling stage 55–95% mortality was reported.

2.5.2 SYMPTOMS

Collar rot is observed at seedling stage in wet soils. Leaves of affected seedlings turn yellow, droops, and dry. The seedlings generally collapsed and show symptoms of rotting at collar region covered with whitish mycelia growth. The same can be observed on taproot of infected plants even after several days of its death. At advance stage of disease white sclerotial initials and mature sclerotia are easily visible.

2.5.3 CAUSAL ORGANISM

Sclerotium rolfsii
Kingdom: *Fungi*
Phylum: *Basidiomycota*
Class: *Agaricomycetes*
Order: *Atheliales*
Family: *Atheliaceae*
Genus: *Sclerotium*
Species: *S. rolfsii*

2.5.4 DISEASE CYCLE AND EPIDEMIOLOGY

Collar rot is generally observed in high soil moisture and warm temperature at the seedling stage. This disease is usually problem in areas where paddy plants were grown before chickpea cultivation. The disease is favored by the presence of undecomposed organic matter on the soil surface and excessive moisture at the time of sowing and at the seedling stage.

2.5.5 INTEGRATED DISEASE MANAGEMENT (IDM)

- Adopt wide row spacing;
- Soil solarization during summer months;
- Avoid high moisture content;
- Remove all undecomposed organic stuff while preparing seed bed;
- Treat seeds with vitavax @ 3 g/kg of seed;
- Apply bio-control agent like *Trichoderma viride* @ 2.5 kg/ha mixed with well rotten FYM or NSKC of about 100 kg.

2.6 RUST

2.6.1 INTRODUCTION

Rust is important disease common in Uttar Pradesh, Punjab, Bihar, and West Bengal.

2.6.2 SYMPTOMS

Characteristic symptom is appearance of uredospores on lower areas of leaves. Rust pustules may also appear in pods, floral parts and stems. Severely affected plants appear rusty and dries up prematurely. In later stage telial stages appears on leaves.

2.6.3 CAUSAL ORGANISM

Uromyces ciceris-arietini
Kingdom: *Fungi*
Phylum: *Basidiomycota*
Class: *Pucciniomycetes*
Order: *Pucciniales*
Family: *Pucciniaceae*
Genus: *Uromyces.*

The pycnial or aecial stages are so far unknown. The uredia are scattered, minute, round, powdery, light brown in color. The uredopsores are globose, loosely echinulate, 20–28 μm in diameter, yellowish brown color and 4–8 germ pores. The teliospores are rounding, ovate or angular with a roundish unthickened apex. They measure 18–30 x 18–24 μm.

2.6.4 DISEASE CYCLE

Due to high temperature in plains during summer, urediospores and teliospores are inactivated resulting to lack of primary source of inoculums. Pathogen survives in the weed *Trigonella polycerata* which may serve as primary source of inoculum while the secondary spread of disease occurs through wind or air. Fungus survives in its uredial stage while role of telial stages are unknown.

2.6.5 EPIDEMIOLOGY

Rust is favored by moderately warm weather. Irrigated crops suffer more than crops grown in the rainfed conditions. Optimum temperature for germination of uredospores is 11–20°C.

2.6.6 INTEGRATED DISEASE MANAGEMENT (IDM)

- Remove collateral hosts like *Trigonella polycerata* which may harbor the rust pathogen.
- Disease may be escaped by early sowing of crops.
- Chickpea lines namely NRC 34, NEC 249, JM 583, JM 2649, HPC 63, HPC 136 and HPC 147 are resistant at Himachal Pradesh.
- Foliar spray with hexaconoazole or propiconazole @ 1 ml/l will help to manage the disease.

2.7 CHICKPEA STUNT

2.7.1 INTRODUCTION/ECONOMIC IMPORTANCE

Disease is reported in chickpea growing areas and is an important virus diseases. Yield loss due to chickpea stunt may be 58.0–89.0%.

2.7.2 SYMPTOMS

Affected plants are diagnosed based on their stunted growth and appearance of yellow, orange, or brown discoloration. Typical symptom of the disease is shortening of internodes. Browning occurs at collar region of infected plant.

Stems may also turn brown in few cases. Tips and margins of leaflets become chlorotic before turning reddish brown. Leaf discoloration symptom is more in desi types (reddish) than in Kabuli types (yellow). The most diagnostic symptom of the disease is phloem browning which appears after removing the bark from collar region.

2.7.3 CAUSAL ORGANISM

Luteo virus and Chickpea Chlorotic Dwarf Virus (Gemini virus) are found to be associated with the disease.

2.7.4 TRANSMISSION AND EPIDEMIOLOGY

Aphids transmit Chickpea Stunt Disease while Chickpea Chlorotic Dwarf Virus is transmitted through leaf hopper. Groundnut and faba bean are its other host where virus can perpetuates. Incidence of Chickpea Stunt is favored in early sown areas and crops having wider spacing in India. The activity of *Aphis craccivora* and *Myzus persicae* influences disease incidence.

2.7.5 MANAGEMENT

- Uproot the affected crop and burn it.
- Insecticides like Imidachloprid 17.8 SL can be sprayed to control vector population.
- Sow the crop when activity of vectors is minimum.

KEYWORDS

- **chickpea stunt**
- **collar rot**
- **dry root rot**
- **farm yard manure**
- **integrated disease management**
- **neem seed kernel cake**

REFERENCES

Abida, A., Muhammad, I. S., Rizwana, A. Q., & Chaudhary, A. R., (2008a). Variability among isolates of *Sclerotium rolfsii* associated with collar rot disease of chickpea in Pakistan. *Mycopath.*, *5*(1), 23–28.

Agarwal, A., & Tripathi, H. S., (1999). Biological and chemical control of botrytis gray mould of chickpea. *J. Mycol. Pl. Pathol.*, *29*, 52–56.

Agarwal, A., Tripathi, H. S., & Rathi, Y. P. S., (1999). Integrated management of gray mould of chickpea. *J. Mycol. Pl. Pathol.*, *29*, 116, 117.

Ahmed, A. U., Bakr, M. A., Hossain, M. S., & Chowdhury, J. A., (2002). Integrated management of botrytis grey mould disease in chickpea. *Bangladesh Journal of Agricultural Research*, *27*, 237–242.

Bakr, M. A., & Ahmed, F., (1992). Botrytis gray mold of chickpea in Bangladesh. In: Haware, M. P., Faris, D. G., & Gowda, C. L. L., (eds.), *'Botrytis Gray Mold of Chickpea. Summary Proceedings of the BARI/ICRISAT Working Group Meeting'* (pp. 10–12). ICRISAT, Patancheru, AP, India.

Bakr, M. A., Hossain, M. S., & Ahmed, A. U., (1997). Research on botrytis gray mold of chickpea in Bangladesh. In: Haware, M. P., Lenne, J. M., & Gowda, C. L. L., (eds.), *Recent Advances in Research on Botrytis Gray Mold of Chickpea* (pp. 15–18). ICRISAT, Patancheru, AP, India.

Butler, D. R., (1993). How important is crop microclimate in chickpea botrytis gray mold? In: Haware, M. P., Gowda, C. L. L., & McDonald, D., (eds.), *Recent Advances in Research on Botrytis Gray Mold of Chickpea* (pp. 7–9). ICRISAT, Patancheru, AP, India.

Chand, H., & Khirbat, S. K., (2009). Chickpea wilt and its management. *Agric. Rev.*, *30*(1), 1–12.

Davidson, J. A., Pande, S., Bretag, T. W., Lindbeck, K. D., & Kishore, G. K., (2004). Biology and management of *Botrytis* spp. in legume crops. In: Elad, Y., Williamson, B., Tudzynski, P., & Delen, N., (eds.), *'Botrytis: Biology, Pathology and Control'* (pp. 295–318). Kluwer Academic Publishers, Netherlands.

Dubey, S. C., Suresh, M., & Singh, B., (2007). Evaluation of *Trichoderma* species against *Fusarium oxysporum* f. sp. *ciceris* for integrated management of chickpea wilt. *Biol. Control*, *40*, 118–127.

Grewal, J. S., & Laha, S. K., (1983). Chemical control of botrytis blight of chickpea. *Indian J. Phytopath.*, *36*, 516–520.

Gurah, S. N., Singh, G., & Sharma, Y. R., (2003). Diseases of chickpea and their management. In: Massod, A., Shiv, K., & Singh, N. B., (eds.), *'Chickpea Research in India'* (pp. 195–227). Indian Institute of Pulses Research: Kanpur, India.

Hanumanthegowda, B., (1999). Studies on stem rot of groundnut caused by *Sclerotium rolfsii* Sacc. *M. Sc. (Agri.) Thesis*. Univ. Agric. Sci., Dharwad.

Haware, M. P., (1998). Diseases of chickpea. In: Allen, D. J., & Lenne, J. M., (eds.), *'The Pathology of Food and Pasture Legumes'* (pp. 473–516). ICARDA, CAB International, Wallingford, UK.

Haware, M. P., Mukherjee, P. K., Lenne, J. M., Jayanthi, S., Tripathi, H. S., & Rathi, Y. P. S., (1999). Integrated biological-chemical control of Botrytis gray mold of chickpea. *Indian Phytopath.*, *52*, 174–176.

Haware, M. P., Narayana, R. J., & Pundir, R. P. S., (1992). Evaluation of wild Cicer species for resistance to four chickpea diseases. *International Chickpea Newsletter*, *27*, 16–17.

Jarvis, W. R., (1980). Epidemiology: In: Coley, S. J. R., Verhoeff, K., &. Jarvis, W. R., (eds.), '*The Biology of Botrytis*' (pp. 219–250). Academic Press, London.
Jiménez-Díaz, R. M., Castillo, P., Del Mar, Jiménez-Gasco, M., Landa, B. B., & Navas-Cortés, J. A., (2015). *Fusarium* wilt of chickpeas: Biology, ecology and management. *Crop Prot.*, *73*, 16–27.
Joshi, M. M., & Singh, R. S., (1969). A Botrytis gray mould of gram. *Indian Phytopath., 22*, 125–128.
Kaur, N. P., & Mukhopadhyay, A. N., (1992). Integrated control of chickpea wilt complex by trichoderma and chemical methods in India. *Trop. Pest Manag.*, *38*, 37–41.
Kishore, G. K., (2005). Cultural, morphological, pathogenic, and genetic variation in Botrytis cinerea, causal agent of gray mold in chickpea. *PhD Thesis*. Jawaharlal Nehru Technological University, Hyderabad, Andhra Pradesh, India.
Landa, B. B., Navas, C. J. A., Hervas, A., & Jimenez-Diaz, R. M., (2001). Influence of temperature and inoculum density of *Fusarium oxysporum* f. sp. *Ciceris* on suppression of *Fusarium* wilt of chickpea by rhizosphere bacteria. *Phytopathol., 91*, 807.
Latha, S. K., & Grewal, J. S., (1983). Botrytis blight of chickpea and its perpetuation through seed. *Indian Phytopath., 36*, 630–634.
Mahmood, M., & Sinha, B. K., (1990). Gray mold disease of Bengal gram in Bihar. *Final Technical Bulletin* (p. 35). Bihar, India.
Mandhare, V. K., & Suryawanshi, A. V., (2008). Efficacy of some botanicals and *Trichoderma* species against soil borne pathogens infecting chickpea. *J. Food Legumes, 21*(2), 122–124.
Meeta, M., Bedi, P. S., & Kumar, K., (1986). Chemical control of gray mold of gram caused by *Botrytis cinerea* in Punjab. *J. Res., 23*, 435–438.
Muhammed, S., Vannia, R. P., & Viswanatha, K. P., (2011). Screening for multiple disease resistance in chickpea. In: "*Nat. Seminar on—Plant Genetic Research for Eastern and North-Eastern India*" (p. 163). Umaim, Meghalaya.
Nene, Y. L., Reddy, M. V., Haware, M. P., Ghanekar, A. M., Amin, K. S., Pande, S., & Sharma, M., (2012). *Field Diagnosis of Chickpea Diseases and Their Control: Information Bulletin No.28* (p. 60). Patancheru, A. P. 502 324, India: International Crops Research Institute for the Semi-Arid Tropics.
Nene, Y. L., Sheila, V. K., & Sharma, S. B., (1984). *A World list of Chickpea (Cicer arietinum L.) and Pigeon Pea (Cajnus cajan (L.) Millsp.) Pathogens* (p. 19). ICRISAT Pulses pathology progress report 32.
Pande, S., Galloway, J., Gaur, P. M., Siddique, K. H. M., Tripathi, H. S., Taylor, P., et al., (2006). Botrytis grey mould of chickpea: A review of biology epidemiology and disease management. *Australian J. Agric. Res., 57*, 1137–1150.
Pande, S., Singh, G., Rao, J. N., Bakr, M. A., Chaurasia, P. C. P., Joshi, S., et al., (2002). Integrated management of botrytis gray mold of chickpea. *Information Bulletin No.61*. ICRISAT, Andhra Pradesh, India.
Ramakrishnan, T. S., (1930). A wilt of zinnia caused by *Sclerotium rolfsii. Madras Agric. J.*, *16*, 511–519.
Rangaswami, G., & Mahadevan, A., (2016). *Diseases of Crop Plants in India* (4th edn., pp. 270–276).
Reddy, M. V., Ghanekar, A. M., Nene, Y. L., Haware, M. P., Tripathi, H. S., & Rathi, Y. P. S., (1993). Effect of vinclozolin spray, plant growth habit and inter-row spacing on Botrytis grey mold and yield of chickpea. *Indian J. Pl. Prot., 21*, 112–113.
Reddy, M. V., Nene, Y. L., Singh, G., & Bashir, M., (1990). Strategies for management of foliar diseases of chickpea. In: Van Rheenen, H. A., Saxena, M. C., Walley, B. J., & Hall, S.

D., (eds.),'*Chickpea in the Nineties: Proceedings of the Second International Workshop on Chickpea Improvement* (pp. 117–127).' ICRISAT, India. ICRISAT, Patancheru, AP, India.

Rolfs, P. H., (1892). Tomato blight some hints. *Bulletin of Florida Agricultural Experimental Station* (p. 18).

Saccardo, P. A., (1911). Notae mycologicae. *Annales Mycologici*, *9*, 249–257.

Sattar, A., (1933). On the occurrence perpetuation and control of blight caused by *Ascochyta rabiei* with special reference to Indian conditions. *Ann. Appl. Biol.*, *20*, 612–632.

Sharma, K. D., & Muehlbauer, F. J., (2007). *Fusarium* wilt of chickpea: Physiological specialization, genetics of resistance and resistance gene tagging. *Euphytica.*, *157*, 1–14.

Shaw, F. J. P., & Ajrekar, S. L., (1915). The genus Rhizoctonia in India. *Mem. Department of Agricultural Indian Bot. Ser.*, *7*, 177–194.

Singh, G., & Bhan, L. K., (1986). Chemical control of gray mold in chickpea. *International Chickpea Newsletter*, *15*, 18–20.

Singh, G., & Kapoor, S., (1984). Role of incubation and photoperiod on the intensity of botrytis gray mold of chickpea. *International Chickpea Newsletter*, *12*, 23, 24.

Singh, M. P., (1989). Studies on survivability of *Botrytis cinerea* Pers. ex. Fr., causal agent of grey mould of chickpea. MSc Thesis. Govind Ballab Pant University of Agriculture and Technology, Pantnagar, UP, India.

Singh, R. S., (2009). *Plant Diseases* (9th edn., pp. 454–457). Oxford & IBH Publishing Co. Pvt. Ltd.

Tripathi, H. S., & Rathi, Y. P. S., (1992). Epidemiology of botrytis gray mold of chickpea. In: Haware, M. P., Faris, D. G., & Gowda, C. L. L., (eds.), '*Botrytis Gray Mold of Chickpea*' (pp. 8, 9). ICRISAT, Patancheru, A. P., India.

Tripathi, H. S., Singh, R. S., & Chaube, H. S., (1987). Effect of Fungicidal seed and foliar application of chickpea blight. *Indian Phytopathol.*, *40*, 63.

Warda, J., Mariem, B., Boukteb, A., Béji, M., & Kharrat, M., (2017). *Fusarium* wilt affecting chickpea crop: A review. *Agriculture*, *7*(23), 2–16.

CHAPTER 3

Important Diseases of Field Pea (*Pisum sativum* var. *Arvense*) and Their Management

A. KHULBE[1] and K. K. SHARMA[2]

[1]*Krishi Bhawan, ICAR, New Delhi, India, E-mail: anjanikhulbe@gmail. com*

[2]*Dr. D. R. Bhumbla Regional Research Station for Kandi Area (PAU), Ballowal Saunkhri, SBS Nagar–144521, Punjab, India, E-mail: kksharma@pau. edu*

Field pea, commonly known as pea, is an important pulse crop having its center of origin in Mediterranean region of Southern Europe and Western Asia, is a winter season crop and is known to be grown worldwide. Canada rank first in area (21%) and production (35%) at Global level. India occupies second position in area (14%) and 4th position in production (7%). According to twelfth plan (2012–2016), a total area of 9.01 lakh hectares and a total production of 8.49 lakh tonnes were recorded. Uttar Pradesh ranked first both in area and production (37.90% and 41.58%) followed by Madhya Pradesh (38.67% and 32.98%) and Jharkhand (3.80% and 4.85%). In case of productivity Rajasthan ranked first with (1867 kg/ha) followed by Punjab (1297 kg/ha) and Jharkhand (1203 kg/ha). The lowest yield was observed in Maharashtra (390 kg/ha) followed by C. G. (437 kg/ha) and Assam (817 kg/ha) (Annual Report, 2016–17, DPD).

It is an important high yielding pulse crop that can be used for vegetable purpose when unripen and the dry seeds are used as dal and also as snacks after boiling. The left over part of the plant is used for feeding the cattle.

Being a winter season crop it requires a cool growing season with moderate temperature throughout the life. High temperature is more injurious to pea crop than frost. Frost can damage the plants during flowering stage. High humidity associated with cloudy weather results into spread of fungal diseases like damping-off and powdery mildew. Optimum monthly temperature suitable for growth is 13–18°C.

3.1 DIFFERENCE BETWEEN FIELD PEA AND VEGETABLE PEA

Field pea is *Pisum sativum* var. *arvense* while vegetable pea is *Pisum sativum* var. *hortense*. The flower color of field pea is purple while that of vegetable pea is white. Former one has smooth seed while wrinkled seed is found in the later one. The field pea is less sweet due to high amount of starch, however, the vegetable pea is sweet due to high sugar content. The plant of may have either single stem or many axillary stems with tap root system and leaves composed of leaflets in pair to several in which lower leaflets are larger than the upper.

Similar to vegetable pea, field pea is also affected by number of diseases such as damping off, rust, powdery mildew, ascochyta blight, wilt, and root knot, etc., which are reported worldwide. These diseases cause high economic losses to the pea growers. For large scale commercial production specific crop varieties play a major role, however, the role of disease management practices with timely diagnosis at appropriate growth stages of the crop is equally important to save the crop from significant yield losses. This chapter deals with various aspects of important diseases of field pea and their integrated management practices.

3.1.1 DISEASES OF MAJOR CONCERN IN FIELD PEA CROPS

- Ascochyta/Mycosphaerella blight;
- Powdery mildew;
- Root rot;
- Seed decay and seedling blight;
- Sclerotinia stem and pod rot;
- Bacterial blight;
- Bacterial rot;
- Nematode diseases;
- Viral diseases.

3.2 FUNGAL DISEASES

3.2.1 ASCOCHYTA/MYCOSPHAERELLA BLIGHT

3.2.1.1 INTRODUCTION AND ECONOMIC IMPORTANCE

Ascochyta blight (synonym: black spot) is one of the most important diseases affecting field peas. It is caused by four pathogens that occur as a complex in the field and cause a single disease. It is one of the destructive diseases of peas distributed worldwide among almost all areas where pea is grown (Bretag et al., 2006).

This disease complex is caused by three different species of fungi, each of which produces a clearly recognizable symptom: *Mycosphaerella pinodes* (Berk. & B lox.) Vestergr (the perfect stage of *Ascochyta pinodes*) which causes blight; *A. pisi* Lib. which causes leaf and pod spot and *Phoma rnedicaginis* var. *pinodella* (Jones) which causes foot rot (Wdlen, 1974; Lawyer, 1984a; Kraft, 1991).

It was reported first time from Europe (1830) by Libert. Variable yield losses were estimated as 10% 60% from Australia (Bretag et al., 2006), 40% in France (Tivoli et al., 1996) and up to 50% from Canada in experimental fields (Xue et al., 1996). In India, it is prevalent in Himalayan region.

3.2.1.2 CAUSAL ORGANISM

Ascochyta pisi, *Phoma medicaginis var. pinodella*, *A. pinodes.*

3.2.1.3 PATHOGEN

A. pisi has light-colored mycelium with abundant carrot-red spores as compare darker mycelium of *A. pinodes* and *Phoma medicaginis* var. *pinodella*. The pycniospores of *A. pisi* and *A. pinodes* are larger than those of *A. pinodella*. The incubation period required to cause the disease is reported to be 2–4 days in case of *A. pinodes* and *Phoma medicaginis var. pinodella* and 6–8 days for *A. pisi*. Air-borne ascospores are produced only by *A. pinodes*. All these three pathogens are able to infect the seed and get carried from one place to another. The pathogens perpetuates through infected seeds and pea straw. The pathogen of ascochyta blight remains in soil till the complete decomposition of infected pea straw. When the weather becomes moist, the pycniospores and ascospores both may be produced on previous season infected pea straw left in the soil. Rain splashes helps in dissemination of Pycniospore to a short distance only up to few plants in patches whereas the a large number of wind-borne ascospores (*A. pinnodes*) get blown up by wind current and spread uniformly in the entire area causing more damage. Under favorable conditions, pycniospores, and ascospores both are repeatedly produced on infected plant parts in field. The weather parameters namely rainfall and dew, humidity, and wind speed are very crucial for further development of ascochyta blight.

3.2.1.4 SYMPTOMS

Symptoms appear on all the aboveground parts of field pea plants as well as the crown region below ground level. In general, planting of seeds infected with Ascochyta species causes weakening or killing of young seedlings due to foot rot near the soil line. Necrotic spots appear on stems, leaves, and pods that coalesce into large lesions. In severe cases root rot may occur. Infection in the early stage may cause death of the seedling.

1. **Symptoms on Leaves:** The spots on leaves are small, irregular, dark-brown, and scattered throughout the leaf, or large, circular, and brown in color. Leaves of diseased plants shrivel and finally dry exhibiting blighted appearance.
2. **Symptoms on Stem:** Lesions are purplish-black in color and may cause streaking of the lower stem. The infection continues on petiole around the nodes where lesions may coalesce leads to girdling of the whole stem which becomes weak and may break easily.
3. **Symptoms on Pods:** Spots on the pods may join and form big, sunken, purple-black spots.
4. **Symptoms on Seeds:** Infected seeds may appear purplish-brown in color.

Ascochyta pisi produces symptoms on above ground parts as slightly sunken, light brown, circular spots on leaves, pods, and elongated on stem with darker border and numerous black pycnidia whereas *A. pinodes* causes irregularly shaped, darker brown spots on leaves pods and cotyledons. *A. pinodella* infection generally affects belowground parts particularly near soil line exhibiting severe foot rot symptom (Linford and Sprague, 1927; Lawyer, 1984). However, it is not happen always because same type of symptoms may also be seen with *A. pinodes* and *A. pinodella* infection. Peever et al. (2007) described above ground symptoms on pea caused by *Ascochyta pisi* as tan colored spots with discrete border whereas brown to tan lesions can be observed on aboveground as well as underground parts of the affected plants although such lesions are lacking aforesaid margins (Figure 3.1).

3.2.1.5 MANAGEMENT

Growers should use healthy and disease free seeds to avoid initial infection. Seed treatment is a good option to reduce the external inoculum present on seed. To get rid of soil borne inoculums one should go for crop rotation with non-host crop for at least 4 or 5 years and it is always advisable that the fields are rotated in a manner so that every 3–4 years a field would rest and be fallow. Infected stubble or vines, an important source of inoculum for the succeeding crop, should be destroyed by plowing deep in soil immediately after harvest and thereafter any cereal crop can be sown. Fungicides like tebuconazole, iprodione, carbendazim, and fludioxonil were found highly effective to control more than 80% disease tested under field situations by Liu et al. (2016).

FIGURE 3.1 Symptoms of ascochyta blight on (A) leaves and (B) stem.

3.2.2 POWDERY MILDEW

3.2.2.1 INTRODUCTION AND ECONOMIC IMPORTANCE

Powdery mildew disease of pea is worldwide distributed and causing huge economic damage to the crop. The disease may appear from early growth stages to pod formation. The disease is caused by a fungus, *Erisiphe pisi* in India while *E. polygoni* and *E. cichoracearum* have also been reported and identified in other countries. The annual losses in yield were estimated about 25–50% (Nisar et al., 2006; Rathi and Tripathi, 1994; Dixon, 1987) and 30–40% (Singh, 1999; Upadhyay and Singh, 1994).

3.2.2.2 CAUSAL ORGANISM

Erisiphe pisi, Erysiphe polygoni, Erysiphe cichoracearum.

3.2.2.3 PATHOGEN

Plant diseases caused by genus *Erysiphe* is commonly known as powdery mildews. Sexual stage of this fungus is closed fruiting body called cleistothecium. It is an obligate parasite and survives and multiplies on the living host tissues (leaves, stem, tendril, and pods) of susceptible plants. The disease is favored by sunny hot days followed by humid nights and average temperature of 20°C at which germination of conidia takes place which lead to aggravating the disease.

3.2.2.4 SYMPTOMS

Symptoms first develop as off-colored spots on the upper surface of the lowest and oldest leaves, then rapidly spread to cover the entire surface of leaves, stems, and pods with a fine, powdery, bluish-white mildew growth. During advanced stage of the disease, other parts are also infected and diseased portions are fully covered with greyish-white powdery patches and become dirty in appearance. The infected plant part show discolored specks.

The severity of disease depends on the environmental conditions. In sever condition of disease, the infected plant parts become malformed and shriveled whereas entire dried and diseased leaves dropped down from the stem (Gritton and Ebert, 1975). The plant growth and development is adversely affected and it leads to development of immature pods. The quality and quantity is also badly affected (Figure 3.2).

3.2.2.5 MANAGEMENT

Farmers are advised to grow either disease resistant or tolerant varieties. Generally crop rotation is advocated to avoid pathogenic population build up and disease development. The pathogen is obligate in nature therefore its management strategy includes removal and destruction of infected plant parts at early stage of the disease. Once the disease acquires severe condition in field, its management is very difficult. However, fungicides like sulfur dusting (25–30 kg/ha) or two to three foliar spray of Karathane (0.2%) are recommended for its management.

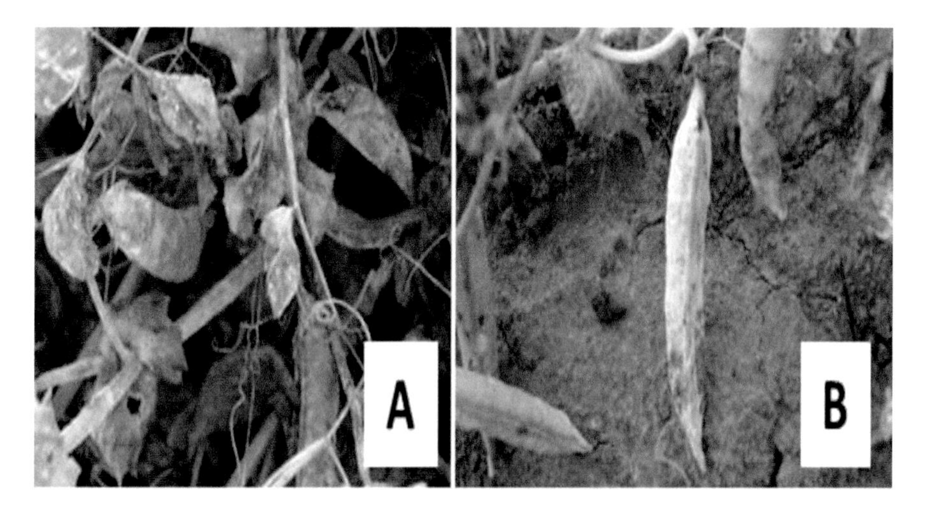

FIGURE 3.2 Powdery mildew symptoms A. on leaves and stem B. on pods.

The management through application of fungicides is an effective way of disease management (Shabeer et al., 2006). Hexaconazole (0.05%) and Penconazole (0.05%) are also highly effective and economical for controlling the disease and for giving better seed yield (Surwase et al., 2009). Some microbial products like Biofungus (*Trichoderma* spp.) and Biofox C (*Fusarium oxysporum*-non pathogenic strain) are also recommended for effective management. In India, microbial product like Kalisena (*Aspergillus niger*), Tri-control (*Trichoderma spp.)*, Ecofit (*Trichoderma virde*), and Basderma (*T. viride*) are advocated for its management.

3.2.3 ROOT ROT COMPLEX

3.2.3.1 INTRODUCTION AND ECONOMIC IMPORTANCE

It is a disease complex in which a group of root invading fungi attack singly or collectively on the outer layer of the root and lower stem or the cortical region. The root rot or damping off disease is considered as one of the most important cause of reduced plant population due to seed rot, root rot, pre-, and post-emergence death of seedlings finally affecting the early crop stand. Almost all the pea varieties get affected with root rot, damping off, and their complex in pea growing areas due to perpetual presence of *Rhizoctonia solani* and, *Fusarium solani,* and *Pythium* spp. (Negi et al., 2005).

3.2.3.2 CAUSAL ORGANISM

Fusarium solani, Rhizoctonia solani, and Pythium aphanidermatum.

The root rot complex is soil-borne in nature and the inoculum perpetuate either in soil or diseased plant debris where it multiplies year after year.

3.2.3.3 SYMPTOMS

The infected seeds start rotting underground before germination due to individual or cumulative effect of attack by different fungi. Effect of the root rot complex generally starts at seedling stage. The affected tissues at collar region develop yellowish-brown, somewhat depressed, and eroded lesions. Later on, it spreads both upward, producing darker brown; sunken; heavily worn lesion leading to girdling of stem, and downward affecting root system. In Fusarium, root rot, pre-emergence damping off: Due to rotting of seeds radical and plumule do not emerge out from infected seeds.

Post-emergence damping off: If seedlings emerge out anyhow above soil surface, they develop water soaked lesions on stem near ground level. Such seedlings collapse at the ground level and die subsequently. Soils with heavy texture and high moisture content enhance the severity of the disease complex (Figure 3.3).

3.2.3.4 MANAGEMENT

Use tolerant or resistant varieties if available. Good agriculture practices (GAP) should be promoted which will discourage the establishment of root rot fungi. To reduce the soil borne initial inoculum collects and destroy the infested crop residues after harvesting, avoid monoculture and adopt crop rotation with non-host crop for 4–5 years. Avoid excessive irrigation in heavy textured soils. Ensure proper drainage to avoid water stagnation in field. Seed treatment with fungicides may certainly protect from seed rot and damping off at initial stage of seedling growth but not beyond that. Seed treatment with Bavistin @ 1 g/kg of seeds or Thiram @ 2 g/kg of seeds is generally adopted. Bio-control agents are effective to manage root rot complex of pea.

FIGURE 3.3 Root rot complex.

3.2.4 WILT

3.2.4.1 INTRODUCTION AND ECONOMIC IMPORTANCE

This disease was first time reported in U. S. A. in 1929 whereas the record of its occurrence in India ways back to 1957 (Sukapure et al. (1957). The wilt disease of pea, caused by a *Fusarium* fungus has been reported from almost all pea-growing countries. This is a very devastating disease as it is associated with significant economic losses. In India, condition early sown crops are more affected by this disease.

3.2.4.2 CAUSAL ORGANISM

Fusarium oxysporium f. sp. *Pisi.*

3.2.4.3 PATHOGEN

The pathogen produces conidia and persists in soil or decaying host tissue for long. Besides, chlamydospores are also produced. Chlamydospores or macroconidia germinate to produce germ tube that infects epidermal as well as cortical cells of the susceptible host. The pathogen colonizes internal plant system by producing hyphae and mycelium and restricts the uptake and translocation of water and minerals in the plant thereby leading to production of disease symptoms. Soil temperature of 23–27°C favors the spread and development of disease.

3.2.4.4 SYMPTOMS

The diseased plant show downward curling of leaves and stipules, thickening of basal internode, brittle leaves, and stem. Slight yellowing of leaves, thickening of lower internode and the stem of affected plant becomes stiff and shriveled to some extent followed by top wilting of the affected plant. Yellowing of the leaves progresses from the base of the plant to the top of the foliage. Unlike the root rot complex, the cortical tissues of the root and stem are generally not attacked and remain healthy. However, saprophytic microbes prefer to grow on roots and stem of wilted plants. At later stage, host plant completely dies and finally rot. Plants often die due to loss of the root system (Haglund, 1984; Tu, 1987a). In roots when cut longitudinally, there is a yellowish, orange color in the vascular tissue, which can extend into the basal area of the stem (Figure 3.4).

3.2.4.5 MANAGEMENT

The cultivation of resistant or tolerant pea varieties is generally recommended. Knowing the race of the region or zone is very important to avoid disease incidence and severity of disease. Early planting should be avoided in severely infested fields to prevent the crop from disease. Sanitation practice is always recommended for its management since sources of disease initiation are infected seed material and contaminated soil on boots and equipment; infected crop debris on trellises and stakes etc. The infected plants debris should be collected properly after harvesting and destroyed carefully. *Fusarium* is a soil-inhabiting fungus, and flourishes in soil with dry conditions. So it is recommended to adopt non-host crop in crop rotation to minimize the disease incidence and yield loss. The fungicides like Bavistin @ 0.1% or Captan @ 2 g or formulation of *Pseudomonas fluorescence* @ 10–15 g as seed treatment is recommended (Dhall, 2017). In severe disease conditions, one or two foliar spray of Bavistin @ 0.2% can be adopted.

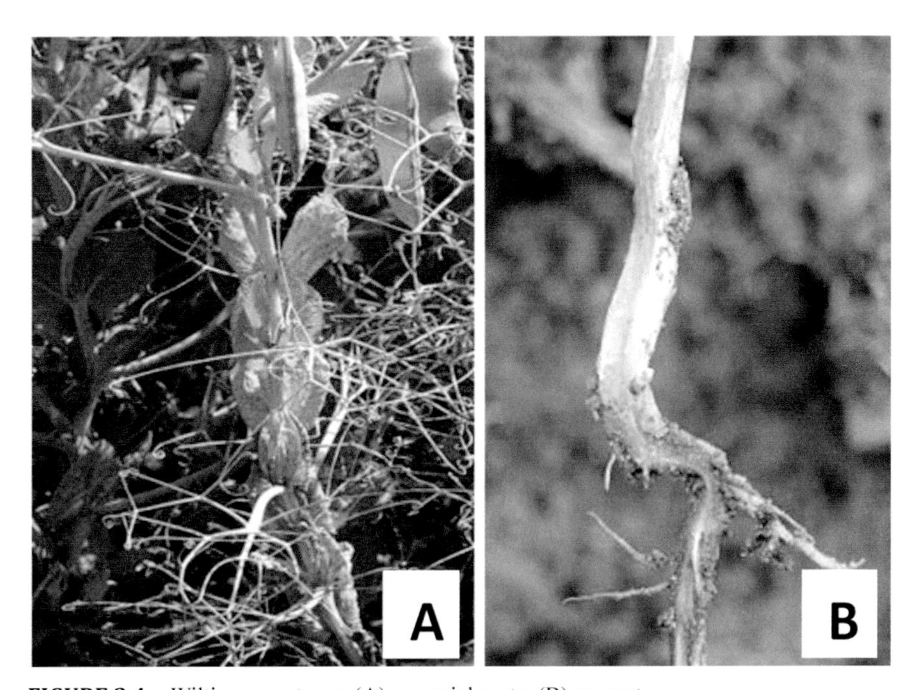

FIGURE 3.4 Wilting symptoms: (A) on aerial parts; (B) on root.

3.2.5 RUST

3.2.5.1 INTRODUCTION AND ECONOMIC IMPORTANCE

The disease is more prevalent in Himalayan region, Tarai region, and Southern part of the country. Upadhyay and Gupta (1998) reported huge yield losses in pea due to pea rust disease in sub-mountainous region of North India. This rust is polycyclic and autoecious in nature (Singh and Tripathi, 2012; Singh et al., 2013), which is associated with huge yield losses within short period of time due to speedy development and spread under favorable conditions. Yield loss was reported 30–40% annually (Singh, 1999; Upadhyay and Singh, 1994).

3.2.5.2 CAUSAL ORGANISM

Uromyces viciae-fabae.

3.2.5.3 PATHOGEN

The rust pathogen is an obligate fungal parasite and autoceious in nature. The pathogen produces different kinds of spores. The infection cycle start with multiplication of aeciospores and causes damage to the crop. The urediospores and teliospores are formed on the surface of host plant and complete its life cycle on the same host. The basidial stage and initiation of primary infection are not fully understood. The disease is favored by high humidity, cloudy and rainy weather condition. Disease development in field is favored between 20 and 22°C (Singh et al., 2013).

3.2.5.4 SYMPTOMS

In Indian conditions, generally two types of symptoms can be seen. Initially round to oval shaped, light yellow to bright orange colored pustules develop on abaxial side of older leaves and form fruiting bodies called aecidia. Under favorable conditions, these pustules act as a source of disease development. Normally all plant parts get infected and the plant appear discolored. Infected parts exhibit restricted growth with shriveled tendrils and immature pods. In severe condition, grain size is reduced significantly which badly affect shine and color of the grain. Besides these, infected plant gets mature earlier with limited flower and pod development (Pfunder and Roy, 2000) (Figure 3.5).

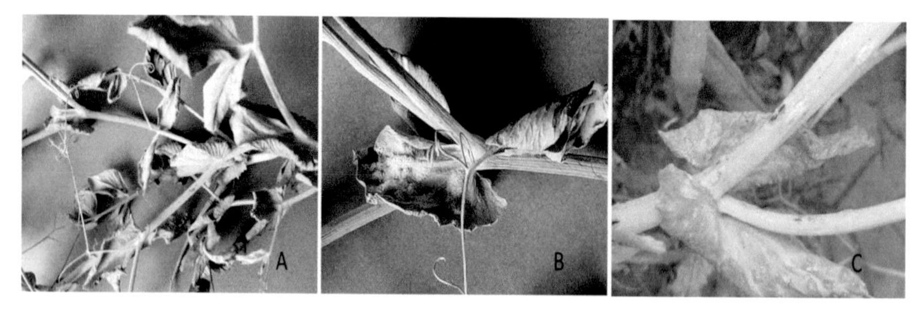

FIGURE 3.5 Pea Rust symptoms on (A) aerial parts, (B) leaf, and (C) stem.

3.2.5.5 MANAGEMENT

The rust pathogen is an obligate parasite and can survive on the susceptible host plant so cultivation of resistant varieties should always be adopted. Crop rotation with non-host crop is also recommended for disease management. Management of disease through application on fungicides *viz.,* Tridemorph (0.1%) and Mancozeb (0.25%) is advocated for its management but once disease acquire sever condition, application of these fungicides become ineffective for its management (Brewer and Larkin, 2005). Besides the application of fungicides, botanicals, and animal origin products are also suggested for its management. In addition to application of non-hazardous products, some microbial-based products of Bacillus and Trichoderma are recommended for its management (Deshmukh et al., 2010; Ragab et al., 2015). In a study conducted by Abhishek and Simon (2017), plant extract Nimbicidin @ 0.3 was also found effective to manage the disease.

3.2.6 DOWNY MILDEW

3.2.6.1 INTRODUCTION AND ECONOMIC IMPORTANCE

Unlike powdery mildew, this disease may occur in severe form only during winters on susceptible to tolerant varieties in North Indian regions of the country, Nepal, and Bangladesh where cool and foggy weather provide conducive environment for the spread and development of downy mildew (Pande et al., 2000). Downy mildew is present during the early part of the growing season and is troublesome in cool and moist season (Hagedorn, 1991).

3.2.6.2 CAUSAL ORGANISM

Peronospors viciae (syn. *P. pisi*).

3.2.6.3 PATHOGEN

Downy mildew disease of pea is caused by a fungus *Peronospora viciae.* The pathogen is obligate parasite of vascular plants. Oospores in the soil can infect pea seedlings systematically. The spores produced in the diseased plants are windblown due to which the disease spread in the field. For the infection of the host plant, it requires wet and cool conditions and night temperature below 10°C and morning time formation promotes disease development.

3.2.6.4 SYMPTOMS

Pathogen is able to infect host at any growth stage but losses is more reported in early stage infection. White fluffy growth is observed underside on leaves at initial infection which exhibits greenish yellow to brown patches on affected portions. In infected plant parts or host plant show retarded plant height and deformed tendrils and pod structure.

3.2.6.5 MANAGEMENT

The approaches for disease management are adoption of resistant cultivars. Follow deep plowing to bury the infected plants in soil. Adopt crop rotation. Foliar application of Dithane M-45 (0.3%), Blitox-50 and Dithane Z-78 (0.3%) or seed treatment with Ridomil MZ 50 WP @ 4 g/kg seed can be followed to manage the disease.

3.2.7 WHITE ROT

3.2.7.1 INTRODUCTION AND ECONOMIC IMPORTANCE

This disease causes comparatively less economic loss if timely managed and it is restricted also to specific areas of Jammu and Kashmir, Tarai region of Uttarakhand and Himachal Pradesh.

3.2.7.2 CAUSAL ORGANISM

Sclerotinia sclerotiorum.

3.2.7.3 SYMPTOMS

At the initial stage of the disease, irregular watery lesions appear on aerial parts of the plant near the soil surface. Later on, white fluffy mycelial growth of the fungus can be easily seen on the infected plant parts when moist and cool weather conditions prevail. Brownish-black to black hard resting structures (sclerotia) is produced within the stem and pods which provide the inoculum for next season. The sclerotial bodies can be easily seen after split open these infected stem and pods. Generally, severity of the disease is higher at flowering and pod formation (Figure 3.6).

FIGURE 3.6 Symptoms of white rot of pea on A. Stem and B. sclerotia inside the stem.

3.2.7.4 *MANAGEMENT*

Either decreasing the chance of buildup of initial inoculum or reducing it if already has built up, growers should avoid to grow pea continuously for several years (monoculture) and pea crop should not also be cultivated if the susceptible crops like brinjal, cauliflower, carrot, and are grown on the same field. The plant debris of the crop should be collected and burnt (Dhall, 2017). Collect and destroy the infected plant residue harvesting of the crop. Maintain proper distance between rows as recommended. Farmers should spray on crop with Bavistin @1 g/l of water as soon as whitish growth appears on aerial parts near soil line. Spray the crop with Dithane M-45 @

2.5 g/l of water at 15–20 days interval which should be started at flowering and continued up to seed formation stage of the crop.

3.3 BACTERIAL DISEASES

3.3.1 BACTERIAL BLIGHT

Bacterial blight is a serious disease of peas throughout the pea growing regions of the world.

3.3.1.1 CAUSAL ORGANISM

Pseudomonas syringae pv. *pisi.*

3.3.1.2 SYMPTOMS

Water-soaked lesions appear initially near the nodes and stipules and are shiny. These lesions spread to other parts like stems, peduncles, and tendrils and pods and later become dark in color. The water-soaked lesions on underside of the leaves and stipules appear dark green to brown on the upper surface. Later the spots become angular with dark margins and light papery in the center. When held up to the light the lesions appear translucent. Watery, dark spots sometimes appear on the seed. The damage is serious when occur at early stage. Seedlings attacked by the bacteria are not able to survive.

3.3.1.3 DISEASE CYCLE

The bacteria are seed-borne in nature. The seed is covered by a dry, white film of bacteria that is invisible with naked eye. The infection in the seed can remain for up to three years. Infection occurs as the plumule contacts the infected seed coat during germination.

Soil moisture is a critical factor in the infection. Optimal temperatures for *P. syringae* pv. *pisi* growth are 26–28°C. The higher the soil moisture, the higher is the rate of infection. Any type of mechanical injury from hail, frost, wind, animals or machinery predisposes the plants to infection. Secondary spread can occurs through rain splash, wind, machinery, animals, and irrigation.

3.3.2 BROWN SPOT

3.3.2.1 CAUSAL ORGANISM

Pseudomonas syringae pv. *syringae.*

3.3.2.2 SYMPTOMS

The symptoms of brown spot can be confused with that of bacterial blight. It is easily distinguishable under laboratory conditions. Initially, the symptoms are small, water-soaked lesions on leaves, petioles, and stems which later on turn tan in color and appear burnt. Infected leaves dry and fall off at later stage. The disease becomes serious at high humidity.

3.3.2.3 DISEASE CYCLE

The disease cycle is similar to *P. syringae* pv. *pisi* and the difference is only that *P. syringae* pv. *syringae* is both seed or soil-borne. It does not survive for long on the surface of seed coat but survive for long in soil. The temperature optimum for this pathogen is 24°C.

3.3.2.4 MANAGEMENT

Use of disease-free seed to avoid the disease. Plant resistant cultivars. Machinery should be disinfected when moved from one field to another.

3.4 VIRAL DISEASES

A number of viruses are known to cause affect field pea. The name some important viruses causing disease in field pea are PEMV-pea enation mosaic virus, AMV-alfalfa mosaic virus, CMV-cucumber mosaic virus, PSbMV-pea seed-borne mosaic virus, BYMV-Bean yellow mosaic virus, BLRV-bean leaf roll virus, BWYV-beet western yellow virus, SCSV-subterranean clover stunt virus, TNV-top necrosis virus. However there are only few that are known to occur in Indian conditions.

3.4.1 PEA ENATION MOSAIC

Pea enation mosaic disease is worldwide in occurrence in all pea-growing countries. It is the only viral disease extensively studied in India. Virus is transmitted by aphid *A. pisum, M euphorbiae,* and *M. persicae* (Smith, 2012), which may also infect field Peas, faba beans, and soybean.

3.4.1.1 CAUSAL AGENT

Pea Enation Mosaic Virus (PEMV-1 & 2)

This disease is caused by two interdependent viruses, PEMV-1 and PEMV-2, which are Enamo virus and Umbra virus, respectively.

3.4.1.2 SYMPTOMS

Symptoms are vein-clearing, mosaic, stunted growth of plant and enation on lower surface of the pea leaves. Infected leaf has show yellowish-white spots and become crinkled which develops small cracks, eruptions, and outgrowths (enations) on the veins apparently visible on undersurface of the leaves. Distorted leaves of the infected plant are unable to perform their normal functions and hence, plant does not attain its proper growth and remain stunted. These infected, stunted plants produce abnormal and infertile flowers in bunches. Pods are generally distorted, smaller, and shriveled with few seeds if developed.

1. **Top Necrosis Virus (TNV):** The infected plants exhibited top necrosis, and greyish-brown discoloration of leaves, petioles, and stems. Severely infected plants died prematurely.
2. **Alfalfa Mosaic Virus (AMV):** Chlorosis in new shoots, necrotic spots on leaves and streaking, necrosis of new or old leaves.
3. **Cucumber Mosaic Virus (CMV):** Chlorosis and yellow tips.
4. **Bean Yellow Mosaic Virus (BYMV):** Symptoms can be variable. In some cases the virus may be symptomless or there can be mosaic symptoms, leaf deformation, severe stunting of the plant and premature senescence.
5. **Pea seed-Borne Mosaic Virus (PSbMV):** In some cases, the virus may show no symptoms or there may be downward curling of leaves and mild to severe plant stunting. The infected plants may produce seeds having brown marks or ring spots.

6. **Beet Western Yellows Virus (BWYV):** Stunting and yellowing of plants.

3.4.1.3 COMMON PRACTICES FOR MANAGEMENT OF VIRAL DISEASES

- Growing virus resistant varieties if available.
- Sowing virus free seeds.
- Remove weeds and avoid self-sown pulses as these can be a source of virus and aphids.
- Monitoring aphids population in the crop particularly at the early stage of growth. If aphids population is high at early stage then it spraying of insecticide is recommended.
- Maintain optimum plant stands with minimal bare.
- Follow crop rotation with cereals.
- Remove diseased plants.

3.5 NEMATODE DISEASES

Pea is reported to be infested with nematodes namely *Heterodera* spp. (cyst), *Meloidogyne* spp. (root-knot) and *Rotylenchulus* spp. (reniform), but *Heterodera goettingiana* may be considered as devastating (Pande et al., 2000). Among various nematodes, root-knot nematode (RKN) cause severe yield losses of up to 20%~33% (Sharma et al., 2006). It affects the crop directly and indirectly by interaction with various soil-borne fungi, bacteria, and viruses.

3.5.1 CAUSAL ORGANISM

The most predominant species of RKNs are *Meloidogyne incognita*, *M. javanica*, *M. arenaria* and *M. hapla*. *Meloidogyne* spp. are sedentary endoparasites and among the most damaging agricultural pests. The female lay eggs in the infested roots and second stage juvenile (J2) hatched from these eggs penetrate the roots and cause infection. The infection induces giant cells in susceptible hosts. From these juvenile, globose adult female develops which lay eggs parthenogenetically in a gelatinous matrix, secreted by their rectal glands, on the surface of galled root. Males are vermiform, wander in soil and are non parasitic (Perry et al., 2009; Jepson, 1987). The whole life cycle is completed in about in about 25 days at 25–30°C, which is optimum for most species.

3.5.2 SYMPTOMS

Typical symptoms of nematode injury can be seen on both aboveground and below ground plant parts. Foliar symptoms of nematode is not distinguishable, although infestation of roots by the nematodes generally appears as yellowing and unthriftiness of leaves, reduced flowering, stunted growth and smaller pods and are similar to those induced by nutrient deficiency and water stress. Damage is most pronounced when infection occurs in the early stage of plant growth. Plants exhibiting stunted or decline symptoms usually occur in patches of non-uniform growth rather than as an overall decline of plants within an entire field. In this case, aboveground symptoms will not always be readily apparent early within crop development, but with time and reduction in root system size and function, symptoms become more pronounced and diagnostic. Below ground symptoms are typical where characteristic 'root gall' or 'knotted roots' are produced. Generally, in the initial stage of plant growth, galls are small. But as the nematode completes one life cycle, reinfection by second generation J-2 leads to formation of more galls, the adjacent galls coalesce to form bigger compound galls, which is easily visible at later stages of crop growth. The intensity of galling and size of the galls are variable depending on RKNs species, nematode population, susceptibility of crop, and age of crop.

3.5.3 MANAGEMENT

The use of cultural control methods to manage RKNs is the most environmentally sustainable and potentially most successful method for limiting RKNs damage. The deep summer plowing at 10–15 days interval during May–June are very effective and killing most of the J-2 in soil due to desiccation and exposure to high temperature. This practice is successfully adapted in tropical and subtropical areas where temperature goes very high during summer. Pea fields infested with RKNs can potentially be planted to a non-host crop such cereal crops like wheat, maize, rice, sorghum, pearl millet, or antagonistic crops like onion, garlic, and sesamum. Prevent the introduction of nematodes in clean fields by thoroughly cleaning equipment between fields. Provide adequate fertilizer and irrigation for crop development to reduce the impact of nematode feeding. Sprinkler irrigation tends to be more effective than furrow irrigation for supplying even water to nematode-damaged plants. Control weeds that may serve as alternate hosts of nematodes. Promptly incorporate crop residues after harvest to limit nematode reproduction (Sasser and Carte, 1985)

RKNs are very difficult to manage because they are soil-borne pathogens with a wide host range and it requires applications of large amounts of chemicals with specialized equipment. Fumigants (such as 1, 3-dichloropropene, methyl bromide and dazomet) are commonly applied as pre-plant treatments to reduce nematode numbers, but they must thoroughly penetrate large soil volumes to be effective. In addition to broad-spectrum fumigants, nervous system toxins (including oxamyl and fenamiphos) have been shown to be extremely effective for controlling RKNs. Currently, these chemicals are the most economically feasible control method for RKNs. Because they are not toxic to plants, they are the only chemical options for established plants.

Organic amendments for root knot nematode infested soil are highly effective due to their nematotoxicity, imparting tolerance to plant and encouraging buildup of natural antagonists. However, requirement of very high dosages and cost considerations deter their use at field scale. Nevertheless, products like neem cake can be used for small fields (Viaene, 1998).

Management measures employing organisms antagonistic to RKNs have been attempted by many researchers. The most commonly used biological control agents are fungi and bacteria. There are many kinds of nematophagous (nematode-feeding) fungi. Oviparasitic fungi, *Paecilomyces lilacinus* and *Pochonia chlamydosporia* and a bacteria parasite, *Pasteuria penetrans* are promising biocontrol agents of root knot nematodes. Some fungi use mycelial traps or sticky spores to capture nematodes, for example, *Arthrobotrys* spp. and *Monacrosporium* spp. Besides, nematode antagonistic bacterium *Pseudomonas fiuorescens* and fungi like *Trichoderma harzianum* and *T. Viride* has also been found effective.

KEYWORDS

- **alfalfa mosaic virus**
- **bean yellow mosaic virus**
- **beet western yellows virus**
- **cucumber mosaic virus**
- **good agriculture practices**
- **pea enation mosaic virus**

REFERENCES

Abhishek, S., & Simon, S., (2017). Eco-friendly management of powdery mildew and rust of garden pea (*Pisum sativum* L.). *Journal of Pharmacognosy and Phytochemistry*, *6*(5), 90–93.

Beckman, C. H., (1987). *The Nature of Wilt Diseases of Plants*. St. Paul, MN: American Phytopathological Society Press.

Bretag, T. W., Keane, P. J., & Price, T. V., (2006). The epidemiology and control of ascochyta blight in field peas: A review. *Aust. J. Agr. Res.*, *57*, 883–902. doi: 10.1071/AR05222.

Caver, T. L. W., & Jones, S. W., (1988). Colony development by *Erysiphe graminis* f. sp. hordei on isolated epidermis of barley coleoptile incubated under continuous light or short-day conditions. *Transactions of the British Mycological Society*, *90*, 114–116.

Charchar, M., & Kraft, J. M., (1989). Response of near-isogenic pea cultivars to infection by *Fusarium oxysporum* f. sp. pisl races 1 and 5. *Can. J. Plant Sci.*, *69*, 1335–1346.

Deshmukh, A. J., Mehta, B. P., & Patil, V. A., (2010). In vitro evaluation of some known bioagents to control *Collectotrichum gloeosporioides* Penz, and Sacc, causing Anthracnose of Indian bean. *Inter. J. Pharma. & Bio. Sci.*, *1*(2), 1–6.

Dhall, R. K., (2017). Pea cultivation. *Bulletin No. PAU/2017/Elec/FB/E/29* (1st edn.). ISBN: 978-93-86267-37-5.

Dixon, G. R., (1987). Powdery mildew of vegetables and allied crops. In: Spencer, D. M., (ed.), *The Powdery Mildew* (p. 565). Acad. Press.

Elgersma, D. M., MacHardy, W. E., & Beckman, C. H., (1972). Growth and distribution of *Fusarium oxysporum* f. sp. lycopersici in nearisogenic lines of tomato resistant or susceptible to wilt. *Phytopathology*, *62*, 1232–1233.

Gritton, E. T., & Ebert, R. D., (1975). Interaction of planting date and powdery mildew on pea plant performance. *Journal of American Society of Horticultural Science*, *100*, 137–142.

Hagedorn, D. J., (1991). *Handbook of Pea Diseases*. Cooperative Extension Publications, University of Wisconsin Extension.

Haglund, W. A., (1984). *Fusarium wilts*. In: Hagedorn, D. J., (ed.), *Compendium of Pea Diseases* (pp. 22–25). The American phytopathological society.

Horticultural Statistics at a Glance, (2017). *Area, Production and Yield of Horticultural Crops* (p. 150). Ministry of agriculture & farmers welfare, Government of India.

Jepson, S. B., (1987). *Identification of Root-Knot Nematodes (Meloidogyne Species)*. CAB International, Wallingford, UK.

Kushwaha, C., Chand, R., & Srivastava, C. P., (2006). Role of aeciospores in outbreak of pea (*Pisum sativum*) rust (*Uromyces fabae*). *Eur. J. Plant Path.*, *115*, 323–330.

Lawyer, A. S., (1984). Disease caused by *Ascochyta* spp. In: Hagedorn, D. J., (ed.), *Compendium of Pea Diseases* (pp. 11–15). The American Phytopathological Society: Minnosota, USA.

Libert, M. A. (1837). *Plantae Cryptogamae, quas in Arduenna Collegit.* Fasc.1(1), 59pp.

Linford, M. B., & Sprague, R., (1927). Species of Ascochyta parasitic on the pea. *Phytopathology*, *17*, 381–398.

Linford, M. B., (1929). Pea disease in the USA in 1928. *Pl. Dis. Rept. Supple*, *67*, 14.

Liu, N., Xu, S., Yao, X., Zhang, G., Mao, W., Hu, Q., Feng, Z., & Gong, Y., (2016). Studies on the control of ascochyta blight in field peas (*Pisum sativum* L.) caused by *Ascochyta pinodes* in Zhejiang province, China. *Front. Microbiol.*, *7*, 481. doi: 10.3389/fmicb.2016.00481.

Maheshwari, S. K., Jhooty, J. S., & Gupta, J. S., (1983). Survey of wilt and root rot of complex of pea in Northern India and the assessment of losses. *Agricultural Science Digest, India*, *3*(3/4), 139–141.

Markell, S., Pasche, J., & Lyndon, P. (2016). *Pea Disease Diagnostic Series*. NDSU Extension Services. https://www. ag. ndsu. edu/publications/crops/pea-disease-diagnostic-series/ pp1790. pdf (accessed on 13 January 2020).

Negi, D. S., Sharma, P. K., & Gupta, R. K., (2014). Management of root-rot complex disease and assessment of plant growth promoting characters in vegetable pea with native and commercial antagonistics through seed biopriming. *International Journal of Recent Scientific Research*, *5*(8), 1416–1421.

Nisar, M., Ghafoor, A., Khan, M. R., &. Qureshi, A. S., (2006). Screening of *Pisum sativum* L. Germplasm against *Erysiphe pisi* Syd. *Acta Biologica Cracoviensia Series Botanica*, *48*(2), 33–37.

Pande, S., Sharma, S. B., & Ramakrishna, A., (2000). *Biotic stresses affecting legumes reduction in the Indo-Gangetic plain.* In: *Legumes in Rice and Wheat Cropping Systems of the Indo-Gangetic Plain-Constraints and Opportunities* (pp. 129–155). International Crops Research Institute for the Semi-Arid Tropics, Patancheru, Andhra Pradesh, India. ISBN: 92-9066-418-5.

Peever, T. L., Barve, M. P., Stone, L. J., & Kaiser, W. J., (2007). Evolutionary relationships among Ascochyta species infecting wild and cultivated hosts in the legume tribe's cicereae and vicieae. *Mycologia*, *99*, 59–77.

Perry, R. N., Moens, M., & Starr, F. J., (2009). *Root-Knot Nematodes*. CAB International, Wallingford, UK.

Pfunder, M., & Roy, B., (2000). Pollinator-mediated interactions between a pathogenic fungus, *Uromyces pisi* (Pucciniaceae), and its host plant, *Euphorbia cyparissias* (Euphorbiaceae). *Amer. J. of Bot.*, *87*(1), 48–55.

Ragab, M. M. M., Abada, K. A., Abd-El-Moneim, Maisa, L., & Abo-Shosha, Yosra, Z., (2015). Effect of different mixtures of some bioagents and *Rhizobium phaseoli* on bean damping-off under field condition. *Inter. J. of Sci. and Eng. Res.*, *6*(7), 1009–1106.

Rathi, A. S., & Tripathi, N. N., (1994). Assessment of growth reduction and yield losses in pea (*Pisum sativum*) due to powdery mildew disease caused by *E. Polygoni* DC. *Crop Research (Hissar)*, *8*, 371–376.

Sasser, J. N., & Carter, C. C., (1985). *An Advanced Treatise on Meloidogyne*: *Biology and Control* (Vol. I,). Department of Plant Pathology and Genetics, North Carolina State University and the United States Agency for International Development, Raleigh, NC.

Shabeer, A., Irfan, U., & Attauddin, (2006). Powdery mildew and its yield loss assessment in pea under natural field conditions. *Sarhad. J. Agric.*, *22*(1), 163–168.

Sharma, A., Haseeb, A., & Abuzar, S., (2006). Screening of field pea (*Pisum sativum*) selections for their reactions to root-knot nematode (*Meloidogyne incognita*). *Journal of Zhejiang University Science B.*, *7*(3), 209–214.

Singh, D., Kumar, A., Singh, A. K., Prajapati, C. R., & Tripathi, H. S., (2013). Studies on survivability of pea rust caused by *Uromyces viciae- fabae* (Pers.) de Bary in Tarai region of Uttarakhand (India). *African Journal of Agricultural Research*, *8*(17), 1617–1622.

Singh, R. S., (1999). *Plant Diseases* (p. 686). Oxford and IBH, New Delhi.

Singh, S. K., & Tripathi, H. S., (2012). Management of rust (*Uromyces fabae*) of pea through fungicides. *Pantnagar Journal of Research*, *10*(1), 51–55.

Sukapure, R. S., Bhide, V. P., & Patel, M. K., (1957). *Fusariums* wilt of garden pea (*Pisum sativum* L.) in Bomby State. *Indian Phytopath.*, *10*, 11–17.

Surwase, A. G., Badgire, D. R., & Suryawanshi, A. P., (2009). Management of pea powdery mildew by fungicides, botanicals, and bio-agents. *Ann. Pl. Protec. Sci.*, *17*(2), 384–388.

Tivoli, B., Beasse, C., Lemarchand, E., & Masson, E., (1996). Effect of Ascochyta blight (*Mycosphaerella pinodes*) on yield components of single pea (*Pisum sativum*) plants under field conditions. *Ann. Appl. Biol., 129*, 207–216. doi: 10.1111/j.1744–7348.1996. tb05745. x.

Tu, J. C., (1987). Etiology and control of *fiisarium* wilt (*Fusarikm oxyspomrn*) and rom rot (*Fusarium salani*) of green pea (*Pisurn sativum*). Mededelingen-van-de-Faculteit – Landbouwwetenschappen. *Rijksuniversiteit. Gent.52*(8), 15–823.

Upadhyay, A. L., & Singh, V. K., (1994). Performance of pea varieties/lines against powdery mildew and rust. *Annals of Agricultural Research*, *7*, 92–93.

Viaene, N. M., (1998). Management of *Meloidogyne hapla* on lettuce in organic soil with Sudan grass as a cover crop. *Plant Disease, 82*, 945–952.

Warkentin, T. D., Rashid, K. Y., & Xue, A. G., (1996). Fungicidal control of powdery mildew in field pea. *Can J. Plant Sci., 76*, 933–935.

Xue, A. G., Warkentin, T. D., & Kenaschuk, E. O., (1996). "Mycosphaerella blight of field pea-potential damage and fungicide control." In: *Proceedings of Manitoba Agri-Forum, Winnipeg (Manitoba)* (pp. 5–6).

CHAPTER 4

Important Diseases of Green Gram (*Vigna radiata* L. Wilczek) and Their Management

JAI SINGH,[1] K. K. MISHRA,[2] and A. K. CHAUBEY[1]

[1]*JNKVV Krishi Vigyan Kendra, Sidhi–48661, Madhya Pradesh, India*

[2]*JNKVV, Zonal Agricultural Research Station, Powarkheda, Hoshangabad, Madhya Pradesh, India*

Pulses had always been the integral part of food since the ancient times (Achaya, 1998). Searches of older documents have revealed that pulses had to have grown in the Indian sub-continent for more than 6000 years. References of most of the pulses cultivated today in India were basically from ancient scriptures (Nene, 2000).

Pulses happen to be the major source of dietary protein for vast population of India. When supplemented with cereals in diet pulses provide a desired mix of vegetable protein of high biological value. Pulses also used to be excellent feed and fodder for domestic animals. The leguminous crops are also bestowed to have addition benefits to the soil by improving organic matter (leaf drop) and fixing atmospheric nitrogen, carbon sequestration and other physiochemical properties as well. Since pulses have resilient to water requirement and temperature regime, they found to be the indispensable component of sustainable crop production system in the arid and semi arid zones. Pulses are often standing good for crop diversification and intensification. The pulses are more significant that to when impending risk of non-availability the nitrogen as fertilizer from petroleum sources.

Pulses occupy 68.32 million ha area and contribute 57.51 million tones to the world food production (Singh et al., 2012). India is the largest producer of pulses in the world, with 24% share in world production. The important pulses are Chickpea (48%), Pigeon pea (15%), Mungbean

(7%), Urdbean (7%), Lentil (5%), and Field pea (5%) (Ali and Gupta, 2012).

Mungbean (*Vigna radiata* (L.) Wilczek) commonly known as green gram or golden gram is an extensively grown pulse crop. Due to suitability to tropical and subtropical region, mungbean is able to withstand temperature as high as 40°C. It is considered as a "poor men's protein" (Mian, 1976). Apart from 26% protein, it also contains 51% carbohydrate, 10% moisture, 4% minerals, and 3% vitamins (Khan, 1981). It is widely cultivated throughout Asia including India, Pakistan, Bangladesh, Sri Lanka, Thailand, Cambodia, Vietnam, Indonesia, Malaysia, and South China (Yadav et al., 2014). It is also grown in parts of East and Central Africa, Australia, United State of America and West Indies. In India, it is grown in almost all parts of the country during summer and rainy season. India is the largest producer of mungbean where it is third most important pulse crop with an area of approximately 3.43 million hectares (about 15% of the national pulse crop area), production 1.71 million tons of grain with productivity of 498 kg/ha (Anonymous, 2012). To meet global demand, it is imperative to improve the current average global productivity (400 kg ha^{-1} as well as to expand the crop into new regions (Nair et al., 2012).

The yield of mungbean is affected by several biotic and abiotic factors (Grewal, 1988). Among the biotic factors, Powdery mildew, Mungbean yellow mosaic virus (MYMV), Cercospora leaf spot (CLS), Anthracnose, Dry root rot, Leaf crinkle virus, Web blight, Rust, etc., are of prime importance in reducing crop yield. The account of diseases and their management practices are mentioned below.

4.1 POWDERY MILDEW

4.1.1 INTRODUCTION

Powdery mildew is a major problem to mungbean cultivation. Recently it's become a limiting factor in Rabi sown mungbean crop in Odisha, Andhra Pradesh, Karnataka, and Tamil Nadu. The disease has been reported from Australia, India, Philippines, Sri Lanka, Taiwan, Colombia, Ethiopia, Thailand, Korea, and the USA. Disease causes 9–50% yield losses at farmers field (AVRDC, 1982).

4.1.2 SYMPTOMS

The disease appears on all above ground part of plant. Initially white powdery patches appear on leaves and other green parts which later become dull colored. These patches gradually increase in size and become circular covering the lower surface also. In case of severe infection, both the surfaces of the leaves are completely covered by whitish powdery growth. The powdery mass consisting of mycelium and conidia eventually turn dirty white. Severely affected parts get shriveled and distorted. In severe infections, foliage becomes yellow causing premature defoliation. The disease also creates forced maturity of the infected plants which results in heavy yield losses.

4.1.3 PATHOGEN

The powdery mildew incited by obligate parasite *Erysiphae polygoni*, the pathogen fungus is ectophytic, spreading on the surface of the leaf, sending haustoria into the epidermal cells to obtain nutrients. Conidiophores arise vertically from the leaf surface, bearing conidia in short chains. Conidia are hyaline, thin walled, elliptical or barrel shaped or cylindrical and single celled. Later in the season, cleistothecia appear as minute, black, globose structures with myceloid appendages. Each cleistothecium contains 4–8 asci and each ascus contains 3–8 ascospores which are elliptical, hyaline, and single celled.

4.1.4 PREDISPOSING CONDITIONS

The pathogen favored by dry weather, temperature range between 22 to 26°C and relative humidity (RH) of 80–85%. Disease incidence was higher on older plant than the 40 days old plants. The disease was more severe in closely planted crop during July than in those with widely spaced late sown crop (August).

4.1.5 DISEASE CYCLE

The fungus is an obligate parasite and survives as cleistothecia in the infected plant debris. Primary infection is usually from ascospores which develop in cleistothecia. The pathogen has a wide host range and survives in oidial form on various hosts viz. *Vigna mungo*, *Vicia faba*, *Lathyrus sativa*, *Vigna ungulate,* etc., in offseason which may also act as reservoirs of the inoculums. The secondary spread is carried out by the air-borne conidia. Rain splash also helps in the spread of the disease.

4.1.6 MANAGEMENT

Destruction of plant debris can minimize the survival habitat of pathogen. The growing powdery mildew resistant varieties/genotype has been considered to be the most effective method of reducing disease incidence, such as TARM-1, TARM-2, TARM-18, COG-4, Pusa-105, Kamdev ML-131, ML-1299 and LGG-460 (Akhtar et al., 2014; Yadav et al., 2014). Disease incidence can be minimized by early sowing with wider spacing. Two foliar spray of Carbendazim (12%) + Mancozeb (63%) –75WP @ 0.2% (Yadav et al., 2014) or Karathane (0.1%) (Suryawanshi et al., 2009) or Wettable sulfur 1.5 kg or Benlet (0.05%) or Tridemorph @ 0.05% or difencoazole@ 0.1% (Rakhonde et al., 2011) or propiconazole @ 0.1% (Akhtar et al., 2014) at the initiation of disease and repeat 15 days later, has been found effective to control the disease under field condition. Spray with NSKE@ 50 g/l or Neem oil 3000 ppm @ 20 ml/l twice at 10 days interval from early onset of disease.

4.2 ANTHRACNOSE

4.2.1 INTRODUCTION

Anthracnose incited by *Colletotricum truncatum* (Schw.) Andrus and Moore is the very important seed-borne fungal disease of mungbean resulted intremendous reduction in seed quality and yield every year with varying magnitude. The disease causes both qualitative and quantitative losses (Sharma et al., 1971). Losses in yield due to anthracnose have been estimated to be in the tune of 24 to 67% (Deeksha and Tripathi, 2002; Shukla et al., 2014). The yield losses caused by mungbean anthracnose are often proportional to the disease severity and vary remarkably depending on the stage of infection of crop, cultivar grown, and environmental factors.

4.2.2 SYMPTOMS

The anthracnose may appear on all above ground part of the plant and at any stage. Often symptom occurs on the ventral surface of the leaf than on the dorsal surface and also occurs son petioles, stem, and pods. Initially water soaked lesions seen on leaves and pods which later become circular, black, sunken spots with dark center and bright red orange margins on leaves and pods. In severe cases, the affected parts wither off. Seedlings get blighted if infection occurs soon after seedling emergence.

4.2.3 PATHOGEN

Infected Moong bean seeds were found to be associated with different species of *Collectotrichum* viz. *Collectotrichum lindemuthianum* Sacc & Marga) Bri & Cav (perfect state *Glomerella lindemuthionum* (Sacc & Margn) Shear); *C. dematium* (Pres. ex Fr.) Grov., *C. trucatum* (Sehw.) Andrus & Moor and *C. graminicola* (Ces. Wilson). Among these, *C. linemuthionum* and *C. capsici* were found wide spread and causes severe infections under conducive environmental conditions.

The mycelium of pathogen is septate, hyaline, and branched. Conidia are produced in acervuli, arise from the stroma beneath the epidermis which later rupture to become erumpent. A few dark colored, septate setae are seen in the acervulus. The conidiophores are hyaline and short and bear oblong or cylindrical, hyaline, thin walled, single celled conidia with oil globules. The sexual stage of the fungus produces perithecia with limited number of asci, which contain typically 8 ascospores either one or two celled with a central oil globule.

4.2.4 PREDISPOSING CONDITIONS

The disease incidence was correlated positively with RH and rainfall, and negatively with maximum and minimum temperature. The disease score were highest at 91–100% RH and 26–29°C, with low rainfall (2.5–53.5 mm). Early sowing (23 June) were more vulnerable than the late sowing (Thakur and Khare, 1992). Intermittent rains at frequent intervals favors the epidemic development of the disease.

4.2.5 DISEASE CYCLE

The fungus is seed-borne and cause primary infection. It also lives in the infected plant debris in soil. The secondary spread by air borne conidia produced on infected plant parts. Spattering rains splash associated with wind current are the main source for the local dissemination of the pathogen.

The pathogen infects several leguminous hosts which may also act as reservoirs of the primary inoculums.

4.2.6 MANAGEMENT

Crop debris removal from the field after harvesting the crop. Hot water treatment at 58°C for 15 minutes proved effective to minimize the seed borne infection and increase proportion of seed germination. Mungbean genotypes/cultivars viz., TM-92-2, TARM-18, MLTG-9 were showed highly resistance reaction against anthracnose disease (Rathaiah and Sharma, 2004; Kulkarni et al., 2009). Seed treatment plays an important role in reducing disease to minimize early establishment of the pathogens. Carbendazim 50 WP @ 2 g/kg seed or Thiram+ Carbendazim (2:1) @ 3 g/kg of seed was found to be effective in eliminating the seed borne infection. Shukla et al. (2014) proved that, two sprays of carbendazim (0.1%) are sufficient to manage the disease and realize good yields. Foliar spray of Mancozeb (@ 0.2%) soon as the symptoms appeared and second spray after 15 days of the first spray is also effective to reduce disease incidence.

4.3 CERCOSPORA LEAF SPOT (CLS)

4.3.1 INTRODUCTION

The CLS occurrence was for the first time reported from Delhi, India (Munjal et al., 1960). Now the disease is prevalent in all part of the country and causes severe leaf spotting and defoliation. Asian Vegetable Research and Development Center, Taiwan considered this disease as one of the most serious diseases since it has been reported to occur there in 1973 (Mew et al., 1975).

Yield losses due to this disease have been observed up to 50–70% (Lal et al., 2001; Chand et al., 2012). Distribution of the disease has been reported from Indonesia, India, Taiwan, Philippines, Thailand, Bangladesh, and west Malaysia (Pandey et al., 2009). In India, disease occurs in most of the mungbean growing states of country particularly, Punjab, Haryana, Uttar Pradesh, Madhya Pradesh, Rajasthan, Himachal Pradesh, Odisha, Assam, and Maharashtra.

4.3.2 SYMPTOMS

Initially disease appears on the lower leaves, as water soaked spot and then spread all over the areal parts of the plant. The characteristic symptom appears as small leaf spots produced numerous in numbers with pale brown center and reddish brown margin. Similar spots may also occur on branches, petioles, and pods. Under congenial environmental conditions, the spots increase in size and if it is at the time of flowering and pod formation may lead to defoliation. Sometimes the leaves may become malformed and wrinkled. Maturity is delayed in the diseased plants resulting poor pod formation and yield.

On the pods symptoms rage from few spot to solid blacking and killing of the entire pod and seeds within the infected pod become shriveled, remain small, and darker in colors (Poehlman, 1991). Five species of Cercospora may have reported to infect mungbean with slight variation in their symptoms.

C. cruenta infected mungbean produces typical spots around flowering stage. The spots on leaves are usually angular being limited by veins, large in size (5–7 mm) than those produced by *C. cenescens* and have tan to grey center surrounded by prominent reddish purple border.

C. canescens produced numerous, small (1–3 mm), circular to semi-circular pale brown spots on leaves with reddish brown border. With the advancing age and the spore formation, the spots assumes silvery to grey color in center and the spots are mainly form on leaves.

C. kikuchii produced sub circular to irregular, pale brown with tan to grey centers and measure 3–6 mm in diameter. The spots are elliptical on stem and pods. *C. dolichi* is prominent on Rabi mungbean crops and develops reddish brown blotches. *C. corocollae* has been found attaching on mungbean.

4.3.3 PATHOGEN

Among these five species of Cercospora viz., *C. cruenta* Sacc (Perfect state: *Mycospharrella cruenta* Latham), *C. canescens* Ell. & Martin, *C. kikuchii* Matsumato & Tomoyassun, *C. dolichi* Ell. & Evt. and *C. corocollae* (Speg.) former two species are frequently observed on mungbean. The fungus produces clusters of dark brown septate conidiophores. The conidia are linear, hyaline, thin walled and 5–6 septate. The conidiophores of *C. canescens* is long, multiseptate, usually straight or slightly curved, light to dark brown, 20–250 x 4.5–5.0 mm; conidia hyline, straight or curved, multiseptate 50–225 x 3.5–5 mm. The conidiophores of *C. cruenta* are short, aseptate, light brown, straight, and rarely branched 10–77 x 3–5 mm in size, conidia subhyline or pale olivaceous, multiseptate, slightly curved, 30–150 x 3–5 mm. Conidiophores are pale olivaceous brown, sparingly septate, straight to curved, 10–45 x 4.5–5.5 mm in *C. dolichi*; conidia olive brown, obclavate, cylindrical, multiseptate, 40–100 x 3.7–4.5 mm. In *C. kikuchii*, conidiophores occur in dense fascicles, medium dark brown, multiseptate, 45–220 x 4–6 mm, conidia hyaline, acicular, multiseptate, 150–300 x 3.7–4.5 mm.

4.3.4 PREDISPOSING CONDITIONS

CLS is most prevalent in rainy season and can cause heavy defoliation in severe conditions on mungbean especially at optimum temperature range 25–30°C with high RH viz.90–100%. Symptoms of this disease appear mainly on leaves but may also occur on stem, branches, and pods. Hossain et al. (2011) described that the disease was mainly observed when crop was at flowering, pod formation and ripening stage. The severity of the disease depends on the humidity, and temperature regime coincided with virulent isolate.

4.3.5 DISEASE CYCLE

The fungus survives on diseased plant debris and on seeds. Infected seed produce seedlings with cotyledonary lesions. Sporulation on the cotyledonary lesions provides primary inoculums for young leaves infection. In humid weather, sporulation is profuse. Conidia disseminated by wind and splash off rain to other leaves and stems and initiate secondary infections that again produce conidia which infect other leaves, stems, and pods during warm and wet conditions. Peak dispersal of conidia occur at dew-off in the morning and at the onset of rainfall.

4.3.6 MANAGEMENT

Different approaches such as sanitation, tillage operations, crop rotation, destruction of infected crop debris, removal of collateral hosts in the vicinity of the crop, host plant resistance, fungicides, and different plant extracts are used to managed CLS of mungbean (Hanson and Panella, 2003; Khan and Smith, 2005; Khan, 2008; Dubey and Singh, 2013).

Mungbean entriesML-5, ML-65, ML-131, ML-392, ML-325, ML-406, Ml-408, ML-537, LGG-460 and ML-1299 showed resistance response to CLS (Mathur et al., 1990; Yadav et al., 2014; Akhtar et al., 2014). Similarly, Haque et al. (1997) reported that 12 genotypes (NH-98, 98-cmg-003, c2/94-4-42, NM-1, NM-2, 98 cmg-018, BRM-188, Co-3, Basanti, PDM-11, BARI mung-2 and VC-3960-88) were found highly resistant to CLS. Agronomic practices, i.e., intercrop the mungbean with tall growing cereals and millets and mulching reduces the disease incidence resulting in increase yield. Two foliar spray of Difenconazole (0.0125%) or Carbendazim (12%) + Mancozeb (63%)–75WP (0.0.2%) or Hexaconazole (5%) + Captan (70%) – 75 WP (0.05%) or Radomil Gold (0.1%) or Propiconazole (0.1%) are effective for minimizing CLS incidence, severity, and increasing yield of mungbean (Kapadiya and Dhruj, 1999; Akhtar et al., 2014; Shahbaz et al., 2014; Yadav et al., 2014). Many researchers diseases (Singh and Dwivedi, 1987; Tariq and Magee, 1990; Lakshmanan et al., 1990). It has been observed that neem leaf extract resulted significant reduction of CLS of mungbean (Uddin et al., 2013).

4.4 RUST

4.4.1 INTRODUCTION

The disease is occurs on mungbean and may pose a threat to the successful cultivation of crop under favorable development of the disease. It has been reported from India and Indonesia, it is usually present in the rainy season.

4.4.2 SYMPTOMS

Initially the disease symptoms appears as circular, reddish brown pustules which appear more common on the underside of the leaves, less abundant on pods and sparingly on stems. When leaves are severely infected, both the surfaces are fully covered by rust pustules. Shriveling followed by defoliation results in yield losses.

4.4.3 PATHOGEN

Rust disease caused by *Uromyces phaseoli typica* (Syn: *Uromyces appendiculatus*), is autoecious, and macrocyclic rust fungi. The uredospores are unicellular, spiny, thin walled, globoid to ellipsoid in shape and yellowish brown. The teliospores are globose to broadly ellipsoid in shape, unicellular, pedicellate, chestnut brown in color with hyline papilla at the top. Yellow colored pycnia appear on the upper surface of leaves. Orange colored cupulate aecia develop later on the lower surface of leaves. The aeciospores are unicellular and elliptical (Souza et al., 2008).

4.4.4 PREDISPOSING CONDITIONS

The temperature range of 21–26°C, availability of free moisture at leaves surface, cloudy humid weather, high RH and night with heavy dews favor infection and subsequent disease development.

4.4.5 DISEASE CYCLE

The pathogen is an obligate parasite and being autoecious in nature. It produces all the five stages, i.e., pyconidial, aecial, uredial, and teial the same host. The uredial stage repeated several times and it is often noticed on species of Dolichos and Vigna. The urediospores build up in masses and spread by wind to causes secondary infections. The fungi can perpetuates through the teliospores in the soil and as well as uredospores in crop derbies (Courtesy: http://www. eagri. org/eagri50/PATH272).

The teliospore upon germination produces basidiospores which in turn produce pycnidia, aecia, and uredia. Pycnia appears in yellowish spots on the upper surface of leaves. Whereas orange colored, aecia are formed on the opposite side. Primary infection is by the sporidia developed from teliospores. Secondary spread is by windborne uredospores. The fungus also survives on other legume hosts (Courtesy: http://www. eagri. org/eagri50/PATH272).

4.4.6 MANAGEMENT

Removal and destruction of plant debris and manipulation in dates of sowing may help in reducing the disease incidence. Spraying of wettable sulfur at the rate of 2 kg/ha or Carbendazime at the rate of 500 g/ha or Propiconazole @ 1.0 l/ha, immediate as the disease appears and repeat after 15 days. Foliar spray of Captan 70% + Hexaconazole 5% (Taqat-75WP) is highly effective to control rust (Adinarayana et al., 2013).

4.5 DRY ROOT ROT

4.5.1 INTRODUCTION

Dry root rot is an important disease of mungbean incited by *Macrophomina phaseolina* (Tassi) Goid, is a serious problem in hampering the production of the mungbean in all growing areas. It was first reported from Jabalpur (M. P.) India (Philip, 168).

4.5.2 SYMPTOMS

Pathogen attacks on all parts of plants, i.e., root, stem, branches, petioles, leaves, pods, and seeds. In pre emergence stage, the fungus causes seed rot and mortality of germinating seedlings. In post emergence stage, seedling gets blighted due to soil or seed born infection. The Fungus attacks the stem at ground level on one-month-old crop and farming localized raised white canker at the base of stem. They gradually enlarge in size and turn brown streaks spreading upwards. The leaves of infected plants are dark green, mottled, and reduced in size. The leaves are dropped suddenly and dry, resulting heavy reduction in flowering and podding. When affected plants are spilt vertically from the collar downwards, reddish discoloration of the internal tissues is clearly visible while the internal root tissues appear white. The affected plants can be easily pulled out leaving dried, rotten root portions in the ground. The rotten tissues of stem and root contain a large number of black minute sclerotia.

4.5.3 PATHOGEN

The disease is caused by *Macrophomina phascolina* (Tassi) Goid [*Rhizoctonia bataticola* (Taub) Butler]. The pathogen invades the mungbean plants both inter and intra-cellularly. It grows fast covering large area of the host tissues and eventually killing them. The fungus produces dark brown, septate mycelium with constrictions at hyphal branches. It also produces numerous sclerotial bodies on the host tissues. Sclerotia are minute, dark to jet-black color, round, and smooth to oblong or irregular and abundance in nature. The fungus also produces dark brown, globose ostiolated pycnidia immersed in the host tissues, are erumpent at maturity. The pycnidiospores are thin walled, hyaline, single celled and elliptical some time curved or irregularly contoured and variable in size, 10–42 x 6–10 mm.

4.5.4 PREDISPOSING CONDITIONS

The fungus reported as a high temperature pathogen. The disease develops rapidly and become sever at high temperature (Day temperature of 30°C) and prolonged water stress (15% RH) followed by irrigation. The disease severity is increases with increasing temperature.

4.5.5 DISEASE CYCLE

The fungus survives in the infected debris and also as facultative parasite in upper layer of the soil. The primary spread is through seed-borne and soil-borne sclerotia, which either do not germinate or produce seedlings that may die soon after emergence. Since the fungus attacks a wide range of hosts, it perpetuates freely and becomes virulent when the optimum pre-disposing conditions in the host exist. The fungus being soil and seed borne in nature, it spread in the field through irrigation water, farm implements, and cultural practices. The secondary spread is through air-borne pycnidiospores and sclerotia and causes further spread of the pathogen.

4.5.6 MANAGEMENT

Follow 2–3 years crop rotation for reducing the inoculums load from soil. Expose the soil through deep summer plowing is effective methods for weed control as well as disease. Destruction of infected plant debries by burning and burying in soil. Seed dressing with Carbendazim @ 2 g/kg seed or *Trichoderma harzianum* or *T. viride* or *G. virense* @ 10 gram per kg of seed were effective in controlling the seed borne infections (Kumar et al., 2012; Deshmukh and Raut, 1992; Manczinger et al., 2002). Foliar spray of Carbendazim and Benomyl @ 500 ppm at 15 days interval reduced Macrophomina leaf blight. Meanwhile, Kumari et al. (2012) reported that integrated management approach showed that vermicompost and carbendazime in combination was more effective in reducing the root rot incidence in pod conditions.

4.6 YELLOW MOSAIC DISEASE

4.6.1 INTRODUCTION

Yellow mosaic disease is the most serious limiting factor in mungbean cultivation. The occurrence of MYMV was first time reported from Delhi, India (Nariani, 1960). Since then, it has been reported from all over India and other countries of the Indian subcontinent (Verma and Malathi, 2003), but it is widely prevalent in Uttar Pradesh, Bihar, Chhattisgarh, Delhi, Punjab, Haryana, Himachal Pradesh, Rajasthan, Odisha, Andhra Pradesh, Tamil Nadu, Karnataka, and Madhya Pradesh.

Yield losses due to this disease vary from 5 to 100% depending upon disease severity, susceptibility of cultivars and population of whitefly (Nene, 1972; Rathi, 2002). The infection not only drastically reduces yield but also delayed in pod maturity (Singh et al., 1982) and reduced number of pod per plant. In mungbean, Yellow mosaic disease incidence in farmer's fields might be as high as 100% (Rishi, 2009).

4.6.2 SYMPTOMS

Initially small yellow patches or spots appear on green lamina of young leaves. Soon it develops into characteristics bright yellow mosaic or golden yellow mosaic symptom. Yellow discoloration slowly increases and leaves turn completely yellow. Maturity is delayed in the diseased plants and flower and pod formation are severely reduced. Seeds that develop on severely infected plants are small and immature. In addition to yellow mosaic to the yellow mottle symptoms, a necrotic center develops in the yellow spots in the some varieties. The pods are small and distorted. Early infection causes death of the plant before seed set.

4.6.3 PATHOGEN

It is caused by Mungbean yellow mosaic India virus (MYMIV) in Northern and Central region and MYMV in western and southern regions. It is a

Begomovirus belonging to the family Gemini viridae. Geminate virus particles, are isometric, paired, 15 x 30 μmin size (Honda et al., 1981), having ssDNA, bipartite genome with two gemonic components DNA-A and DNA-B. These typse of DNA were identified as DNA-A and DNA-B by nucleic acid hybridization tests (Verma et al., 1991) The virus particles are limited to phloem-associated elements in infected plants (Srivastava et al., 1985).

4.6.4 PREDISPOSING CONDITIONS

The incidence and spread of yellow mosaic disease are generally high in summer crops, as compared with rainy or winter season crops. Weed serves as reservoir of the virus, and is a source of primary inoculums. The virus starts spreading with the onset of the monsoon in India (Grewal, 1988), reaching a high incidence in areas where the temperature ranges between 31 and 35°C and RH is about 70% (Sharma et al., 1993). These conditions favor the disease development and multiplication of the vector *Bemisia tabaci.*

4.6.5 DISEASE CYCLE

It is transmitted by whitefly, *Bemisia tabaci* and grafting under favorable conditions in a circulative, non-propogative manner and by grafting. It is not seed or soil borne or sap transmissible. Transmission of the virus is discussed detail by Singh et al. (1998). Disease spreads by feeding of plants by viruliferous whiteflies. Summer sown crops are highly susceptible. Weedhosts viz., *Croton sparsiflorus, Acalypha indica, Eclipta alba,* and other legume hosts serve as reservoir for inoculums.

4.6.6 MANAGEMENT

The most effective way to minimize the impact of MYMV to grow resistant varieties, i.e., Pant Mung-3, Pant Mung-4, Pant Mung-5, Pant Mung-6, Pusa Vishal, Basanti, ML-267, ML-337, PDM-54, PDM 139, LGG-407, LGG-460, Narendra mung-5, HUM-1, TM-99-37, HUM-16, TARM-1, Pusa-95-31, Sweta, and Samrat have been identified. Manipulation in sowing dates, plant density and intercropping with two rows of maize (60 x 30 cm) or Sorghum (45 x 15 cm) for every 15 rows of greengram are inhibit the congenial atmosphere for vector and reduced the movement of vector. The incidence of MYMV in mungbean was the lowest in crop raised from the seeds treated with Thiamethoxam @ 4 g/kg seed (Ganapathy and Karuppiah, 2004). Thiamethoxam as seed treatment (4 g/kg seed) providing production to healthy seedling from MYMV, and its subsequent sprays (0.02%) (Dubey and Singh, 2013) or seed treatment with Imidachloprid-70WS @ 4 g/kg seed and two subsequent spray at 21 and 35 days protect the crop up to the pod formation stage.

4.7 LEAF CRINKLE DISEASE

4.7.1 INTRODUCTION

Leaf crinkle, caused by urdbean leaf crinkle virus (ULCV), is a limiting factor for successful cultivation of mungbean (Vigna radiata (L.) Wilczek) in India. ULCV has been reported in natural infections of mungbean in India by Singh et al., (1979). ULCV has the potential to cause heavy yield losses in mungbean. In early-infected plants seed production may fail completely. A field experiment conducted over 3 years showed losses in grain yield ranging from 2 to 95% in both mungbean and urdbean, depending on the plant age at symptom onset (after 55 days and before 24 days old respectively) (Kadian, 1982a). The virus also reduced yield components such as plant height, root length, nodulation, pods/plant, pod length, seeds/pod and weight of 100 seeds, but it increases the leaf area and dry weight of roots and shoots (Singh, 1980).

4.7.2 SYMPTOMS

The initial symptoms appear on younger leaves as chlorosis around some lateral veins and its branches near the margin. The leaves show downward curling of margin. Some of the leaves show twisting. The veins show reddish brown discoloration on the under surface which also extended to petioles. Plants show symptoms within 5 weeks after sowing, invariably remain stunted and majority of these die due to top nerosis within a week or two. Plants infected in late stages of growth do not show severe curling and twisting of the leaves but show conspicuous veinal chlorosis anywhere on the leaf lamina. Petioles as well as internodes are shortened. Infected plant gives a stunted and bushy appearance. Flowering is delayed, inflorescence, if formed, are malformed with small size flower buds and fail to open. Pollen fertility and pod formation is severely reduced on infected plants which decreases the yield drastically.

4.7.3 PATHOGEN

The 'Urdbean leaf Crinkle disease caused by Leaf Crinkle Virus.' The virus particles are spherical and approximately 25–30 nm in size. Dubey et al. (1983) reported that ULCV particle purified from urdbean were 32 nm in diameter. In the sections of urdbean cells, particles measured 25 to 30 nm in diameter (Bhaktavatsalam et al., 1983). The A_{260}/A_{280} ratio was 1.51 (Bhaktavatsalam et al., 1983). No serological relationships were found between ULCV and black gram mottle Carmovirus, broad bean mottle bromovirus or bean pod mottle comovirus (Dubey et al., 1983).

4.7.4 TRANSMISSION

The ULCV is transmitted through sap, grafting, insect vectors and seeds. Beniwal et al. (1980) reported it has been seed transmitted to an extant of 15% in mungbean. Many workers have reported its transmission by various insects viz., Aphid, *Aphis craccivora* and *A. gossypii* (Bhardwaj and Dubey, 1986; Dhingra, 1975), beetle, *Henosepilachna dodecastigma* (Beniwal and Bharathan, 1980) and white fly, *Bemisia tabaci* (Narayanasami and Jaganathan, 1974). The virus is also sap transmissible.

4.7.5 PREDISPOSING CONDITIONS

The presence of weed hosts like *Aristolochi abracteata* and *Digera arvensis*. Kharif season crop is highly susceptible. Continuous cropping of other legume crops which also harbor the virus. Environmental factors showed positive interaction in case of maximum and minimum temperature but in case of RH the interaction was negative. With an increase the temperature, the infection rate increased and with increase in RH the infection rate decreased (Binyamin et al., 2011). Ashfaq (2007) reported that maximum disease incidence was developed at maximum temperature of 35–42°C.

4.7.6 DISEASE CYCLE

The virus is seed-borne and primary infection occurs through infected seeds and presence of weed hosts like *Aristolochia bracteata* and *Digera arvensis.* Kharif season crop and continuous cropping of other legumes, i.e., Urdbean (*Vigna mungo*), Cowpea, Pigeon pea, Taparybean (*Vigna aconitifolicus*) (Kolte and Nene, 1975; Beniwal, 1983) serve as source of inoculum. White fly, aphids, and beetle help in the secondary spread of virus.

4.7.7 MANAGEMENT

To avoid early infection into the crop, planting of disease free seeds are an important measures of management of disease. Removed collateral weed hosts around the field and rogue out the diseased plants could reduce the chance of further spread and reduced production of more infected seed. Hot water seed treatment at 52°C for 20–30 min reduced seed borne infection. Cultivation of resistant mungbean entries also decreased diseased incidence. Foliar spray of Thiomethaxam 0.02% at 15 days after showing, decreased whitefly population, and incidence of leaf crinkle (Ganapathy and Karuppiah, 2004). Binyamin et al. (2011) reported that 2% neem and akk extract reduced the severity of ULCV infection, while Zeshan et al. (2012) have reported that nitrogen, phosphorus, potash, zinc, boron, and naphthalene acetic acid (NAA) were effectively reducing ULCV infection in different varieties of mungbean by the same order from 65 to 58.57%.

4.8 WEB BLIGHT

4.8.1 INTRODUCTION

Web blight is one of the major constraints in the production of many pulses in warm humid tropical zones of the world. On Mungbean Rhizoctonia, blight was reported for the first time from Philippines (Nacien et al., 1924). In the year 1985, Alam et al., reported occurrence of web blight of mungbean in Pakistan.

In India first report of its occurrence on mungbean was given by Dwivedi and Saksena (1974) from Kanpur, Uttar Pradesh subsequently this disease had also been reported from Assam (Saikia, 1976), Punjab (Bains et al., 1988), Madhya Pradesh (Tiwari and Khare, 1998), Bihar, Rajasthan, Haryana, Himachal Pradesh, and Jammu & Kashmir (Anonymous, 2004).

The disease was observed to reduce 33 to 40% grain yield and 28.6% decrease in 1000-grain weight at different level of disease severity with different variety of mungbean (Gupta et al., 2010; Singh et al., 2012). Since then the web blight of mungbean has become one of the most serious problems of this crop in Northern India causing extensive damage to mungbean.

4.8.2 SYMPTOMS

The symptoms of web blight on mungbean occur on roots, stems, petioles, and pods but the disease is the most destructive on foliage. The pathogen is soil borne, seed borne, and air borne in nature. During second to third week of plants growth, it causes seedling mortality. Seed decay, pre, and post emergence mortality also occurs. The first symptoms appear as small circular brown spots on the primary leaves. These spots enlarge, often show concentric banding and surrounded by irregular water soaked areas. The lesion expands and coalesces and white mycelia fungal growth can be seen under surface of infected leaves and young branches. The mycelium on infected leaves appears as spider web thus suggested the name web blight disease (Dwivedi and Saksena, 1974; Singh, 2006).

On young pods the spots are light tan and irregular in shape but generally on mature pods, they are dark brown and sunken. Spots may coalesce to cover entire pod under favorable conditions (Galvez et al., 1979).

4.8.3 PATHOGEN

The pathogen has been reported on mungbean under the name *Thanatephorus cucumeris* (frank) Donk (Dwivedi and Saksena, 1974; Singh, 2006). Web blight pathogen *R. solani* has been characterized as:

- Colony color pale to dark brown.
- Rapidly growing mycelium of relatively large diameter (>10 μm).
- Branching near the distal septum of hyphal cell, often nearly at right angles in older hyphae.
- Production of moniloid cells (Barrel shaped cells or chlamydospores) in chain or aggregates.
- Production of sclerotia nearly uniform texture, varying in size (<1 mm to several cm) and shape (round to thin crust).

- Pathogenic to vide range of host (142 crops belonging to 32 plant families) and producing in a variety of symptoms (Damping off, Root Rot, stem rot, Blighting of stems and leaves, Decay of fruits and seeds).
- On the host, the sclerotia are superficial, variable in shape and size, dark brown to black, scab like and accompanied with dark, short celled, abundantly branched, stout mycelium without clamp connections (Saikia and Roy, 1976; Singh, 2006).

The perfect state *T. cucumeris* develops on healthy tissue adjacent to the lesion on green gram leaves as well as *in vitro* (Tiwari and Khare, 1998). The sexual stage is induced on the host under warm and humid conditions with heavy dew formation during the night. Fructifications are resupinate, creamy to greyish white, loosely attached to the substratum, composed of arachnoid, repent hyphae giving rise to thin hypochnoid or sub membranous fertile patches. Cluster of basidia are produced terminally in a discontinuous hymenium. Basidia are mostly 9–25 x 5–12.4 μm in size but sometimes short, wider, barrel shaped to sub cylindrical without a median construction. Sterigmata are straight, attenuate towards the apex usually bearing 4 basidium, rarely 2, 3 or up to 7 in number. Basiospore are hyaline smooth thin walled, oblong to ellipsoid and dorsally fattened or broad ovoid and commonly widest at the distal end form secondary basidiospores (Saksena, 1973).

The pathogen *R. solani* is cosmopolitan with a very wide host range and attacks large number of crop plants and weeds. It is claimed that there is hardly any plant species which cannot be infected by the pathogen (Ogoshi, 1987). As a plant pathogen *R. solani* regarded as unspecialized fungus and composed of a 14 complex assemblage of anastomosis groups (AGs) (AG-1 to AG-13 and AG BI) that are genetically distinct nonmating populations have a wide variation in their morphological, cultural, pathological, and physiological characteristics (Ogoshi, 1987;Sneh et al., 1991; Kuninga et al., 1997). Among them, isolates of AG 1 have been divided three sub groups; AG 1-IA, AG 1-IB and AG 1-IC (Sneh et al., 1991) and web blight of mungbean in India caused by AG 1-IB (Singh, 2006). The web blight isolates of *R. solani* from mungbean (Tiwari and Khare, 1998) and soybean (Sharma et al., 2013) are varying in their cultural, morphological appearance, pathology, anastomosis, and physiology.

4.8.4 PREDISPOSING CONDITIONS

Environmental factors play a vital role in the development of web blight disease caused by *R. solani*. Higher aerial temperature (26 to 32°C) RH near 100% and soil temperature 30–33°C favored the development of high disease severity. Rainfall (91–97 mm) had a significant role in severe development of web blight during early stage of crop (Sharma and Thripathi, 2001; Gupta et al., 2010). The maximum basidiospores production and their discharge occur during mid night and early morning before sunrise (Saksena and Dwivedi, 1973). The spore production is favored by night temperature below 24°C, RH above 95% and rate of evaporation below 1.5 mm per day. In nature, basidiospore production starts in the mid August in north India. The basidiospores germinate in two hours and penetrate leaf through the intact surface by formation of infection cushion or direct penetration through stomatal openings. Host susceptibility is also important in disease incidence and severity. In case of web blight of legumes, the plant are susceptible seedling stage until maturity and the severity is maximum in 30–70 days old plants probably because of dense canopy of the crop which facilitates the early spread of the pathogen through contact of plants and leaves with one another, forming mycelial bridges.

4.8.5 DISEASE CYCLE

The pathogen is soil borne, seed borne and airborne in nature. Epidemiology of Rhizoctonia blight may be divided into two phases, one before and the after canopy closures. The first phase is soil borne and the second is leaf borne (Yang, 1990). Though, the pathogen *R. solani* has been isolated from seeds of mungbean, it is primarily soil borne and can survive for many years by producing sclerotia in soil and on plant tissue/seed coat. *R. solani* also survive as mycelium by colonizing soil organic matter as saprophytes. The highest inoculum potential is noted in the top 10 cm soil and no inoculum is found below 40 cm (Kaiser et al., 1970). It germinates to produce vegetative threads (hyphae) of the fungus that can attack mungbean leaves during heavy rains. The collateral weed hosts play an important role in initiation and early spread of the disease to the main host (Saksena, 1985). Because of their proximity to the soil carrying primary inoculum and their special microclimate, the weed hosts are first to take the infection and facilitate the production of the basidiospores of the pathogen. There after the pathogen become air borne to causes leaf, stem or pod infection and produces fresh crops of basidiospores on the main host. In *R. solani* secondary spread via sclerotic through wind and rain splash is also reported. Cobweb like mycelial growth of pathogen also gets transmitted from diseased to healthy leaves in the web blight of legumes (Saksena, 1979; Verma and Thapliyal, 1976).

4.8.6 MANAGEMENT

Plant resistance is one of the most attractive approaches to suppressing plant disease. It is not only compatible with other disease management techniques but eco-friendly also. Singh et al. (2007) reported that out of 85 mungbean entries, only 7 viz., NDM-92-2, Ml-406, EC-27130, EC-1243-1, EC-5551, IC-73362, and IC-39338 showed resistance reaction against web blight disease. Disease injury can be reduced by crop rotation, adjusting time of sowing (early sowing) to avoid periods of low temperature with wider row spacing (50 cm) and draining wet areas of the field. Seed dressing with Carbendazim and Thiophante methyl were found effective in eliminating the seed borne infection of *Thanatephorus cucumeris* causing web blight of mungbean (Singh et al., 1995).

Trichoderma virens and *T. viride* were found effective against *R. solani* (Dubey, 2003). Seed treatment *T. viride*/*T. virens* + Carboxin + Rhizobium gave maximum seed germination, plant vigor, number of root nodules and grain yield with minimum seedling mortality and disease intensity (Dubey, 2003). Integration of soil application of *P. glabra* cake (200 kg/ha), seed treatment with T. viride (2 g/kg seed) + Carboxin (1 g/kg seed) + Rhizobium sp. (25 g/kg seed) and foliar spray of *P. glabra* leaf extract (10%) suppressed disease severity to a significant extent (92.7%). This treatment also increased seed germination (32.4%), improved plant vigor, and enhanced production (49.2%). The same combination excluding carboxin was also effective and could be an option for organic production of mungbean. The integration of any two modes of applications of treatments was superior to any single mode of application (Dubey (2006).

4.9 BACTERIAL LEAF BLIGHT

4.9.1 INTRODUCTION

Disease has been reported from Punjab, Haryana, Rajasthan, Uttar Pradesh, Madhya Pradesh, Maharashtra, Andhra Pradesh, and Tamil Nadu.

4.9.2 SYMPTOMS

The disease Bacterial leaf spot was first time reported from China (Fang et al., 1964). In India Patel and Jindal (1972) reported two new disease on mungbean, i.e., Bacterial leaf spot and Halo blight during 1968 and 1969. This is characterized by brown, circular to irregular, dry, raised spots on leaves and stem. Initially it appears as superficial eruptions and gradually invades the tissue giving corky or rough appearance. The lesions on petiole are brown, flat, or raised and occur as long streaks without defoliations. In severe infections, dark brown longitudinal split or canker develops on stem. When the disease is sever, the leaves become yellow and fall off prematurely.

4.9.3 CASUAL AGENT

Bacterial leaf blight is incited by *Xanthomonas campestris* pv. *Vignaeradiate* [(Sabet, Ishaq, and Khalil) Dye]. The bacterium is a gram negative, rod shaped and having single polar flagellum. The pathogen can also infect *Phaseolus vulgaris*, *P. hontus*, *P. braclantus, Dolichos lablab,* and *Lens culineris*.

4.9.4 PREDISPOSING CONDITIONS

The disease develops rapidly under warm and wet weathers. The optimum temperature for the growth of bacterium is lies between 30 and 35°C.

4.9.5 DISEASE CYCLE

The bacterium survived on seeds, diseased plant debris as well as other host plants. In seed borne condition, the bacteria develops in seed coat of the germinating seeds and infects the surface of expended cotyledon, penetrating through rifts in the cuticle and grow intercellurly until the vascular system is reached. It may then precede as vascular pathogen throughout the large xylem vessels of the plant. Both leaf lesions and stem cankers may arise from systemic infections. Primary local infection may also take place through stomata and produced systems on cotyledons. Secondary spread of inoculums from primary lesions is carried out by splashing water or windblown rain, implements, men, and animals.

4.9.6 MANAGEMENT

Disease transmission through seeds could be reduced by treatment with Streptomycin sulfate (@ 500 ppm), bleaching powder (@0.025%), Vitavax

(@0.2%) Captan (@0.3%). Three protective sprays of Streptomycin (@ 100 ppm), Zineb (@0.3%), Blitox-50 (@0.25%), or Benomyl (@0.25%) were found quite effective to reduced secondary infections. Bora et al. (1993) reported that application of *Eriwinia herbicola* 48 h before inoculation and bacterization of seeds of Pusa Baisakhi with Bacillus sp. reduced seedling infection (Borah et al., 2000). Number of cultivars, i.e., Jalgaon 781, ML-1, ML-2, ML-3, ML-8, ML-9, ML-10, Yellow mung and P70-68 showed resistance (Chand et al., 1977). The Resistance of bacterial leaf spot is govern by a single dominant gene (Singh, 1972).

KEYWORDS

- **anastomosis groups**
- **bacterization**
- **cercospora leaf spot**
- **mungbean yellow mosaic virus**
- **naphthalene acetic acid**
- **Urdbean leaf crinkle virus**

REFERENCES

Achaya, K. T., (1998). *Indian Food: A Historical Companion* (pp. 322). Oxford University Press, Delhi, India.

Adinarayana, M., Mahalakshmi, M. S., & Rao, Y. K., (2013). Field efficacy of new fungicide Taqat-75 WP against foliar fungal disease of blackgram. *J. Biopesticides*, *6*(1), 46–48.

Akhtar, J., Lal, H. C., Kumar, Y., Singh, P. K., Ghosh, J., Khan, Z., & Gautam, N. K., (2014). Multiple disease resistance in green gram and blackgram germplasm and management through chemicals under rain fed conditions. *Legume Res.*, *37*(1), 101–109.

Alam, S. S., Qureshi, S. H., & Bashir, M., (1985). A report on web blight of mungbean in Pakistan. *Pakistan J. Bot.*, *17*(1), 165.

Ali, M., & Gupta, S., (2012). Carrying capacity of Indian agriculture: Pulse crops. *Current Sci.*, *102*, 874–881.

Anonymous, (2004). *Annual Report (Kharif)* (p. 112). All India Coordinated research project on Mulla RP (ICAR) IIPR, Kanpur.

Anonymous, (2012). *Project Coordinator's Report AICRP on MULLARP Crops*. IIPR, Kanpur.

Ashfaq, M., (2007). Characterization of epidemiological and biochemical factors in relation to resistance against Urdbean leaf crinkle virus (ULCV) and its management. PhD Thesis (p. 37). University of agriculture, Faisalabad, Pakistan.

AVRDC, (1982). Asian vegetable research and development center, Shanhua Taiwan. *Center Point, 2*(1).

Bains, S. S., Dhaliwal, H. S., & Basandrai, A. K., (1988). A new blight of mung and mash in Punjab. *Ann. Biol. Ludhiana, 4*(1/2), 113, 114.

BARI, (2007). *Research Report for 2006–2007*. Bangladesh Agricultural Research Institute, Jodhpur, Ghazipur.

Beniwal, S. P. S., (1983). Urdbean leaf crinkle virus. In: Boswell, K. F., & Gibbs, A. J., (eds.), *Viruses of Legumes* (pp. 85, 86). Descriptions and keys from VIDE. Austral. Natl. Univ., Canberra.

Beniwal, S. P. S., Chaubey, S. N., & Bharathan, N., (1980). Presence of Urdbean leaf crinkle virus in seeds of mungbean germplasm. *Indian Phytopathol., 33*, 360.

Bhaktavasalam, G., Nene, Y. L., & Beniwal, S. P. S., (1983a). Ultrastructure changes in Urdbean infected by Urdbean leaf crinkle virus. *Indian Phytopathol., 36*, 228.

Bhaktavatsalam, G., Nene, Y. L., & Beniwal, S. P. S., (1983b). Influence of certain physio-chemical factors on the infectivity and stability of Urdbean leaf crinkle virus. *Indian Phytopathol., 36*, 489–493.

Bharadwaj, S. V., & Dubey, G. S., (1986). Studies on the relationship of Urdbean leaf crinkle virus and its vectors, *Aphis craccivora* and *Acyrthosiphom pisum. J. Phytopathol., 83*, 115.

Binyamin, R., Khan, M. A., Ahmad, N., & Ali, S., (2011). Relationship of epidemiological factors with Urdbean leaf crinkle virus disease and its management using plant extracts. *Int. J. Agric. Biol., 13*, 411–414.

Bora, L. C., Gangopadhyay, G., & Chand, J. N., (1993). Biological control of bacterial leaf spot of mungbean with phylloplane antagonists. *Indian J. Mycol. Pl. Pathol., 23*, 162–168.

Borah, P. K., Jindal, J. K., & Verma, J. P., (2000). Biological management of bacterial leaf spot of mungbean caused by *Xanthomonas axopodis* pv. *vignaeradiate. Indian Phytopath., 53*, 384–394.

Chand, J. N., Yadav, O. P., & Yadav, H. C., (1977). Reaction of genetic stock of mungbean to bacterial blight. *Indian Phytopath., 30*, 565.

Chand, R. Singh, V., Pal, C., Kumar, P., & Kumar, M., (2012). First report of new pathogenic variant of *Cercospora canescens* on mungbean (*Vigna radiata*) from India. *New Disease Reports, 26*, 6.

Chauhan, M. P., & Gupta, R. P., (2004). Genetics of *Cercospora* leaf spot disease resistance in mungbean [*Vigna radiata* (L.) Wilczek]. *Legume Res., 27*, 155, 156.

Deeksha, J., & Thripathi, H. S., (2002). Cultural, biological, and chemical control of anthracnose of Urdbean. *J. Mycol. Pl. Path., 32*, 52–55.

Deshmukh, P. P., & Raut, J. G., (1992). Antagonism by trichoderma sp. on the five plant pathogenic fungi. *New Agriculturist, 3*, 127–130.

Dhingra, K. L., (1975). Transmission of Urdbean leaf crinkle virus by two aphid species. *Indian Phytopathol., 28*, 80.

Dubey, G. S., Sharma, I., & Prakash, N., (1983). Some properties of Urdbean leaf crinkle virus. *Indian Phytopathol., 36*, 762.

Dubey, S. C., & Singh, B., (2013). Integrated management of major diseases of mungbean by seed treatment and foliar application of insecticides, fungicides, and bioagent. *Crop Protection, 47*, 55–60.

Dubey, S. C., (2003). Integrated management of web blight of URD and mungbean. *Indian Phytopath., 56*, 413–417.

Dubey, S. C., (2006). Integrating bioagent with plant extract, oil cake, and fungicide in various mode of application for better management of web blight of Urdbean. *Archives of Phytopathology and Plant Protection*, *39*, 341–351.
Dwivedi, R. P., & Saksena, H. K., (1974). Occurrence of web blight caused by *Thanatephorus cucumeris* on mungbean. *Indian J. Farm Sci*., *2*, 100.
Fang, C. T., Chen, H. Y., & Chu, C. I., (1964). A comparative study of the species of the genus *Xanthonomas* from leguminous plants. *Acta Phytopath Simica*, *7*, 21–31.
Galvez, G. E., Guzman, P., & Castano, M., (1979). In: Schwariz, H. F., & Galvez, G. E., (eds.), *Web Blight in Bean Production Problems, Disease, Insect, Soil and Climatic Constraints of Phaseolus Vulgaris* (pp. 101–110). Cali, Colombia, Centro Inter. De Agric. Trop.
Ganapathy, T., & Karuppiah, R., (2004). Evaluation of new insecticides for the management of white fly (*Bemisia tabaci* Genn), mungbean yellow mosaic virus (MYMV) and Urdbean leaf crinkle virus (ULCV) diseased in mungbean (*Vigna radiata* (L.) *wilezek*). *Indian J. Pl. Prot., 32*, 35–38.
Grewal, J. S., (1988). Diseases of pulse crops: An overview. *Indian Phytopath*., *41*, 1–14.
Gupta, R. P., Singh, S. K., & Singh, R. V., (2010). Assessment of losses due to web blight and weather effects on disease development in mungbean. *Indian Phytopath., 63*(1), 108, 109.
Gupta, S., (2011). Indian Institute of Pulses Research (IIPR) Vision (2030). In: *Indian Institute of Pulses Research (ICAR)* (p. 42). Kanpur, India.
Hanson, L. E., & Panella, L., (2003). Disease control in sugerbeet. *Int. Suger. J., 105*, 60–68.
Haque, M. F., Mukherjee, A. K., Mahto, R. N., Jha, D. K., Chakraborty, M., Srivastava, G. P., & Prasad, D., (1997). Birsa-Urd-1 new variety for Chotanagpur region of Bihar. *J. Res., 9*, 177, 178.
Honda, Y. M., Twaki, Y., Saito, P., Thangmeearkom, P., Kittisak, K., & Deema, N., (1981). Mechanical transmission, purification and some properties of whitefly transmitted mungbean yellow mosaic virus in Thailand. *Plant Disease, 65*, 801–804.
Honda, Y., & Ikegami, M., (1986). Mungbean yellow mosaic virus: *AAB Descriptions of Plant Viruses 323*, 4.
Hossain, A. M. M. Z., Chatterjee, D. D., Howlader, B., Kanti, K., & Jalauddin, (2011). Incidence and severities of *Cercospora* leaf spot on different strain of growing mungbean. *Bangladesh Res. Pub. J*., *5*, 1–6.
Kadian, A. O. P., & Verma, J. P., (1981). Some properties of virus causing leaf crinkle disease of Urdbean and mungbean in India in *Absrt.5th Int. Congress Virology. Absrt.,* p. 22/06.
Kadian, O. P., (1980). Studies on leaf crinkle disease of Urdbean (*Vigna mungo* (L.) *Hepper*) mungbean (*V. radiate* (L.) *Wilczek*) and its control. PhD Thesis (p. 177). Department of Plant Pathology, Haryana Agriculture University, Hisar, India.
Kadian, O. P., (1982a). Screening for resistance to leaf crinkle in mungbean. *Tropical Pest Management, 28*, 436.
Kadian, O. P., (1982b). Yield loss in mungbean and Urdbean due to leaf crinkle virus. *Indian Phytopathol., 35*, 642–644.
Kaiser, W. J., (1970). Rhizoctonia stem canker disease of mungbean in Iran. *Plant Dis. Rep., 54*, 240–250.
Kaiser, W. J., Mossahebi, G. H., & Okhorat, M., (1970). Occurrence, pathogenicity and distribution in soil of *Rhizoctonia solani* inciting a stem canker disease of mungbean (*Phaseolus aureus*) in Iran. *Iranian Journal of Plant Pathology*, *6*, 17–25.
Kapadiya, H. J., & Dhruj, I. U., (1999). Management of mungbean *Cercospora* leaf spot through fungicides. *Indian Phytopath., 32*(1), 96–97.

Khan, M. F. R., & Smith, L. J., (2005). Evaluating fungicides for controlling *Cercospora* leaf spot on sugar beet. *Crop Prot.*, *24*, 79–86.

Khan, M. F., (2008). Evaluating strategies for managing *Cercospora* leaf spot of sugerbeet. *Phytopathol.*, *98*, 80.

Khan, M. R. I., (1981). Nutritional quality characters in pulses. In: *Proceedings of National Workshop on Pulses* (pp. 199–206). BARI, Gazipur.

Kolte, S. J., & Nene, Y. L., (1975). Host range and properties of Urdbean leaf crinkle virus. *Indian Phytopathol.*, *28*, 430, 431.

Kulkarni, S., Benagi, V. I., Patil, P. V., Hegde, Y., Konda, C. R., & Deshpande, V. K., (2009). Sources of resistance to anthracnose in greengram and biochemical parameters for resistance. *Karanataka J. Agric. Sci.*, *22*(5), 1123–1125.

Kumari, R., Shekhawat, K. S., Gupta, R., & Khokhar, M. K., (2012). Integrated management against root rot of mungbean [*Vigna radiata* (L.) *Wilczek*] incited by *Macrophomina phaseolina. J. Plant Pathol. & Microb.*, *3*, 136. doi: 10.4172/2157–7471.1000136.

Kuninaga, S., Natsuaki, T., Takeuchi, T., & Yokosawa, R., (1997). Sequence variation of the rDNA ITS regions within and between anastomosis groups in *Rhizoctonia solani*. *Current Genetics*, *32*(3), 237–243.

Lakshmanan, P., Mohan, S., & Jeyarajan, R., (1990). Antifungal properties of some plant extracts against *Thanatephorus cucumeris*, the causal agent of color rot disease of *Phaseolus aureum*. *Madras Agric. J.*, *77*(1), 1–4.

Lal, G., Kim, D., Shanmugasundaram, S., & Kalb, T., (2001). *Mungbean Production by AVRDC* (p. 6). World Vegetable Center, Taiwan, Shanhua: AVRDC-The World Vegetable Center.

Manczinger, L., Antal, Z., & Kredics, L., (2002). Ecophysiology and breeding of mycoparasitic trichoderma strains: A review. *Acta Microbiol. Immunol. Hung.*, *49*, 1–14.

Mathur, A. K., Chitale, K., Gaur, V. K., & Tyagi, R. N. S., (1990). Response of mungbean cultivars to *cercospora* leaf spot in Rajasthan. *Indian J. Mycol. Pl. Pathol.*, *19*, 213, 214.

Mew, I., Pin, C., Wang, T. C., & Mew, T. W., (1975). Inoculum production and evaluation of mungbean varieties for resistance to *Cercospora canescens*. *Plant Dis. Rep.*, *59*, 397–401.

Mian, A. L., (1976). *Grow More Pulse to Keep Your Pulse Well: An Essay of Bangladesh Pulse* (pp. 11–15). Department of Agronomy, Bangladesh Agriculture University, Mymensingh.

Munjal, R. I., Lal, G., & Chona, B. L., (1960). Some *Cercospora* species from India-IV. *Indian Phytopathol.*, *13*, 144–149.

Nacien, C. C., (1924). Studies on rhizoctonia blight of beans. *Philippine Agriculturist*, *8*, 315–321.

Nair, R. M., Schafleitner, R., Kenyon, L., Srinivasan, R., Easdown, W., Ebert, A., & Hanson, P., (2012). Genetic improvement of mungbean productivity. In: *Proc. of the 12th SABRAO Congress on Plant Breeding Towards (2025): Challenges in a Rapidly Changing World* (pp. 27–28). Chiang Mai, Thailand.

Narayansami, P., & Jaganathan, T., (1974). Characterization of blackgram leaf crinkle virus. *Madras Agric. J.*, *16*, 979.

Nene, Y. L., (1972). A survey of viral diseases of pulse crops in Uttar Pradesh. *Final Tech. Report, Res. Bull. No.4* (pp. 1–91). U. P. Agricultural University, Pantnagar, India.

Nene, Y. L., (2000). A new vision of pulse pathology research in India. *Indian J. Pulses Res.*, *13*(1), 1–10.

Nene, Y. L., Naresh, J. S., & Nair, N. G., (1971). Additional hosts of mungbean yellow mosaic virus. *Indian Phytopathology*, *24*, 415–417.

Ogoshi, A., (1987). Ecology and pathogenesity of anastomosis and intraspecific groups of *Rhizoctonia solani* Kuhn. *Ann. Rev. Phytopath.*, *25*, 125–143.
Pandey, S., Sharma, M., Kumari, S., Gaur, P. M., Chen, W., Kaur, I., et al., (2009). Integrated foliar disease management of legumes. In: Masood, A. et al., (eds.), *Grain Legumes: Genetic Improvement, Management and Trade* (pp. 143–161). Indian Society of Pulses Research and Development, Indian Institute of Pulse Research, Kanpur, India.
Patel, P. N., & Jindal, J. K., (1972). Bacterial leaf spot and halo blight diseases of mungbean and other legumes in India. *Indian Phytopath.*, *25*, 517–525.
Poehlman, J. M., (1991). *The Mungbean* (pp. 169–274). Westview Press. Boulder.
Rakhonde, P. N., Koche, M. D., & Harne, A. D., (2011). Management of powdery mildew of green gram. *J. Food Legumes.*, *24*(2), 120–122.
Rathaiah, Y., & Sharma, S. K., (2004). A new leaf spot disease on mungbean caused by *Colletotrichum truncatum*. *J. Mycol. Pl. Path.*, *34*, 176–178.
Rathi, Y. P. S., (2002). Epidemiology, yield losses and management of major diseases of *Kharif* pulses in India. In*: Plant Pathology and Asian Congress of Mycology and Plant Pathology*. University of Mysore, Mysore, India.
Rishi, N., (2009). Significant plant virus diseases in India and a glimpse of modern disease management technology. *J. Gen. Plant Pathol.*, *75*, 1–18.
Saikia, U. N., & Roy, A. K., (1976). Natural occurrence of *Corticium sasakii* on few plants in Jorhat. Assam. *Sci. Cul.*, *42*, 228, 229.
Saikia, U. N., (1976). Blight of mung caused by *Corticium sasakii* a new disease recorded from Assam. *Indian Phytopath.*, *29*, 61, 62.
Saksena, H. K., & Dwivedi, R. P., (1973). Web blight of black gram caused by *Thanatephorus cucumeris*. *Indian Journal of Farm Sciences*, *1*(1), 58–61.
Saksena, H. K., (1973). Banded blight disease of paddy in north India. *Proceedings, International Rice Research Conference.* International IRRI, Los Banos, Philippines.
Saksena, H. K., (1979). Epidemiology of diseases caused by *Rhizoctonia* spp. In: *Proceedings of Consultants Group: Discussion on the Resistance to Soil Borne Diseases of Legumes* (pp. 59–64). ICRISAT. Hyderabad.
Saksena, H. K., (1985). Relationship of environment and rhizoctonia aerial blight. *Indian Phytopathology*, *38*, 584.
Shahbaz, M. U., Iqbal, M. A., Rafiq, M., Batool, A., & Kamran, M., (2014). Efficacy of different protective fungicides against *Cercospora* leaf spot of mungbean [*Vigna radiata* (L.) Wilczek]. *Pak. J. Phytopathol.*, *26*, 187–191.
Sharma, H. C., Khare, M. N., Joshi, L. K., & Kumar, S. M., (1971). *Efficacy of Fungicides in the Control of Diseases of Kharif Pulses Mung and Urd* (p. 2). All India Workshop on kharif Pulses.
Sharma, J., & Tripathi, H. S., (2001). Influence of environmental factors on web blight development in Urdbean [*Vignamungo* (L.) *Hepper*]. *J. Mycol. Pl. Pathol.*, *31*, 54–58.
Sharma, L., Goswami, S., & Narrale, D. T., (2013). Cultural and physiological variability in *Rhizoctonia solani*; responsible for foliar and lesion on aerial part of soybean. *J. Applied and Natural Sci.*, *5*, 41–46.
Sharma, S., Yadav, O. P., Thareja, R. K., & Kaushik, J. C., (1993). Nature and extent of losses due to mungbean yellow mosaic virus and its epidemiology. *HAU J. Res.*, *23*, 51–53.
Shukla, V., Bhagel, S., Maravi, K., & Singh, S. K., (2014). Yield loss assessment in mungbean [*Vigna radiata* (L.) *WilczekI*] caused by anthracnose [*Colletotrichum truncatum* (Schw.) Andrus and Moore]. *The Bioscan, 9*(3), 1233–1235.

Singh, B. R., Singh, M., Yadav, M. D., & Dingra, S. M., (1982). Yield loss in mungbean due to yellow mosaic. *Sci. and Culture, 48*, 435–436.
Singh, D., (1972). Screening and inheritance of resistance to bacterial blight in cowpea and bacterial leaf spot of mungbean. PhD Thesis (p. 72). IARI, New Delhi.
Singh, J. P., (1980). Effect of virus diseases on growth component and yield of mungbean (*Vigna radiata*) and Urdbean (*Vigna mungo*). *Indian Phytopathology, 8*, 405–408.
Singh, J. P., Kadian, O. P., & Verma, J. P., (1979). Survey of virus diseases of mungbean and Urdbean in Haryana. *Haryana Agric. Univ. J. Res., 9*, 345.
Singh, J., (2006). Occurrence, variability in a management of web blight pathogen (*Rhizoctonia solani Kühn*) of mungbean [*Vigna radiata* (L.) *Wilczek*]. PhD Thesis. B. H. U., Varanasi (U. P.).
Singh, J., Singh, R. B., & Balai, L. P., (2012a). Grain yield loss in mungbean due to web blight. *Trends in Biosciences, 5*(2), 147, 148.
Singh, J., Singh, R. B., & Balai, L. P., (2012b). Manipulation of sowing dates and inter-rows spacing and eco friendly management (*Vigna radiata*) web blight. *Trends in Biosciences, 5*(2), 152, 153.
Singh, J., Singh, S. K., & Singh, R. B., (2007). Resistance of mungbean accessions against web blight caused by *Rhizoctonia solani* Kühn. *Journal of Food Legume, 21*(2), 149, 150.
Singh, R. A., Gurha, S. N., & Ghos, A., (1998). Diseases of mungbean and Urdbean and their management. In: Thind, T. S., (eds.), *Diseases of Field Crops and Their Management* (pp. 179–204). National agricultural information center, Ludhiana, India.
Singh, R. K., & Dwivedi, R. S., (1987). Fungitoxicity of different plants. *National Academy Sci., 10*(3), 89–91.
Singh, R., Singh, H., & Dodan, D. S., (1995). Sensitivity of *Rhizoctonia solani* isolates causing seedling mortality of mung to different fungicides. *Plant Dis. Res., 10*, 41–43.
Sneh, B., Burpee, L., & Ogoshi, A., (1996). *Identification of Rhizoctonia Species* (p. 33). APS Press, St. Paul.
Souza, T. L. P. O., Marin, A. L. A., Faleiro, F. G., & Barros, E.g., D., (2008). Pathosystem common bean *Uromyces appendiculatus*: Host resistance, pathogen specialization, and breeding for rust resistance. *Pest Technology, 2*(2), 56–69.
Srivastava, K. M., Raizada, R. K., & Singh, B. P., (1985). *Gemini Viruses: Perspective in Plant Viroology* (Vol.1, pp. 153–188). Print House (India) Lucknow.
Suryawanshi, A. P., Wadje, A. G., Gawade, D. B., Kadam, T. S., & Pawar, A. K., (2009). Field evaluation of fungicides and botanicals against powdery mildew of mungbean. *Agric. Sci. Digest., 29*(3), 209–211.
Tariq, V. N., & Magee, A. C., (1990). Effect of volatiles from garlic bulb extracts on *Fusarium oxysporum* fsp. *lycopersici. Mycological Res., 94*(5), 617–620.
Taunk, J., Bharti, A. B., Yadav, N. R., Yadav, R. C., & Yadav, R. K., (2012). Genetic diversity among green gram [*Vigna radiata* (L.) *Wilczek*] genotypes varying in micronutrient (Fe and Zn) content using RAPD markers. *Indian Journal of Biotechnology, 11*, 48–53.
Thakur, M. P., & Khare, M. N., (1992). Epidemiology of mungbean anthracnose. *Indian J. Pulse Res., 5*, 49–52.
Tiwari, A., & Khare, M. N., (1998). Variability among isolates of *Rhizoctonia solani* infecting mungbean. *Indian Phytopath., 51*, 334–337.
Uddin, N. M., Bakr, M. A., Islam, M. R., Hossain, M. I., & Hossain, A., (2013). Bioefficacy of plant extracts to control *Cercospora* leaf spot of mungbean (*Vigna radiata*). *Int. J. Agril. Res. Innov. and Tech., 3*(1), 60–65.

Verma, A., Dhar, A. K., & Malathi, V. C., (1991). Cloning and restriction analysis of mungbean yellow mosaic virus. In: *International Conf. Virology in the Tropics Lucknow* (p. 14). India.

Verma, H. S., & Thapliyal, P. N., (1976). Rhizoctonia aerial blight of soybean. *Indian Phytopath., 29*, 389–391.

Yadav, D. L., Pandey, R. N., Jaisani, P., & Gohel, N. M., (2014). Sources of resistance in mungbean genotypes to *Cercospora* leaf spot disease and its management. *African J. Agril. Res*., *9*(41), 3111–3114.

Yadav, D. L., Pratik, J., & Pandey, R. N., (2014). Identification of source of resistance in mungbean genotypes and influence of fungicidal application to powdery mildew epidemics. *International Journal of Current Microbiology and Applied Sciences*, *3*(2), 513–519.

Yang, X. B., Snow, J. P., & Berggren, G. T., (1990). Analysis of epidemics of rhizoctonia aerial blight of soybean in Louisiana. *Phytopathology, 80*, 386–392.

Zeshan, A. M., Ali, S., Khan, M. A., & Shahi, S. T., (2012). Role of nutrients and naphthalene acetic acid in the management of Urdbean leaf crinkle virus. *Pak. J. Phytopathol*., *24*, 79–81.

CHAPTER 5

Integrated Disease Management of Horse Gram (*Macrotyloma uniflorum*)

UPMA DUTTA, [1] SACHIN GUPTA, [2] ANAMIKA JAMWAL, [3] and SONIKA JAMWAL[4]

[1]*Division of Microbiolgy SKUAST-J, Jammu, Jammu & Kashmir, India*

[2]*Division of Plant Pathology, SKUAST-J, Jammu, Jammu & Kashmir, India*

[3]*KVK Kathua, SKUAST-J, Jammu, Jammu & Kashmir, India*

[4]*ACRA, SKUAST-J, Jammu, Jammu & Kashmir, India*

Horsegram (*Macrotyloma uniflorum*) (Lam.) Verdc (Family: Fabaceae) is one of the lesser known beans. It is known as poor men's and a crop of poor resource is widely grown in India in almost 200–700 mm rainfall situations at a temperature range of 20–35°C. The seeds of horsegram are utilized as cattle feed and the stems are slightly hairy. It has trifoliate leaves with each leaflet growing between 2.5 cm and 5 cm in breadth. These plants are indigenous the southeastern regions of India. This crop is also found in other parts of the world including Africa, Malaysia, Australia, and the West Indies. They are generally grown from the seeds. The plants can adapt to a wide range of soil types, tropic, and sub-tropic climates for their growth. They can tolerate saline soils with the preferred pH range being 6.0 to 7.5. These plants do not survive in frost and extremely cold weather. The seeds of the horsegram can be harvested by hand or by a harvester. This crop can be consumed as a whole seed, as sprouts, or as whole meal in India, popular especially in southern Indian states. Horsegram is grown in almost all states of India and grown in an area of 0.51 M ha, but its 90–95% area is confined to five major states of Orissa (16.0%), Tamil Nadu (18.0%), Karnataka, (34.0%), Maharashtra (18.0%), and Andhra Pradesh (16.0%).

Horse gram is an important arid legume and generally grown without agronomic and plant protection inputs. However, due to somewhat congenial situations with high humidity and appropriate temperature prevailing in northern hilly tracts and southern high rainfall zone, diseases may appear. Hence, some important diseases which damage horse gram have been discussed below.

5.1 ANTHRACNOSE DISEASE

5.1.1 ECONOMIC IMPORTANCE

Anthracnose is one of the most important diseases in various leguminous crops, viz., cowpea, soybean, French bean, faba bean, etc., including horsegram. The pathogen is known to infect leaves, stems, and pods and can cause considerable yield losses. The incidence of anthracnose infection is 68% in horsegram.

5.1.2 CASUAL ORGANISMS

Colletotrichum lindemuthianum (Sacc. & Mogn.) Bri & Cay.

5.1.3 SYMPTOMS (FIGURE 5.1)

- Most prominent symptom is a characteristic spotting on the pods.
- Firstly, water soaked lesions appear on the pods, later becoming brown and enlarging to form circular spots of varying size.
- Spots are usually depressed with dark centers, and bright red, yellow, or orange margins.
- Spots may occur more often on the under surface of the leaf than on the upper surface and also occur on the petioles and stems.
- When the infections are severe on the leaf petiole and stem the affected parts may wither off. Often the seedlings are blighted due to infection soon after the seeds germinate.

FIGURE 5.1 Anthracnose disease symptoms on pods of Horse gram (*Macrotyloma uniflorum*) (Lam.)

5.1.4 DISEASE CYCLE

Colletotrichum lindemuthianum deploys a complex life cycle which has various development phases (Figure 5.2).

In the imperfect form of *C. lindemuthianum* the reproduction is asexual, the spores are produced inside acervulus and immerse in water-soluble pre-formed mucilage (O'Connell, 1996). The development of fungal spore shows a biphasic behavior which means two life styles, as a saprophyte and biotroph; therefore, the fungus has been classified as hemibiotroph. In life style saprophytic the fungus growth in any carbon source including crystalline cellulose which may be easily converted into molecules fuel by extracellular lytic enzymes. On the other hand, as a biotroph fungus has the ability to feeding of nutrients outright of living plants. As a saprophyte fungus, the spore germination process begins with the spore adhesion to the plant surface under adequate humidity conditions; specifically, correct aqueous content in the spore envelope (mucilage). At this level, the oval spores of the fungus round off by water absorption and active growth. Later, the germinating tube is formed (germinule phase), and the hyphae elongate to colonize the substrate. The aerial mycelia appear; then the fungal reproductive structures are formed where the spores are storage. Finally, their life cycle is completed and it starts all over again.

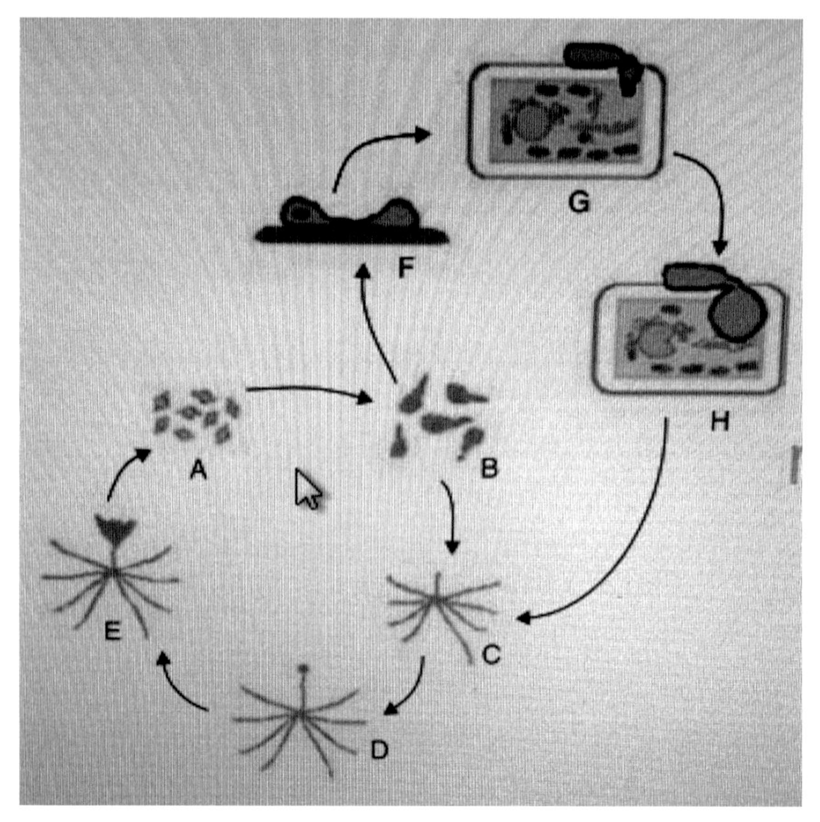

FIGURE 5.2 *Colletotrichum lindemuthianum* disease cycle (A: Spores; B: Germination; C: Vegetative growth; D: Aerial structures; E: Sporogenesis; F: Spore adhesion G: Penetration; H: Primary hyphae formation).

5.1.5 MANAGEMENT

- Crop rotation and field sanitation are convenient approaches to manage the disease.
- Seed treatment with carbendazim @ 2 g/kg of seed will able to manage the primary inoculums.
- Spray with 0.1% carbendazim can managed the disease incidence in the field.

5.2 POWDERY MILDEW

5.2.1 ECONOMIC IMPORTANCE

Powdery mildew is an important disease causing considerable yield loss proportional to the disease severity and the stage at which it occurs. Yield reduction is characterized by the reduction in photosynthetic activity and physiological changes. Severity of disease leads to potential decrease in seed yield.

5.2.2 CASUAL ORGANISMS

Leveillula taurica (Lev.) Arnaudi.

5.2.3 SYMPTOMS (FIGURE 5.3)

- A white powdery mass grow on the leaf surface, spreading to cover the stem and other plant parts.
- Infection are severe when the plants are flowering and it persists until harvest.
- During extreme infection, the entire plant becomes affected and wilted.

5.2.4 DISEASE CYCLE

Infection and colonization of *L. taurica* was studied and described as under:

- After infection to the host leaf, the Conidia start germinate within 2 hours and will reach maximum germination rate after 3 hours.

FIGURE 5.3 Powdery mildew disease symptoms on leaves of Horse gram (*Macrotyloma uniflorum*) (Lam.).

- Germ tubes develop from base of conidium and each conidium produces single germ tube that grows in the direction and orientation of the conidia.
- The germ tube may pass over a stomata and can enter stoma without appressorium and within six hours a lobed appressorium is formed at the apex of the germ tube.
- An infection hypha emerges from the appressorium and grows through a stoma into the leaf tissue.
- The infection hypha elongates into the mesophyll tissue via intercelluar growth and enters into the spongy parenchyma tissue to form haustoria.

- Penetration peg arise from intercellular hyphae and penetrate the host's cell walls with papillae forming against the outer cell wall at the site of penetration.

5.2.5 *MANAGEMENT*

- Grow plant resistant varieties to combat the powdery mildew disease.
- Two spray of calixin (0.05%) were found effective in controlling powdery mildew of horse gram with higher cost: benefit ratio.

5.3 DRY ROOT ROT

5.3.1 *ECONOMIC IMPORTANCE*

Dry root rot can infect the horse gram at the time of flowering and pod formation stages. This disease can be seen in other crops like cowpea, black gram and green gram, etc.

5.3.2 *CAUSAL ORGANISMS*

Macrophomina phaseolina (Tassi) Goud.

5.3.3 *SYMPTOMS*

- First outward symptom of the disease is yellowing of the leaves. Within three or four days they drop off.
- Affected plants wilt and die within a week.
- Bark of the root and basal stem are fibrous and are found associated with black powdery mass of sclerotia of the fungus.
- Plants bear pods with partially filled seeds. The disease appears in patches and becomes severe during dry periods.

5.3.4 *DISEASE CYCLE*

M. phaseolina has a monocyclic disease cycle.

5.3.5 SURVIVAL

- The *M. phaseolina* fungus has aggregates of hyphal cells, which form microsclerotia.
- Microsclerotia overwinter in the soil and crop residue during unfavorable conditions and during spring these microsclerotia become the primary source of inoculum.
- They can survive in the soil for up to three years. They are black, spherical, or oblong structures that allow the persistence of the fungus under poor conditions, such as low soil nutrient levels and temperatures above 30°C. However, in wet soils, microsclerotia survival is significantly lower; often surviving no more than 7 to 8 weeks, and mycelium cannot survive more than 7 days. Additionally, infected seeds can carry the fungus in their seed coats. These infected seeds either do not germinate or produce seedlings that die soon after emergence.

5.3.6 INFECTION

M. phaseolina produce large quantity of microsclerotia under low water potential and high temperatures. When conditions are favorable, hyphae germinate from these microsclerotia. Microsclerotia germinate on the root's surface, and germ tubes on the end of the microsclerotia form appresoria that penetrate the host's epidermal cell walls using turgor pressure or through natural openings.

The hyphae enter the cortical tissue and grow intercellularly, then infect the roots and the vascular tissue. The mycelia and sclerotia are produced within the vascular tissue and plug the vessels and cause the greyish-black color in plants, and it also prevents water and nutrients from being transported from the roots to the upper parts of the plant. Thus, due to this systemic infection, diseased plants often wilt and die prematurely.

5.3.7 MANAGEMENT

- Application of organic amendments like farmyard manure, neem cake and seed treatment with *Trichoderma viride* @ 4 g/1 kg of seed and *Pseudomonas fluorescens* @ 10 g/kg of seed can reduce the disease Incidence.

- Enhanced soil moisture can reduce the *M. phaseolina* population.

5.4 RUST

5.4.1 ECONOMIC IMPORTANCE

Horse gram rust is of great economic importance as they cause huge damage to the crops. Disease is reported all growing countries in world and also in India. Yield reduction is characterized by the reduction in photosynthetic activity and physiological changes. Severity of disease leads to potential decrease in seed yield.

5.4.2 CASUAL ORGANISMS

Uromyces phaseoli typica Arth. (Syn. U. appendiculatus (Pers) Fries.

5.4.3 SYMPTOMS

- Fungus produces characteristic rust pustules on the plant;
- Pustules arc mostly found on the leaf blade;
- They are more conspicuous on the under surface of the leaves;
- Infection may spread to young stem also;
- Plants with heavy rust infection will give a brown tinge when looked from a distance;
- In advanced stage of infection the leaf may wither resulting in considerable damage to the crop.

5.4.4 DISEASE CYCLE

It is autoecious, long cycle rust and all the spore stages occur on the same host. The uredospores are unicellular, globose or ellipsoid, yellowish brown with echinulations. The teliospores are globose or elliptical, unicellular, pedicellate, chestnut brown in color with warty papillae at the top. Yellow colored pycnia appear on the upper surface of leaves. Orange colored cupulate aecia develop later on the lower surface of leaves. The aeciospores are unicellular and elliptical.

5.4.5 FAVORABLE CONDITIONS

- Cloudy and humid weather with temperature of 21–26°C.
- Heavy dews in nights.

5.4.6 MODE OF SPREAD AND SURVIVAL

The pathogen survives in the soil by teliospores and uredospores in plant debris. Primary infection is by sporidia which is developed from teliospores. Secondary infection is by wind-borne uredospores. The fungus also survives on other hosts.

5.4.7 MANAGEMENT

- Development of rust resistant varieties and adjusting the sowing dates of the crop to avoid severe infections of the plant in the field may help considerably in reaping better harvest.

5.5 AERIAL BLIGHT

5.5.1 ECONOMIC IMPORTANCE

Aerial blight of horse gram cause extensive yield losses. The fungus survives in the soil and on plant debris, as well as on certain weed hosts. Infection usually begins at flowering during prolonged periods of high humidity and warm temperatures. The disease can cause 10–60% loss depending upon the severity of the disease.

5.5.2 CASUAL ORGANISMS

Rhizoctonia solani Kuhn.

5.5.3 SYMPTOMS

- Aerial blight is seen on the foliage as irregular water soaked area.
- Under high atmospheric humidity the spots coalesce rapidly and cover a large part of the leaf lamina.

- White mycelial growth seen on the affected area.
- Severely affected leaves shed in large number.

5.5.4 DISEASE CYCLE

R. solani can survive for many years in the soil and on plant tissues by producing small (1 to 3-mm diameter), irregular-shaped, brown to black structures called sclerotia. The fungus is attracted to the plant by chemical stimuli released by a growing plant and/or decomposing plant residue. The process of penetration of a host can be accomplished in a number of ways. Entry can occur through direct penetration of the plant cuticle/epidermis or by means of natural openings in the plant. Hyphae come in contact with the plant and attach to the plant by which through growth they begin to produce an aspersorium which penetrates the plant cell and allows for the pathogen to obtain nutrients from the plant cell. The pathogen can also release enzymes that break down plant cell walls, and continue to colonize and grow inside dead tissue. This breakdown of the cell walls and colonization of the pathogen within the host forms the sclerotia. New inoculums are produced on or within the host tissue, and a new cycle is repeated when new plants become available. The disease cycle begins as such:

5.5.5 MANAGEMENT

- Soil application of *Trichoderma viride*, *T. harzanium,* and *Gliocladium virens* significantly reduce the mycelial and sclerotial production of *R. solani.*
- Use of resistant varieties will be the only economical control measure.

5.6 CERCOSPORA LEAF SPOT (CLS)

5.6.1 ECONOMIC IMPORTANCE

Pathogenis widespread in warmer regions, occurring on various legumes. They can cause considerable leaf spotting of horse gram after flowering when defoliation can lead to Yield losses of up to 20%.

5.6.2 CASUAL ORGANISMS

Cercospora dolichi EIL & EVr.

5.6.3 SYMPTOMS

- Water soaked lesions appear on the leaf blade and soon the affected tissues turn brown to reddish brown.
- Spots are bound by veins with purplish border developed, the center of may turn gray in color.
- Powdery growth, due to the sporulation of the fungus, may also be seen at the center of the spots.
- Spotting is mostly confined to leaf blades, but occasionally may occur on the fruit and floral parts also.
- In some cases, the affected tissue is killed and the dead portion drops out, leaving short hole symptoms on the leaves.

5.6.4 DISEASE CYCLE

The fungus survives in infected seeds and infested crop debris. Infected seeds produce seedlings with cotyledonary lesions. Sporulations on the cotyledonary lesions provide primary inoculum for infecting young leaves. In warm humid weather, sporulation is profuse in the resulting lesions. Conidia dissemination by wind and splashing rain to another leaves and stem to initiate secondary infection that produce conidia which again infect another leaves, stems, and pods during warm and wet conditions. Peak dispersal period for conidia occur at dew dry-off in the morning and at the onset of rainfall.

Besides, the pathogen, being capable of infecting several hosts, which are grown round the year, can easily survive to become more active when climatic conditions are favorable.

5.6.5 MANAGEMENT

- Seed treatment with Captan 3 g/kg of seed can be used.
- Spraying the infected plant with Mancozeb (0.2%) can reduce the incidence of the disease.

5.7 YELLOW MOSAIC DISEASE

Mungbean yellow mosaic Gemini virus.

5.7.1 SYMPTOMS

- Yellow patches or spots appear on the green lamina.
- Young leaves first show the symptoms.
- Later, the area of yellow discoloration increases in new growth and completely turns yellow.
- Infected plants mature later and bear very few flowers and pods.
- Pods are small and distorted. The seeds are also reduced in size and are shriveled.

5.7.2 MANAGEMENT

- Use of resistant cultivars for cultivation is one of the best strategies of disease management.
- Removal of the infected plants and periodical spray with systemic insecticide Monocrotophos (0.1%) or neem oil soap emulsion (3% spray) and early sowing can minimize the disease incidence.

KEYWORDS

- **aerial blight**
- **cercospora leaf spot**
- **globose**
- **rust**
- **teliospores**
- **uredospores**

REFERENCES

Aghakhani, M., & Dubey, S. C., (2009). Morphological and pathogenic variation among isolates of *Rhizoctonia bataticola* causing dry root rot of chickpea. *Indian Phytopath.*, *62*(2), 183–189.

Chahota, R. K., Sharma, T. R., Dhiman, K. C., & Kishore, N., (2005). Characterization and evaluation of Horsegram (*Macrotylo mauniflorum* Roxb.) germplasm from Himachal Pradesh. *Indian J. Plant Genet. Resour.*, *18*, 221–223.

Khan, R. A., Bhat, T. A., & Kumar, K., (2012). Management of chickpea (*Cicer arietinum* L.) dry root rot caused by *Rhizoctonia bataticola* (Taub.) Butler. *International Journal of Research in Pharmaceutical and Biomedical Sciences*, *3*(4), 1539–1548.

Kumar, D., (2007). In: Kumar, D., (ed.), *Production Technology for Horse Gram in India* (pp. 1–15).

Monger W. A., Harju, V., Nixon, T., Bennett, S., Reeder, R., Kelly, P., & Ariyarathne, H. M., (2010). First report of Horsegram yellow mosaic virus infecting *Phaseolus vulgaris*in Sri Lanka. *New Disease Reports* (p. 16).

Murthy, R., (1997). Studies on seed borne aspects and control of anthracnose of horsegram (*Macrotylo mauniflorum*) caused by *Colletotrichum dematium* (Pers. ex. Fr.) groove. MSc Thesis. University of Agricultural Sciences, Bangalore, Karnataka, India.

Singh, R. A., Gurha, S. N., & Ghosh, A., (2005). Diseases of chickpea and their management In: Thind, T. S., (ed.), *Diseases of Field Crop and Their Management* (pp. 179–204).

Udaya, S. A., Anitha, K., Sivara, J. N., Meena, K. K. V. S., Sunil, N., & Chakrabarty, S. K., (2015). Screening of horsegram germplasm collected from Andhra Pradesh against anthracnose. *Legume Research*, *38*(6), 753–757.

CHAPTER 6

Diseases of Lentil (*Lens esculentus* Moench) and Their Management

RANGANATHSWAMY MATH,[1] A. K. PATIBANDA,[2] and J. N. SRIVASTAVA[3]

[1]*Department of Plant Pathology, College of Agriculture, Jabugam, Anand Agricultural University, Gujarat, India, E-mail: rangu. math@gmail. com*

[2]*Department of Plant Pathology, Agriculture College, ANGRAU, Bapatla, Andhra Pradesh, India*

[3]*Department of Plant Pathology, Bihar Agricultural University, Sabour–813210, Bhagalpur, Bihar, India*

6.1 INTRODUCTION

Lentil (*Lens culinaris* Medic.) is one of the protein-rich food legumes grown in India. It is cultivated in Northern and Central India. The area under lentil in India is around 1.59 m. ha with a production of 0.94 m. t and productivity 697 kg/ha (Anonymous, 2011). Owing to biotic and abiotic stresses, the crop yield is below attainable levels. Among the biotic factors, diseases are serious threat to lentil production. Lentil suffers from a number of diseases which are caused by fungi, bacteria, viruses, nematodes, and plant parasites. Among different diseases, the major diseases are discussed in this chapter.

6.2 ASCOCHYTA BLIGHT

6.2.1 INTRODUCTION/ECONOMIC IMPORTANCE

Ascochyta blight of lentil was first observed at Palampur, India in 1934. It is present in from Argentina, Australia, Brazil, Canada, Greece, Italy, Jordan, New Zealand, Pakistan, Russia, Syria, Turkey, and the USA. This disease mainly affect seed and there by its marketability.

6.2.2 SYMPTOMS

Disease symptoms appear on all the above ground parts of the plant. Initially small circular, pale green lesions appear on leaves. Lesions soon enlarge coalesce and turn to light tan with age. Small black, pin-prick sized pycnidia develop and become visible within lesion centers as they mature, particularly on the leaves. On stem lesions are elongated and brown. In severe infection, stem breakage can occur via girdling at the point of infection and subsequent death of all tissues above the lesion. Under severe infection, the leaves become blighted and pods become undersized. Seeds from the affected pods may be shriveled and discolored. Affected crops under high incidence give blighted appearance.

FIGURE 6.1 Aschochyta blight infected leaves

6.2.3 CAUSAL ORGANISM

Ascochyta lentis Bond. & Vassil.
Kingdom: Fungi
Phylum: Ascomycota
Class: Dothideomycetes

Order: Pleosporales
Family: Incertae sedis
Genus: *Ascochyta*
Species: *lentis*

It produces septate branched hyphae. It forms asexual fruiting body, i.e., pycnidia which are brown, gregarious, superficial, ostiolate, depressed, and globose. Conidiophores are cylindrical, straight, or rarely curved, rounded at the ends and hyaline. Conidiospores are generally uni- and bi-septate, with a small percentage triseptate.

6.2.4 DISEASE CYCLE AND EPIDEMIOLOGY

The fungus is both seed and soil borne. It survives in the infected plant residues as well as with the seed. Seedling is infected by windblown conidia produced on diseased residues or from infected seed itself. Secondary spread of the disease takes place through conidia produced in the infected lesions. The pathogen produces pseudothecia that over-winter on lentil debris. Pseudothecia erupt from the tissue surface under cool and moist conditions and release ascospores.

The pathogen favors cool, moist conditions for spread of the disease. An extended period of leaf wetness is required for disease development, with maximum disease developing occurring after 24 to 48 hours of leaf wetness. Rain accompanied by wind is more helpful in dispersal of pathogen spores.

6.2.5 MANAGEMENT

The most economical and sustainable strategies to control ascochyta blight are through resistance breeding along with cultural practices.

- Use disease-free seeds for sowing.
- Deep plowing during summer will reduce soil borne inoculum.
- A minimum of two-year non-host crop rotation will help to manage the disease.
- Early sowing of the crop will help to escape from the disease.
- Take foliar spray of benomyl or carbendazim or ipodion @ 0.1% at the initial appearance of the disease.

6.3 ANTHRACNOSE

6.3.1 INTRODUCTION/ECONOMIC IMPORTANCE

Anthracnose of lentil is caused by *Colletotrichum truncatum* (Schw.) Andrus and Moore. The disease is most serious in western Canada where it was first discovered in 1987 in the province of Manitoba and three years later in Saskatchewan. Disease is present in most of the lentil growing countries such as USA, Bulgaria, Brazil, Pakistan, and Bangladesh.

6.3.2 SYMPTOMS

Symptoms can be seen on all the aboveground plant parts. The initial symptoms of the disease can be observed on lower leaflets when the plants have between 8 and 12 nodes. Initially water soaked lesions appear on the leaves. Under severe infection foliage wilts and falls to the ground. On stem lesion are elongated and dark brown. Lesions soon encircle, girdle the stem resulting in lodging of plant. Acervuli containing conidia and setae develop in lesions on leaflets and stems. Black, pinhead sized microsclerotia can be noticed on the infected stem. Defoliation and stem girdling reduces absorption of mineral nutrients leading to drying of plants. Under severe infection, the entire field shows patches of brown appearance. On pods, small brown necrotic lesions appear and seeds from such infected pods are discolored and greatly shriveled.

6.3.3 CASUAL ORGANISM

Colletotrichum truncatum (Schw.) Andrus and Moore.

The pathogen produces asexual fruiting body, i.e., Acervuli in the infected plant parts. Acervuli contain conidia and black setae. The conidia are single-celled, hyaline, and slightly falcate. Dark brown setae protrude above the conidial masses helps in dispersal of conidia. Pathogen also produces microsclerotia on the infected stem which are hardy structures. Perithecia with mature asci developed after 10–14 days. Perithecia are obpyriform having long neck. The asci measured 53–142 μm x 5–14 μm and each contained eight single-celled ascospores, 12–20 μm x 5–8 μm.

Kingdom: Fungi
Phylum: Ascomycota

Class: Sordariomycetes
Order: Glomerellales
Family: *Glomerellaceae*
Genus: *Colletotrichum*
Species: *truncatum*

6.3.4 DISEASE CYCLE AND EPIDEMIOLOGY

Pathogen survives as microsclerotia on infested plant debris. Microsclerotia can survive for 3 years without losing viability. Microsclerotia surviving in the soil on plant debries initiate the disease in the next growing season. Lentil plants are vulnerable to infection at 30–45 days old. Secondary spread of the disease takes place through conidia carried to the neighboring plants by rain droplets. Under optimum environmental conditions, pathogen completes several infection cycles leading to spread of the disease very quickly. The disease is favored by longer leaf wetness period with moderate temperature of 20 and 25°C.

6.3.5 MANAGEMENT

- Do not take lentil crop on the infested field.
- Rotate the lentil crop with other non-host crop.
- Select healthy disease free seeds for sowing.
- Seed treatment with Captan or thiram @ 3 g/kg of seed will protect the plant at initial stages.
- Take foliar spray with chlorothalonil @ 0.2% when the initial symptoms appear.

6.4 WILT

6.4.1 INTRODUCTION/ECONOMIC IMPORTANCE

Wilt is one of the major diseases of lentil affecting the productivity worldwide. Wilt disease has been reported from Argentina, Bangladesh, Brazil, Canada, Egypt, the USA, Pakistan, Nepal, and India. In India, lentil wilt was first reported from Bengal in 1934. In India, it has been reported from New Delhi, UP, Madhya Pradesh, Bihar, West Bengal, Rajasthan, Haryana, and Punjab. Yield losses vary with time of infection. Infection at early stage results in 100% loss. Yield loss due to wilt ranges between 25–70%. Wilt disease cause more damage at flowering and pod formation stage. In Madhya Pradesh, wilt incidence of 50–78% is reported.

6.4.2 SYMPTOMS

Infection of lentil can occur at different stages. Infection at pre-emergence stage can cause seed rotting and failure of seedling emergence. In post-emergence, infection on seedling cause sudden drooping, followed by drying of leaves and seedling death.

FIGURE 6.2 Wilting symptoms of crop plants: field view in spite of Wilting of plants

Infection at pre flowering to pod setting stage cause drooping and wilting of the uppermost leaflets followed by shrinking and curling of leaves that starts from the lower part of the plants and progressively moves up the stem. Infected root develops poorly and turn yellowish brown to dark brown. The taproot is completely killed. Transverse cutting of infected stem shows brown streaks in the vascular region. Sometimes white fungal hyphal growth can also be observed. Seeds from plants affected in mid-pod-fill to late pod-fill are often shriveled.

6.4.3 CASUAL ORGANISM

Fusarium oxysporum f. sp. *lentis*

Pathogen produces septate hyaline-branched mycelium. It produces micro and macro conidia as well as chlamydospores. The microconidia are hyaline, single celled and ovoid in shape. Macroconidia are long 2–7 septate. The chlamydospores are oval in shape, one celled and formed on apical or intercalary in the hyphae.
Kingdom: Fungi
Phylum: Ascomycota
Class: Sordariomycetes
Order: Hypocreales
Family: Nectriaceae
Genus: *Fusarium*
Species: *Oxysporum* f. sp. *lentis*

6.4.4 DISEASE CYCLE AND EPIDEMIOLOGY

The pathogen is soil as well as seed borne. It survives through chlamydospores in dead plant debries in soil. It can survive in the soil for six years. The primary infection is from dormant mycelium or from chlamydospores. Infection spread within the growing season by micro and macro conidia. Key factors in determining symptom expression and fungal growth rates are temperature and soil moisture (Dhingra et al., 1974; Saxena and Khare, 1988). Disease will be more severe at 20–30°C soil temperature with 30% soil water holding. Disease will be severe in light soil as compared to heavy soil. Wilt incidence decreases with increase in soil moisture.

6.4.5 MANAGEMENT

- Use resistant cultivars like Pant L-4, Pant L-6, Pant L-8 and Noori.
- Summer plowing or soil solarization during summer is effective in reducing wilt.
- Treat the seeds with benomyl or carbendazim @ 2 g/kg of seed reduces wilt incidence.
- Rotate the lentil with cereal crops.

- Adjusting the sowing time, late sowing will reduce the wilt incidence compared to early sowing (October).
- Seed treatment with *Trichoderma harzianum* or *T. viride* @ 4 g/kg seed.
- Soil application of *T. viride* (1 kg) mixed with well rotten FYM (25 kg/acre) before sowing.
- Removal and destruction of severely infected plants will reduce the further spread of the disease.
- Application of Zinc @ 10 kg/ha will also help to control the wilt disease.

6.5 BOTRYTIS GREY MOLD (BGM)

6.5.1 INTRODUCTION/ECONOMIC IMPORTANCE

Botrytis grey mold (BGM) of lentil is reported from Australia, Argentina, Nepal, Myanmar, Bangladesh, and Pakistan. The pathogen causes heavy losses particularly in North Indian conditions. Under severe infection, the yield loss may reach up to 70–80%.

6.5.2 SYMPTOMS

On the lower leaves initially small dark green lesions appear. Lesions soon turn to dark brown enlarge and coalesce. Infected leaves can become covered with a fuzzy layer of grey mold near the base of the plant. Under severe infection leaves dry and are shed. Under humid conditions, light brown lesions develop on stem which soon covered with fungal growth. Conidia develop on the infected plant parts. Infected stem girdle and fall down before setting the pods. Infected pods will fail to fill properly. Infected seeds may be discolored and shriveled.

6.5.3 CAUSAL ORGANISM

Botrytis cinerea
Kingdom: Fungi
Phylum: Ascomycota
Subphylum: Pezizomycotina
Class: Leotiomycetes
Order: Helotiales

Family: Sclerotiniaceae
Genus: *Botrytis*
Species: *Cinerea*

Colonies of *B. cinerea* are white and cottony and turn grey or greyish-brown with age. The hyphae are septate and brown. Conidiophores are septate, brown, erect producing conidia in clusters. Conidia are hyaline, single-celled, ovoid or spherical and thin walled. The fungus produces sclerotia which are black and small.

6.5.4 DISEASE CYCLE AND EPIDEMIOLOGY

The fungus is both soil and seed borne. The pathogen perennates as sclerotia in the infected plant debries. The primary infection takes place through conidia produced on previously infected plant residues. Under humid conditions, pathogen grows on the infected plant and produce masses of conidia. Conidia carried by wind and spread the disease in the entire field. High humidity and moderate temperatures with high moisture favors the disease. Environmental conditions and canopy density influence the development and outbreak of disease in lentil. Temperatures between 15–25°C with high relative humidity (RH) at flowering and after canopy closure favors disease outbreak. Excessive vegetative growth, heavy rains, dense planting and humid conditions favor the disease.

6.5.5 MANAGEMENT

- Use clean and healthy seeds for cultivation.
- Cultivation of resistant genotypes like Pant L-639 and Pant L-406.
- Treat the seeds with benomyl, carboxin or chlorothalonil @ 0.1% before sowing.
- Keep the field weed free and remove the severely diseased plants.
- Use optimum fertilizer dose.
- Avoid excess application of nitrogen which helps to prevent development of dense canopies.
- Destruction of infected debris through burning or burying residues.
- Foliar application of *Pseudomonas fluorescens* @10 ml/lit or *Trichoderma* spp. @ 6 gm/lit of water.

- Take foliar spray with carbendazim or benomyl @ 0.1%.
- Rotate the lentil with other non hosts to break the infection cycle of the pathogen.

6.6 POWDERY MILDEW

6.6.1 INTRODUCTION/ECONOMIC IMPORTANCE

It is an important disease of lentil reported from Argentina, Chile, India, Italy, Jordan, Mexico, Romania, Sudan, and Tanzania. The disease is caused by powdery mildew fungi in two genera, *Erysiphe*, and *Leveillula*. *Erysiphe pisi* on lentil has been reported from various countries including India.

6.6.2 SYMPTOMS

Initially small white spots appear on leaf surfaces. Later lesions expand to cover the entire leaf surfaces and pods. Under heavy infection, leaves become chlorotic, then curled and necrotic prior to abscission. Yields can decline and plants sometimes die.

6.6.3 CAUSAL ORGANISM (ERYSIPHE PISI)

Erysiphe pisi produces ellipsoid-cylindrical conidia on conidiophores. Each chasmothecium contains 4–8 ellipsoid asci. Each ascus contains 3–6 ellipsoid-ovoid ascospores. *L. taurica* produces endophytic mycelium, with conidiophores arising in groups from stomata. It also produces distinctive dimorphic conidia.

Kingdom: Fungi
Phylum: Ascomycota
Class: Leotiomycetes
Order: Erysiphales
Family: Erysiphaceae
Genus: *Erysiphe*
Species: *pisi*

6.6.4 DISEASE CYCLE AND EPIDEMIOLOGY

Life cycles are inadequately investigated and very little is known about the biology of powdery mildew species on lentil. In some areas, the chasmothecium survives on infected plant debries. Ascospores produced in the chasmothecium acts as primary source of inoculum. Secondary spread of the disease takes place through windblown conidia. *Erysiphe pisi* is also a seed borne. Infected seeds themselves act as primary source of inoculum. Disease is favored by moderate temperatures and shady conditions.

6.6.5 MANAGEMENT

Foliar spray with benomyl or tridemorph @ 0.1% may prevent disease development.

6.7 RUST

6.7.1 INTRODUCTION/ECONOMIC IMPORTANCE

Rust disease is a limiting factor in successful cultivation of lentil in many regions. It is reported from Algeria, Argentina, Bulgaria, Chile, Colombia, Cyprus, Hungary, India, Iran, Israel, Italy, Jordan, Palestine, Peru, Portugal, Sicily, Syria, and Turkey. In India is most economically important foliar disease affecting lentil. The disease can result in complete crop failure although yield losses on a plot basis vary from 30 to 60% depending on cultivar and disease severity. The yield losses depend upon the stage at which the crop is infected. Rust disease is a potential threat to lentil cultivation and causes substantial yield losses ranging from 60–69% (Sepulveda, 1985). Complete yield losses up to 100% was reported during 2008–09 in Uttarakhand state.

6.7.2 SYMPTOMS

Symptoms can be noticed on all the above-ground parts of lentil *vz.,* leaf, leaf blade, petiole, stem, and pods. Symptoms of the disease start with appearance of yellowish-white pycnia on the leaves. The yellowish-white aecia appear first on lower surface of leaves, stem, and petioles. Formation of aecial stage is preceded by brown uredial stage. The uredo pustules are more or less circular and brown in appearance. Uredo pustules coalesce with each other and form large patches. The black colored teleuto pustules occur in the same mycelium at the end of the crop growth stage. Severe infections can cause defoliation and deformity of stem and plant dies. Infected plants dries forming only small shriveled seeds.

6.7.3 CAUSAL ORGANISM

Uromyces viciae-fabae
Phylum: Basidiomycota
Sub Phylum: Pucciniomycotina
Class: Pucciniomycetes
Order: Pucciniales
Family: Pucciniaceae
Genus: *Uromyces*
Species: *Viciae-fabae*

Rust pathogen is autoecious and forming all the spores on lentil. Pycnial and aecial stages are dominating on the lentil crop. Spermagonia are subepidermal and globoid. Aecia and uredia are subepidermal in origin, erumpent late. Aeciospores are elliptical, yellowish-brown. Uredospores are brown, echinulated and formed singly on pedicels. Telia are subepidermal in origin, then erumpent on leaves. Teliospores are globose and single celled

6.7.4 DISEASE CYCLE AND EPIDEMIOLOGY

Rust pathogen survives on infected plant debris or on seed as contaminant. Teliospores act as a primary source of inoculum. Teliospores are resting structure of the pathogen which can survive for 2 years without losing viability. Secondary spread of the disease takes place through aeciospores and urediospores. Both aeciospores and urediospores readily germinate on plant surface under suitable weather conditions and cause repeated infections. High RH, Cloudy or drizzling weather with temperature of 20 to 28C favors development of the disease. Mittal (1997) reported that atmospheric temperature around 20°C (maximum) and 5°C (minimum) with high RH (60–70% mean weekly) and light shower or drizzle favor the development and spread of rust in lentil.

6.7.5 MANAGEMENT

- Use clean, healthy seeds without concomitant contamination.
- Cultivate disease tolerant cultivars like Bombay 18, Pant L 236, Pusa 10, Pant L-639, Pant L-406, Pant L-6, pant L-7 and Pant L-8.
- Destruct the diseased plant debries after harvesting.
- Adjusting time of sowing.
- Seed treatment with Diclobutazole or Vitavax power @ 2.5 g/kg seed.
- Three sprays of Mancozeb @ 0.2% or Tebuconazole @ 0.1% or Bayleton @ 0.05% at 10 days interval.
- Lal (2006) studied the effect of fungicides both as seed treatment and as foliar spray against rust of lentil under field conditions and reported that seed treatment with propiconazole @ 1.0 ml/kg seed followed by foliar spray of propiconazole @1 ml/l was the most effective in reducing the disease.

6.8 STEMPHYLIUM BLIGHT

6.8.1 INTRODUCTION/ECONOMIC IMPORTANCE

Stemphylium blight also known as Stemphylium leaf spot disease is caused by *Stemphylium botryosum*. The disease is most devastating in Bangladesh and the northeastern state of Bihar in India where yield losses of 80% and higher have been recorded. The disease is also present in Hungary, Egypt, Syria, and the USA.

6.8.2 SYMPTOMS

Initial symptoms on leaves and stems consist of small light beige lesions that enlarge and coalesce. They are often found first in the upper canopy from where the disease spreads to lower plant parts. Eventually, entire branches become necrotic resulting in the characteristic blighted appearance of lentil plants. Often, plants experience severe leaf drop leaving only terminal leaves on branches. The fungus also infects pedicels and flowers causing flower abortion. Severe infection can significantly reduce plant biomass, lower seed yield, decrease seed size and result in seed staining and low germination rates.

6.8.3 CAUSAL ORGANISM

Stemphylium botryosum Wallr.

The fungus produces very characteristic light brown to black conidia that are conspicuously constricted at the median transverse septum. Conidia are further partitioned through secondary trans- and longisepta.

Kingdom: Fungi
Phylum: Ascomycota
Class: Dothideomycetes
Subclass: Pleosporomycetidae
Order: Pleosporales
Family: Pleosporaceae
Genus: *Stemphylium*
Species: *Botryosum*

6.8.4 DISEASE CYCLE AND EPIDEMIOLOGY

Pathogen survives on debris of lentil or alternative host plants. Conidia of *S. botryosum* are air-borne and violent spore release from the conidiophores. Conidia germinate with several germ tubes and cause primary infection. Conidial germination and infection of lentil can occur over a wide range of temperatures from 5–30°C in the presence of free water. High RH of more than 85% in the morning and more than 50% in the afternoon, less than 8 hours of sunshine per day, more than 30 foggy and cloudy days, and an average mean temperature of 18 ± 2°C during the critical period from December to February are highly favorable for outbreak of disease resulting in epiphytotics in Indian conditions.

6.8.5 MANAGEMENT

- Three applications of iprodione or mancozeb @ 0.2% sprayed at 7-day intervals.
- Late sowing of lentil reduces disease severity.
- Removal and destruction of infected plant parts reduces the carryover of the inoculum.

KEYWORDS

- **anthracnose**
- **ascospores**
- **botrytis grey mold**
- **powdery mildew**
- **stemphylium blight**
- **wilt**

REFERENCES

Ahmed, S., Akem, C., Bayaa, B., & Erskine, W., (2002). Integrating host resistance with planting date and funigicide seed treatment to manage *Fusarium* wilt and so increase lentil yield. *Intern. J. Pest Manage, 48*, 121–125.

Akita, G., Vijay, K., & Tripathi, H. S., (2014). Control of wilt disease of lentil through biocontrol agents and organic amendments in Tarai region of Uttarakhand, India. *Journal of Environmental Biology, 35*, 1067–1070.

Ankita, G., Santosh, K., Mehi, L., & Vivek, S., (2013). Major diseases of lentil: Epidemiology and disease management-A review. *Agriways, 1*(1), 62–64.

Anonymous, (2011–2012). *Project Coordinator's Report (Rabi Crops)*. All India Coordinated Research Project on MULLaRP. Indian Institute of Pulses Research Kanpur 208 024.

Bakr, M. A., & Ahmed, F., (1992). Development of stemphylium blight of lentil and its chemical control. *Bangladesh J. Plant Pathol., 8*, 39, 40.

Beniwal, S. P. S., Bayaa, B., Weigand, S., Makkouk, K. H., & Saxena, M. L., (1993). *Field Guide to Lentil Diseases and Insect Pests*. ICARDA, Aleppo, Syria.

Bhalla, M. K., Nozolillo, C., & Schneider, E., (1992). Observations on the responses of lentil root cells to hypha of *Fusarium oxysporum. J. Phytopathol., 135*, 335–341.

Buchwaldt, L., Anderson, K. L., Morrall, R. A. A., Gossen, B. D., & Bernier, C. C., (2004). Identification of lentil germplasm resistant to *Colletotrichum truncatum* and characterization of two pathogen races. *Phytopathology, 94*, 236–243.

Buchwaldt, L., Morrall, R. A. A., Chongo, G., & Bernier, C. C., (1996). Wind borne dispersal of *Colletotrichum truncatum* and survival in infested lentil debris. *Phytopathology, 86*, 1193–1198.

Chahota, P. K., Gupta, V. P., & Sharma, S. K., (2002). Inheritance of rust resistance in lentil. *Indian Journal of Genetics and Plant Breeding, 62*, 226, 227.

Chen, W., Basandrai, A. K., Basandrai, D., Banniza, B., Buchwaldt, L., Davidson, J., Larsen, R., Rubiales, D., & Taylor, P. W. J., (2009). Diseases and their management. In: Erskine, W., Muehlbauer, F. J., Sarker, A., & Sharma, B., (eds.), *The Lentil: Botany, Production and Uses* (pp. 262–281). Common Wealth Agricultural Bureau International, U. K.

Chongo, G., Bernier, C. C., & Buchwaldt, L., (1999). Control of anthracnose in lentil using partial resistance and fungicide application. *Can. J. Plant Pathol., 21*, 16–22.

Davidson, J. A., Pande, S., Bretag, T. W., Lindbeck, K. D., & Krishna-Kishore, G., (2004). Biology and management of *Botrytis* spp. in legume crops. In: Elad, Y., Williamson, B., Tudzynski, P., & Delen, N., (eds.), *Botrytis: Biology, Pathology and Control* (pp. 295–318). Kluwer Academic Publishers, Dordrecht, Netherlands.

Haware, M. P., & Donald, D., (1992). *Integrated Management of Botrytis Gray Mold.* Summary proceedings of the BARI/ICRISAT working group meeting to discuss collaborative research on botrytis gray mold of chickpea. International Crops Research Institute for the Semi-Arid Tropics Patancheru, Andhra Pradesh 502 324, India.

Kaiser, W. J., & Hannan, R. M., (1987). Seed-treatment fungicides for control of *Ascochytalentis* on lentil. *Plant Dis., 71*, 58–62.

Kaiser, W. J., Hannan, R. M., & Rogers, J. D., (1994). Factors affecting growth and sporulation of *Ascochytafabae* f. sp. *lentis. Plant Dis., 78*, 374–379.

Kannaiyan, J., & Nene, Y. L., (1975). Note on the effect of the sowing date on the reaction of 12 lentil varieties to wilt disease. *Madras Agricultural Journal, 62*, 240–242.

Khare, M. N., & Agrawal, S. C., (1978). Lentil rust severity in Madhya Pradesh. *Proceeding of All India Pulse Workshops* (p. 3). Baroda.

Khare, M. N., Agrawal, S. C., & Jain, A. C., (1979). Lentil diseases and their control. *Technical Bulletin* (p. 29). Jawaharlal Nehru Krishi Vishwa Vidyalaya, Jabalpur, India.

Lal, H. C., Upadhyay, J. P., Jha, A. K., & Kumar, A., (2007). Survey and surveillance of lentil rust and its cross infectivity on different host. *Journal of Research (BAU), 19*(1), 111–113.

Mehrotra, R. S., & Claudius, G. R., (1972). Biological control of root rots and wilts diseases of *Lens culinaris* Medic. *Plant and Soil, 39*, 657–664.

Muehlbauer, F. J., Cubero, J. L., & Summerfield, R. J., (1985). Lentil (*Lens culinaris* Medik.). In: Summerfield, R. J., & Roberts, E. H., (eds.), *Grain Legume Crops* (pp. 266–311). Collins, London, U. K.

Pedersen, E. A., & Morrall, R. A. A., (1994). Effect of cultivar, leaf wetness duration, temperature and growth stage on infection and development of Ascochyta blight of lentil. *Phytopathology, 84*, 1024–1030.

Plant Health Australia, (2009). *Contingency Plan—Field Pea and Lentil Rusts* (pp. 1–35). *Uromycespisi & U. viciae-fabae.*

Prasada, R., & Verma, U. N., (1948). Studies of lentil rust, *Uromyces fabae* (Pres.) De Bary in India. *Indian Phytopathology, 1*, 142–146.

Reddy, R. R., & Khare, M. N., (1984). Further studies on factors influencing the mechanism of resistance to lentil (*Lens culinaris* M.) to rust (*Uromycesfabae* (Pers.) de Bary). *LENS Newsletter, 11*, 29–32.

Saxena, D. R., & Khare, M. N., (1988). Factors influencing vascular wilt of lentil. *Indian Phytopathology, 41*, 69–74.

Sepulveda, R. P., (1985). Effect of rust caused by *Uromyces fabae* (Pers) de Bary on the yield of lentil. *Agric. Technol., 45*, 335–339.

Shiv, K., Kumar, J., Singh, S., Ahmed, S., Chaudhary, R. G., & Sarker, A., (2010). Vascular Wilt disease of lentil: A review. *J. Lentil Res., 4*, 1–14.

Sinha, J. N., & Singh, A. P., (1993). Effect of environment on the development and spread of *stemphylium blight* of lentil. *Indian Phytopath., 46*, 252, 253.

CHAPTER 7

Diseases of Pigeon Pea (*Cajanus cajan* L. Millsp.) and Their Management

MANOJ KUMAR KALITA

Biswanath College of Agriculture, Assam Agricultural University, Biswanath Chariali, Assam, India, E-mail: manojpathoaau@gmail. com

Pigeon pea (*Cajanus cajan*) belongs to the family *Fabaceae* is an important pulse crop commercially grown in India. It is two to four meter tall depending upon variety and lives for about five years. Pigeon pea is believed to have originated in India. Besides food crop, it is also cultivated as a green manure crop, cover crop, inter crop, etc., in many sustainable farming systems. Pigeon pea is susceptible to a number of dreaded diseases suffering heavy losses in yield. The major diseases of pigeon pea are as under.

Sl. No.	Name of the Disease	Causal Organism
1.	Wilt	*Fusarium udum* Butler.
2.	Stem rot/Stem blight	*Phytophthora dreschlera var. cajani*
3.	Stem canker	*Diplodia cajani, Macrophomina cajanicola, Colletotrichum cajani*
4.	Powdery mildew	*Leveillula taurica*
5.	Rust	*Uredo cajani*
6.	Cercospora leaf spot	*Cercospora cajani*
7.	Halo blight	*Pseudomonas amygdali* pv. *phaseolicola*
8.	Leaf spots and stem canker	*Xanthomonas campestris* pv. *cajani*
9.	Pigeon pea Sterility mosaic	Sterility mosaic virus
10.	Pigeon pea mosaic	Pigeon pea mosaic virus
11.	Pigeon pea yellow mosaic	Mungbean yellow mosaic virus

7.1 WILT OF PIGEON PEA

7.1.1 INTRODUCTION/ECONOMIC IMPORTANCE

Among the various diseases of pigeon pea is one of the serious diseases in the major *pigeonpea* growing states in India such as UP, Bihar, Madhya Pradesh and Maharashtra causing heavy loss in yield. The disease was first described by Butler from India in 1906 (Butler, 1906, 1910). Later on wilt disease was also reported from Bangladesh, Ghana, Indonesia, Kenya, Malawi, Mauritious, Tenzania, Thailand, Trinidad, Uganda, and Zambia (Nane, 1980). Kannaiyan et al. (1984) reported that in India, the pigeon pea wilt incidence was varied from 0.1% (Rajasthan) to 22.6% (Maharashtra). In severe conditions, the disease may cause up to 50% loss in yield.

7.1.2 SYMPTOMS

The disease generally attacks young plants of five to six weeks old. Wilting of seedlings and adult plants is the characteristic symptom of the disease. Premature yellowing of the leaves is the typical and first symptom of the disease. This is followed by wilting or withering of the leaves of the affected plants. Leaf infections gradually progress from the bottom towards the top of the plant. Gradual or sudden yellowing may be seen in the infected plants or the affected plants may show, withering, and drying of leaves first and then drying of entire plants. Sometimes wilting symptoms also developed in some branches of the affected plant sand some branches may remain healthy. The fungus *Fusarium udum* Butler develops dense masses of mycelial hyphae which plug the vascular tissue of the stem and roots of the infected plants, preventing free flow of water upward to the leaves which lead to develop the wilting symptom. In case of infection in the early stages of the plant growth, the main root and the base of the stem show blackening symptoms. The roots exhibit dark brown discoloration and devoid of rootlets and show a black streak below the bark. The diseased plants are seen scattered in patches throughout the field. The disease becomes more severe if pigeon pea crops are infected with nematode like *Heterodera cajani* and *Meloidogyne*.

Sometimes partial wilting of the *arhar* plants infected with the wilt pathogen is also seen which may be due to partial plugging of the conductive tissues of the plant or may be due to inadequate amount of inoculums density in soil.

7.1.3 CAUSAL ORGANISM

Fusarium udum Butler is the causal organism of the disease and *Gibberella indica* Rai and Upadhyay is the perfect stage of the fungus. The fungus develops hyaline mycelium and produce three types of spores, *viz.* microconidia, macroconidia, and chlamydospores. Microconidia are small in size, unicellular, elliptical or curved in shape with 1–2 septa and 5–15 x 2–4 micrometer in size. Macroconidia are curved, long 3–4 septate bearing a prominent apical hook. They measures about 15–50 x 3–5 micrometer in size. The chlamydospores are thick walled spores developed in the somatic hyphae, oval to spherical and may produce as single spore or in long chains terminally or as intercalary. In perfect stage, the fungus produces perithecia and within perithecia, ascospores are produced. Ascospores are hyaline, ellipsoidal to ovate in shape and 10–17 x 5–7 micrometer in size.

7.1.4 DISEASE CYCLE

The pathogen is soil borne in nature and can remain in soil for more than 10 years in absence of pigeon pea host. The fungus also has the capacity to live saprophytically in soil as but the saprophytic survival continues till there is presence of dead host. Chaudhury et al. (2001) reported that in case of first crop, the top 15 cm of the soil profile contain maximum population of the fungus and if a second crop is grown after the first crop maximum population of the fungus is present at 15–30 cm under the soil. Apart from survival as saprophytes, the chlamydospore is also important for the survival of the fungus in dormant stage even longer periods. Conidia produced by the fungus also survive for long period of time. The soil borne inoculums are disseminated from plot to plot by various agents like farm implements, farm animals, and irrigation water. It has been reported that the infected seeds also acts as a means of dissemination of the fungus from one area to other.

7.1.5 EPIDEMIOLOGY

A wide range of temperature ranging from 17°C to 29°C is favorable for the disease development. As the activities of the antagonistic are higher at high temperature as such high temperature does not favor disease development. Although, *Fusarium udum* can withstand a soil pH ranging from 4.6 to 9.0 but a neutral to slightly acidic soil is favorable for the disease development.

7.1.6 MANAGEMENT

- Follow three to four years crop rotation of pigeon pea with tobacco (Bose, 1939).
- Madhukeshwara et al. (2003) recommended growing of mixed crop of sorghum or pearl millet with pigeon pea in heavily infested soil to reduce the disease incidence.
- Collection and burning of the plant debris after harvest is very much beneficial to reduce the inocullums.
- Grow disease resistant genotypes like C-11, C-28, F-18, KPL 43, ICPL 96047, ICPL 96061 etc.
- Application of zinc sulfate @ 25.0 kg/ha was reported to suppress the disease.
- Direct soil application and seed coating with culture of *Bacillus subtilis*, neem oil with *Trichoderma viridae* or *T. harzianum* was successful in suppressing the disease (Singh, 2000).
- Application of *Trichoderma harzianum* as seed treatment @ 10–20 g/kg seed is beneficial in managing the disease.
- Siddiqui and Shakeel (2009) advocated application of *Pseudomonas fluorescens* Pf736 along with *Rhizobium* for managing of wilt disease complex of pigeon pea.

7.2 STEM ROT/STEM BLIGHT

7.2.1 INTRODUCTION/ECONOMIC IMPORTANCE

First report of this disease was made from India in 1966. Following that the disease occurrence has been reported from many parts of the country and causing heavy loss in yield.

7.2.2 SYMPTOMS

The infection of this disease is manifested by the development of circular to irregular shaped water soaked lesion on the primary leaves. The size of the spots increases up to 1 cm in diameter and become necrotic. Later on, development of brown to brownish black colored lesions is seen on the stems just above the soil surface. Above portion of the stem and the branches also develop similar type of lesions. Finally, whole stem and the branches are girdles by the lesion leading to drying of above the infected portion which easily break at the infection point due to wind. Patches of disease infected plants are seen scattered over the field. The disease can be easily separated from the wilt disease by examining the roots which remain healthy in case of stem rot. The stem rot affected plants cannot be uprooted easily but break at the affected position whereas plants affected by wilt are easily uprooted.

7.2.3 CAUSAL ORGANISM

Phytophthora dreschlera var. *cajani.*

The stem rot disease is caused by *Phytophthora dreschlera* var. *cajani.* The fungus produces coenocytic hyaline hyphae 3–6 micron in diameter. Ovate to pyriform sporangia measuring 41–78 micro meter x 28–45 micro meter in size, are produced on the sporangiophores which germinate to release reniform, biflagellate, hyaline zoospores. The zoo spores swim in water for 2–5 hours after that they become non motile and form a spherical cyst. The cyst germinates to produce single or multiple germ tubes or sometimes with a terminal microsporangium

For survival during off-season, the fungus produces the oospores and chlamydospores. Oospores are hyaline when young but become purple yellow to brown colored at maturity containing a thick wall around it. Oospores are generally spherical, smooth, and measuring about 23.5–37.0 micrometers in size.

7.2.4 DISEASE CYCLE

Phytophthora dreschlera var. *cajani* penrenates in soil through the production of oospores and chlamydospores and also in the infected plant residues (Bisht and Nane, 1990). Agarwal and Khare (1988) reported that the fungus survives better in 5–15 cm soil depth as compared to the soil surface. Under favorable conditions, these perennating spores germinate start infecting the seedlings as primary source of inoculums. Large number of sporangia produced after the primary infection which acts as source of secondary infection. External agencies like wind, rain drop splashes, water, and movement of soil from infected field to new areas help in spreading the disease to new areas.

7.2.5 EPIDEMIOLOGY

High humidity and moderate temperature ranging from 28°C to 32°C favors the disease development. Optimum temperature for growth, sporangia formation and zoospore germination is around 25–30°C. High level of nitrogenous fertilizer is reported to favor the disease where as higher doses of potassic fertilizer helps in reducing the disease incidence (Pal and Grewal, 1984).

7.2.6 MANAGEMENT

- Cultural practices like crop rotation, growing of pigeon pea in ridges with wider inter row spacing and inter cropping with mungbean is helpful in reducing the disease incidence. Irrigation water should be prevented from flowing through the infected field.
- Seed treatment with *Trichoderma viride, T. hamatum*, strains of *Pseudomonas fluorescens* are beneficial. Treatment of seeds with *P. fluorescens* with metalaxyl + mancozeb was reported to reduce the disease incidence.
- Chemical seed treatment with metalaxyl alone or metalaxyl + mancozeb is also reported to be effective in protecting the seedlings from the early infection.

7.3 STEM CANKER

7.3.1 SYMPTOM

Several types of canker are found on the pigeon pea plant. Diseased pants due to *Diplodia cajani* produce grey elliptical lesions with dark edges on the stem in the collar region. These lesions later on turn into deep seated canker. The cankers spread and girdle the stem and plants collapse.

7.3.2 CAUSAL ORGANISM

Diplodia cajani, Macrophomina cajanicola, Colletotrichum cajani.

7.3.3 MANAGEMENT

- Appropriate care should be taken so as to cause minimum injury to the plant at collar region.
- Follow suitable crop rotation if the canker is a problem.
- Spray Indofil M-45 @ 2.5 g/l of water when disease infection is severe.

7.4 POWDERY MILDEW

7.4.1 SYMPTOM

Patches of white powdery growth of the fungus are seen on the under surface of the leaves. The corresponding upper portion just above patches becomes pale and brown. In case of heavy infection, the infected leaves shed prematurely.

7.4.2 CAUSAL ORGANISM (LEVEILLULA TAURICA)

The cause of the fungus *Leveillula taurica.* It is intercellular in nature and produces haustoria to absorb its nutrition from the host. The conidiophores are arised through the stomata which are long, hyaline in color. They are on septate and slender; rarely branched and bear single conidium at the tip the conidiophore. The fungus produces single celled hyalineconidia, which are elliptical or clavate shaped. Cleistothecia are black and globose bear simple myceloid appendages. Each cleistothecium contains cylindrical asci approximately 9–20 in number. About 3–5 unicellular hyaline ascospores are borned within each ascus.

7.4.3 EPIDEMIOLOGY

Cool and dry weather favors development of the disease.

7.4.4 MANAGEMENT

- Pigeon pea genotypes namely ICPL 96047, ICPL 96061 and ICPL 99046 were reported as tolerant to Powdery mildew of pigeon pea at ICISAT (Saifulla and Byre, 2002). These lines can be utilized for future breeding programmes for new variety development.

- The infected crop can be Sprayed with sulfex or solbar @ 1.5 g/lit. of water to manage the disease.

7.5 CERCOSPORA LEAF SPOT (CLS)

7.5.1 INTRODUCTION/ECONOMIC IMPORTANCE

Cercospora leaf spot (CLS) disease of pigeon pea is a serious disease in the humid regions. Under very severe attack this disease can reduce the yield up to 85%.

7.5.2 SYMPTOMS

Small circular to irregular necrotic spots or lesions appeared first usually on the older leaves. Gradually these lesions enlarged in size and coalesce together causing leaf blight and finally heavy defoliation occurred. In the advanced stage of the disease development lesions appeared on the young branches. Due to this, the tip of the affected branches dry and die back occurred. A fluffy mycelial growth is also seen on the lesions.

7.5.3 CAUSAL ORGANISM

Cercospora cajani

The disease is caused by the fungus *Cercospora cajani.* It survives in the seed and act as Primary source inoculum. Beside this, plant volunteers and infected plant refuge are act as the source of inoculums. Secondary spread occurs through conidia which are disseminated by air currents.

7.5.4 EPIDEMIOLOGY

The disease is favored by cool and humid weather.

7.5.5 MANAGEMENT

- Collection and planting of healthy seed.
- Spraying of crop with Indifil M 45 @ 0.25% or Benlate @0.2% at 7–24 day interval.

7.6 STERILITY MOSAIC OF PIGEON PEA

7.6.1 INTRODUCTION/ECONOMIC IMPORTANCE

It is one of the major diseases of *arhar*, which is mainly prevalent in India, Bangladesh, Nepal, and Myanmar. In India this is one of the dreaded diseases of pigeon pea in the states of UP, Bihar, Gujarat, Tamil Nadu, and Karnataka as it may cause up to 90%loss in yield. According to Singh et al. (1999) reported the pigeon pea sterility mosaic incidence in selected states of India as 21.4% in Bihar, 12.2% in Gujarat, 9.8% in Karnataka, 12.8% in TN, and 15.4% in UP.

7.6.2 SYMPTOM

The leaves in the infected plants turn pale yellow, which are shorter in size and develop mosaic pattern. Such plants remain stunted in growth and branch profusely. Plants suffer from mild infection of the disease may come into flowering and bear fruits but the severely infected plants remain completely sterile without any flowers and fruits. Sometimes only, a few branches are affected while other parts remain healthy in the same plant.

7.6.3 CAUSAL ORGANISM

Pigeon Pea Sterility Mosaic Virus (PPSMV)

The PPSMV is responsible for the disease. It is a RNA virus having the shape of flexuous rod, 3–10 nm in width and with undefined length (Lav Kumar et al., 2003). The RNA genome is divided in 5–7 RNA species.

The eriophyid mite *Aceria cajani* (Acari: Arthopoda) transmit the virus from disease plant to healthy plant. A single mite may transmit the disease with an acquisition feeding period of at least 15 min and then a inoculation feeding period of 90 min. Percentage of disease transmission increases as the number of viruliferous mite per plant increases. Starvation of the mite vectors before acquisition or after acquisition period reduces the inoculation period to 10 min and 60 min respectively. No report of transovarial transmission of this virus is available and also the virus has no latent period (Kulkarni et al., 2002).

7.6.4 MANAGEMENT

- Pigeon pea genotypes like ICPL 88046, ICPL 96047, ICPL 96061, ICPL 93001 Bahar, DA-35, KPL 43, etc., were reported to be resistant against the Virus (Singh et al., 1999). Saifulla and Byre (2002) reported six ICPL lines namely, 93001, 96047, 96061, 99046, 99055 and 87119 as resistant against Fusarium wilt and sterility mosaic disease. These genotypes may be utilized for breeding for resistant varieties to be cultivated in the endemic areas.
- The vector can be controlled by Spraying Metasystox (0.1%) starting at the first sight of the disease appearance.

KEYWORDS

- ***Cajanus cajan***
- **cercospora leaf spot**
- ***Gibberella indica***
- **pigeon pea sterility mosaic virus**
- **powdery mildew**
- **stem canker**

REFERENCES

Agarwal, S. C., & Khare, M. N., (1988). Survival of *Phytophthora drechsleri* f. sp. *cajani* causing stem blight of pigeon pea. *Indian Phytopath., 41*, 250.

Bisht, V. S., & Nene, Y. L., (1990). Studies on survival and dispersal of pigeon pea *Phytophthora. Indian Phytopath., 43*, 375.

Bose, R. D., (1939). The rotation of tobacco for prevention of wilt disease of pigeon pea. *Agric. Livestock, India, 6*, 653.

Butler, E. J., (1906). The wilt disease of pigeon pea and pepper. *Agric. J. India., 1*, 25–36.

Butler, E. J., (1910). The wilt disease of pigeon pea and parasitism of *Neocosmospora vasinfecta* Smith. *Mem. Dep. Agric. India, Bot. Ser., 2*, 1–64.

Chaudhury, R. G., Kumar, K., & Dhar, V., (2001). Influence of dates of sowing and moisture regimes in pigeon pea on population dynamics of *Fusarium udum* at different soil strata. *Indian Phytopath., 54*, 44.

Kannaiyan, J., Nene, Y. L., Reddy, M. V., Ryan, J. G., & Raju, T. N., (1984). Prevalence of pigeon pea diseases and associated crops in Asia, Africa, and the Americas. *Trop. Pest Management, 30*, 62.

Kulkarni, N. K., Lav, K. P., Muniyappa, V., Jones, A. T., & Reddy, D. V. R., (2002). Transmission of pigeon pea sterility mosaic virus by eriophyid mite, *Aceria cajani* (Acari: Arthopoda). *Plant Dis*., *86*(12), 1297.

Lav Kumar, P., Jones, A. T., & Reddy, D. V. R., (2003). A noval mite transmitted virus with a divided RNA genome closely associated with pigeon pea sterility mosaic disease. *Phytopathology*., *93*(1), 71.

Lav Kumar, P., Jones, A. T., Sreenivasulu, P., & Reddy, D. V. R., (2000). Breakthrough in the identification of the causal virus of Pigeon pea sterility mosaic disease. *J. Mycol. Pl. Pathol., 30*(2), 249.

Lav, K. P., Jones, A. T., & Reddy, D. V. R., (2000). Mechanical transmission of pigeon pea sterility mosaic virus. *J. Mycol. Pl. Pathol*., *32*(1), 88.

Madhukeshwara, S. S., Shankaralingappa, B. C., Mantur, S. G., & Anilkumar, T. B., (2003). Effects of intercrops and fertility levels on wilt incidence of pigeon pea. *Indian Phytopath*., *56*(1), 88.

Nane, Y. L., (1980). A world list of pigeon pea and chickpea pathogens. ICRISAT *Pulse Pathol. Prog. Rep*., *8*, 14.

Pal, M., & Grewal, J. S., (1976). Effect of NPK fertilizers on the phytophthora blight of pigeon pea. *Indian J. Agric. Sci*., *46*, 32.

Prasad, R. D., Rangeswaran, R., Hedge, S. V., & Anuroop, C. P., (2002). Effect of soil and seed application of *Triciderma harzianum* on pigeon pea wilt caused by *Fusarium udum* under the field conditions. *Crop Protection*, *21*(4), 293.

Rai, B., & Upadhay, R. S., (1982). *Gibberella indica*, the perfect state of *Fusarium udum*. *Mycologia., 74*, 343.

Reddy, M. V., Raju, T. N., Sharma, S. B., Nene, Y. L., & McDonald, D., (1993). Handbook of pigeon pea diseases. *Information Bull.42.* International Crop Research Institute for the semi Arid Tropics. Patancheru, AP.

Saifulla, U. N., & Byre, (2002). Screening of pigeon pea genotypes against multiple disease resistance. *Indian Phytopath*., *55*(3), 367.

Siddiqui, Z. A., & Shakeel, U., (2009). Biocontrol of wilt disease complex of pigeon pea (*Cajanus cajan* (L.) Millsp.) by isolates of *Pseudomonas* spp. *African Journal of Plant Science*, *3*(1), 001–012.

Singh, A. K., Agarwal, A. C., & Rathi, Y. P. S., (1999). Occurrence and sources of resistance to sterility mosaic of pigeon pea. *Indian Phytopath*., *52*, 156.

Singh, R. H., (2000). Effect of biocontrols and biopesticides on wilt (*Fusarium udum*) of pigeon pea. *Indian Phytopath., 53*(3).

Singh, R. S., (2006). *Plant Diseases* (8th edn.). Oxford and IBH, New Delhi.

CHAPTER 8

Integrated Disease Management in Pigeon Pea

SANJEEV KUMAR, S. N. SINGH, U. K. KHARE, USHA BHALE, JAYANT BHATT, and M. S. BHALE

Department of Plant Pathology, College of Agriculture, Jawaharlal Nehru Krishi Vishwa Vidyalaya-Jabalpur, Madhya Pradesh, India, E-mail: sanjeevcoa@gmail. com (S. Kumar)

Pigeon pea (*Cajanus cajan* (L.) Millsp.) is one of the most important legume crops of India. It is know as red gram, arhar, and tur in the country. It occupies 4.9 m ha area with 3.1 mt production which accounts for a productivity of 1145 kg/ha; ranking ninth in the world. The major pigeon pea growing states are Maharashtra, Uttar Pradesh, Madhya Pradesh, Karnataka, Gujarat, Andhra Pradesh, Tamil Nadu, Bihar, and Chhattisgarh. It is an important source of proteins (22%) along with carbohydrates, fiber, certain minerals viz., iron, calcium, magnesium, zinc, iodine, potassium, and phosphorus and 'B' complex vitamins. Pigeon pea stalks are also a chief source of fire-wood and livestock feed. This pulse crop is grown generally as an intercrop between cereals crops and plays an exceptional role in enriching the soil, by adding 40–90 kilogram nitrogen per hectare over a given season. It has the capacity to resist drought and to add large quantities of biomass to the soil in addition to nitrogen fixation makes it a good choice for rainfed as well irrigated production systems. The deep root system of the crop helps to recycle plant nutrients from deeper layers, and the acid secretion from its root increase the availability of phosphorus in the soil. Constraint to the increasing productivity of pigeo pea is also due to abotic and biotic stresses prevalent across the pulse growing regions. Among biotic stresses diseases *viz.,* Fusarium wilt, Phytophthora Blight, sterility mosaic, and foliar diseases lead to 10–30% yield losses (Table 8.1). The diagnostic symptoms of major diseases of pigeon pea and their integrated management are as follows.

TABLE 8.1 Diseases of National and Regional Significance

S. No.	Name of Disease	Name of Pathogen	Distribution
		National Significance	
1.	Wilt	*Fusarium udum* Butler	All over India
2.	Stem blight	*Phytophthora dreschalari f.* sp. *cajani*	All over India
3.	Sterility mosaic	Pigeon pea sterility mosaic virus	All over India
		Regional Significance	
4.	Yellow disease	Mungbean yellow mosaic virus	Tamil Nadu, Utter Pradesh, Gujarat, and Odisha
5.	Dry root rot	*Macrophomina phaseolina*	Uttar Pradesh, Tamil Nadu, Karnataka, Madhya Pradesh, Maharashtra, and Delhi
6.	Cercospora leaf spot	*Cercospora cajani* Hennings	Uttar Pradesh and Bihar
7.	Powdery mildew	*Leveillula taurica* (Lev/)	Uttar Pradesh, Maharashtra, and Tamil Nadu

8.1 MAJOR DISEASES OF PIGEON PEA

8.1.1 PHYTOPHTHORA BLIGHT

8.1.1.1 CAUSAL ORGANISM

Phytophthora drechsleri Tucker f. sp. cajani.

8.1.1.2 DIAGNOSTIC SYMPTOM

Blight causes rapid wilting of plants, desiccation, and upward rolling of leaflets followed by withering of petioles and small stems. Infected plants have brown water soaked circular or irregular lesions on leaves which become necrotic afterward. Affected plants also show brown to dark brown slightly sunken marked lesions on their stems near soil surface and on above ground part of stem. The lesions enlarge in size and girdle the stems which break at this point. In advanced stages, the stem is commonly swollen into cankerous structures near the lesions. The seedlings die suddenly due to infection. In severe cases, the whole foliage becomes blighted. White-pink fungal growth appears on the blighted area under congenial weather conditions.

8.1.2 ***FUSARIUM WILT***

8.1.2.1 CAUSAL ORGANISM

Fusarium udum Butler.

8.1.2.2 DIAGNOSTIC SYMPTOM

It is the most destructive disease of pigeon pea all over India. The main symptoms are wilting of seedlings and adult plants. The wilting starts gradually showing yellowing and drying of leaves following by wilting of whole infected plant. The affected plants can easily be recognized in patches in the field. Wilt appears on the young seedlings but mainly observed during flowering and podding stage. Brown or dark purple colored band are found on the surface of stem which start from base to several feet above ground level. Drying of plants may be partial or complete. The branches arising from discolored parts show the witling symptoms first. The wilting may be partial, as the branches on one side will show wilting while on the other side they remain healthy.

8.1.3 ***POWDERY MILDEW***

8.1.3.1 CAUSAL ORGANISM

Leveillula taurica.

8.1.3.2 DIAGNOSTIC SYMPTOM

White powdery patches are found on both surfaces causing premature defoliation. Severe infection cause yellowing of leaves with crinkling. The fungus is airborne, spreading from one field of pigeon pea to another in the vicinity.

8.1.4 ***ALTERNARIA BLIGHT***

8.1.4.1 CAUSAL ORGANISM

Alternaria alternata.

8.1.4.2 DIAGNOSTIC SYMPTOM

The symptom appears in the form of light to dark prominent brown small necrotic spots on the leaves and pods. Severe infections cause defoliation and destruction of crop. Disease is mostly confined to older leaves but may infect new leaves in post rainy season.

8.1.5 CERCOSPORA LEAF SPOT (CLS)

8.1.5.1 CAUSAL ORGANISM

Cercospora indica.

8.1.5.2 DIAGNOSTIC SYMPTOM

The symptom first appears as minute circular to irregular brown nectotic spots, one to two mm in diameter, mainly on lower surface of older leaves. Several lesions coalesce causing leaf blight and premature defoliation. Sometime, lesions appear on petiole, young branches and cause their tips to dry and die back. Fluffy mycelial growth or concentric zonations on lesions are also seen.

8.1.6 STERILITY MOSAIC

8.1.6.1 CAUSAL ORGANISM

Pigeon pea sterility mosaic virus (PPSMV).

8.1.6.2 TRANSMISSION

This is transmitted by Eriophid mite *Aceria cajani.*

8.1.6.3 DIAGNOSTIC SYMPTOM

The disease is one the most damaging disease of pigeon pea in India subcontinent. Infection by the pathogen in plants when they are less than 45 days old results in 95–100% losses, while older plants suffer 26–97% losses. The disease is characterized by malformed, crinkled leaves, mosaic symptoms, reduction in leaf size and ring spots on leaflets are common. Plants present pale green and bushy appearance without flowers and pods. Sterility of plants can be partial or full. Partially sterile plants produce discolored and shriveled seeds. The plants infected with sterility mosaic remain stunted. The leaves show mosaic symptoms and the symptoms may develop on all the leaves of infected plants. The flowering is partially or completely stopped and a few flowers, which develop are sterile.

8.2 IDM COMPONENT

It includes:

- Sowing a crop variety with resistance to a disease.
- Modified farm management practices that result in reduction of disease loss.
- Enhancement of natural control processes.
- Need based application of pesticides.

Based on the above components, a package has been developed for the management of pigeon pea diseases depicted below:

1. **Summer Plowing:** Due to summer plowing most of the crop residue which harbor the dormant mycelium/spores of the pathogen get exposed to the temperature above 40°C for several days. Thus most of the incoulums of soil borne pathogens (*Fusarium* spp., *Rhizoctonia* spp., *phytophthora* spp., and nematodes) automatically get killed under hot condition and their numbers are reduced.
2. **Soil Solarization:** In this practice, plowed fields are covered with a thin polythene sheet for a period of 6–7 weeks during April–May, which results in rising of the soil temperature up to 60°C. The resultant high temperature is lethal to soil inhabiting pathogen, nematode, and insects. This practice is very effective in controlling the *Heterodera cajani* and *Fusarium udum* provided the soil is irrigated before polythene covering.
3. **Plant Nutrients:** Pigeon pea is a legume crop so it is believed that it requires less amount of nitrogen. Therefore, di-ammonium phosphate (DAP) @ 100 kg/ha is applied in the field which provides nitrogen and phosphorus. Application of 20 kg/ha reduces the incidence of

Phytophthora blight. Manures and oil cakes reduces wilt incidence. It increases the population of beneficial microorganism and reduces the population of fungal and bacterial propagules.

4. **Varieties:** The improved pigeon pea varieties having good yield potential as well as resistant/tolerant to diseases are mentioned in Table 8.2.

TABLE 8.2 Disease Resistant/Tolerant Varieties of Pigeon pea

Disease	Resistant/Tolerant varieties
Fusarium wilt	H 76-65, ICP 8863 (Maruti), ICP 9145, ICPL 267, Mukta, AL 1, BDN 2, Birsa Arhar1, DL 82, H 76-11, H 76-44, H 76-51
Phytophthora blight	ICPL 304, KPBR 80-1-4, KPBR Hy 4, ICPL 150, ICPL 288
Sterility mosaic	ICPL 87051, MA 165, MA 166, PDA 2, PDA 10, Rampur Rahar, Prabhat, Sharda, TT 5, TT 6, Bageshwari, Bahar, DA-13
Alternaria blight	DA 2, MA 128-1, MA 128-2, 20-105
Wilt and Sterility mosaic	Aaha, MA 3, BSMR 736, Narendra Arhar-1

5. **Seed Treatment:** with Carbendazim + Thiram (1:2) @ 2.5 g/kg of seed reduces the incidence of wilt and *Alternaria* blight immensely. In *Phytophthora* prone area, metalaxyl should be used for seed treatment @ 2.5 g/kg of seed. Seed treatment with *Trichoderma* formulations @ 5 g /kg of seed reduces the incidence of wilt considerably.
6. **Sowing Time and Season:** Date of sowing and season also plays an important role in disease management *Phytophthora* blight occurs less in pre monsoon sown crops as compared to late sown crops of which early plant stage is caught by cloudy weather and continuous rains. The disease occurs more on plants at the age of 15 to 20 days and the susceptibility decreases with advancing age, being minimum at 120 days. The pre Rabi pigeon pea crop escapes the attack of phytophthora blight as well as wilt. On the contrary, *Alternaria* blight is serious on pre Rabi crops as compared to rainy season crop. Early sowing of pigeon pea deduces the sterility mosaic incidence.
7. **Method of Sowing:** Ridge sowing helps in the better crop establishment than the flat bed sowing method. Ridges not only provide proper drainage but also avoid the water to come in contact of plants for the longer duration. Otherwise, water stagnated in the field facilitates caongenial condition for sporangium formation, germination, and liberation of zoospores. It also helps in free swimming of zoospore and infection of seedling to cause *phytophthora* blight disease.

8. **Crop Geometry:** The crop geometry influences the crop growth as well as microclimate of the area. The closely spaced plant is more weak and prone to attack of *phytophthora* blight and stems rot disease, while widely spaced plant suffer less. The optimum row-to-row spacing recommended for pigeon pea is 40 cm, 50–60 cm and 75 cm for early, medium, and late maturing varieties.
9. **Inter/Mixed Cropping:** It provides diversity in the field and ultimately creates sequential and temporal discontinuity to the pathogen. Crops like sorghum, urdbean, sunhemp, mungbean, and sesame are grown as inter/mixed crop during cultivation. These crop are of short duration and mature at different intervals and perform better under such conditions. Roots of sorghum, exudates hydrogen cyanide (HCN), which inhibit the growth and reproduction of the pathogen.
10. **Crop Rotation:** with non-host crop helps in reduction of soil borne disease. Three to four-year rotations with tobacco/sorghum and cereal crops reduces intensity of wilt and *Phytophthora* blight disease
11. **Sanitation:** The pathogens *phytophthora, Fusarium, Sclerotium, and* several others survive in the field on the infected plant debris. Therefore, removal, and destruction of the infected plant debries after harvesting or at the time of field preparation, helps in reducing the severity of diseases. The mite vector as well as the virus of SDM survives on the perennial plants of pigeon pea growing in the field bund or near the house. Such plant should be uprooted and destroyed to eliminate the inoculums. In the standing crop also, the diseased plants may be destroyed as soon as they appear I n the field.
12 **Chemical Management:** Drench the soil with carbendazim 50 WP @ 1.0 g/l of water in case of severe disease incidence of wilt in the field. Control Alternaria blight, Cercospora blight and powdery mildew by foliar spray at fortnightly interval with carbendazim (0.1%). In sterility mosaic prone areas, soil application of phorate@1 kg a. i/ha at the time of sowing can protect the pigeon pea plant from infection of sterility mosaic disease for the period of 75 days after sowing (DAS). Control of vector mites by foliar spray of chemical insecticides e.g., dicofol 18.5 EC (2.5 ml/l) or dimethoate 30 EC (1.7 ml/l) reduces the spread of the disease considerably
13. **Integrated Disease Management (IDM) Packages:**
    - Selection of disease-free fields;
    - Cultivation of disease tolerant varieties such as Pusa-9, NP 38, AL 1430, which are resistant or relatively less susceptible;

- Soil solarization or summer plowing;
- Cultivation of pigonpea on ridges with proper drainage system;
- Soil amendment with Trichoderma @ 1.0 kg + 100 kg FYM at the time of field preparation to reduce the incidence of wilt disease;
- Seed treatment with hexaconazole + captan @ 2.5 g/kg seed or metalaxyl @ 2.0 g per kg of seeds (or) carbendazim or thiram @ 2 g/kg of seed 24 hours before sowing (or) with talc formulation of *Trichoderma viride* @ 4 g/kg of seed (or) *Pseudomnas fluorescens* @ 10 g/kg seed;
- Long crop rotation for 3–4 year with non host crop like, tobacco, sorghum, pearl millet, cotton;
- Wide row interspacing;
- Mixed cropping with sorghum;
- Amendment of soil with oil cakes, appliances of trace elements such as boron, zinc, and manganese;
- Preventive sprays of mancozeb, wettable sulfur or metalaxyl at 15–20 days interval starting from 15 days after germination reduces the incidences of diseases;
- Eradication of self-sown plants in and around pigeon pea fields is very helpful to manager sterility mosaic.

8.3 DO'S AND DON'TS IN IDM

Sl. No.	Do's	Don'ts
1.	Deep plowing is to be done on bright sunny days during the months of May and June. The filed should be kept exposed to sunlight at least for 2–3 weeks.	Don not plant or irrigate the field after plowing, at least for 2–3 weeks, to allow desiccation of weed's bulbs and/or rhizomens of perennial weeds.
2.	Adopt crop rotation.	Avoid mono cropping
3.	Grow only recommended varieties.	Do not grow varieties not suitable for the season or the region.
4.	Sow early in the season	Avoid late sowing as this may lead to reduced yield and incidence of white grubs and diseases.

Sl. No.	Do's	Don'ts
5.	Always treat the seeds with approved biopesticides/ chemicals for the control of seed borne diseases.	Do not seeds without seed treatment with biopesticides/chemicals.
6.	Sow in rows at optimum depths under proper moisture conditions for better establishment.	Do not sow seeds beyond 5–7 cm depth.
7.	Use NPK fertilizers as per the soil test recommendation.	Avoid imbalanced use of fertilizers.
8.	Apply short persistent pesticides to avoid pesticide residue in the soil and produce.	Do not apply pesticides during preceding 7 days before harvest.

8.4 SUMMARY

Of late, pigeon pea production has shifted from a largely subsistence farming to market oriented commercial production. However, it has not been possible to exploit the full genetic potential of high yielding pigeon pea varieties because of immense losses due to diseases and pests. Management of one single disease in isolation would prove counterproductive; hence, it calls for the management of all diseases in integrated manner. An integrated disease management (IDM) to combat the disease is the right approach.

KEYWORDS

- ***Cercospora* Leaf Spot**
- **di-ammonium phosphate**
- **hydrogen cyanide**
- **pigeon pea sterility mosaic virus**
- **solarization**
- **sterility mosaic**

REFERENCES

Butler, E. J., (1906). The wilt disease of pigeon pea and pepper. *Agricultural Journal of India, 1*, 25–36.

Butler, E. J., (1910). The wilt disease of pigeon pea and the parasitism of *Neocosmospora vasinfecta* Smith. *Memoirs of the Department of Agriculture in India, Botanical Series, 2*, 1–64.

Chauhan, Y. S., (1990). Pigeon pea: Optimum agronomic management. In: Nene, Y. L., Hall, S. D., & Sheila, V. K., (eds.), *The Pigeon Pea* (pp. 257–278). CAB International Wallingford, U. K.

FAI, (1999). *Fertilization Statistics 1998–1999* (pp. III 46–51). The Fertilizer Association of India, New Delhi.

Gouder, S. B., & Kulkarni, S., (1999). Compatibility effect of antagonists and seed dressers against *Fusarium udum*: The causal agent of pigeon pea wilt. *Karnataka Journal of Agricultural Sciences, 12*(1–4), 197–199.

Kannaiyan, J., & Nene, Y. L., (1984). Efficacy of metalaxyl for control of phytophthora blight of pigeon pea. *Indian Phytopathology, 37*, 506–510.

Kannaiyan, J., Nene, Y. L., Raju, T. N., & Sheila, V. K., (1981). Screening for resistance to phytophthora blight of pigeon pea. *Plant Disease, 65*, 61–62.

Kannaiyan, J., Nene, Y. L., Reddy, M. V., Ryan, J. G., & Raju, T. N., (1984). Prevalence of pigeon pea diseases and associated crop losses in Asia, Africa, and the Americas. *Tropical Pest Management, 30*, 62–71.

Karimi, R., Owuoche, J. O., & Silim, S. N., (2012). Importance and management of *Fusarium* wilt (*Fusarium udum* Butler) of pigeon pea. *International Journal of Agronomy and Agricultural Research, 2*(1), 1–14.

Khan, J., Ooka, J. J., Miller, S. A., Madden, L. V., & Hoitink, H. A. J., (2004). Systemic resistance induced by *Trichoderma hamatum* in cucumber against phytophthora crown rot and leaf blight. *Plant Disease, 88*, 280–286.

Khare, D., Satpute, R. G., & Tiwari, A. S., (1975). Present state of the wilt and sterility mosaic disease of pigeon pea. *Indian Journal of Genetics, 54*, 331–346.

Kulka, M., & Von Schmeling, B., (1987). In: *Modern Selective Fungicides-Properties, Mechanism of Action* (p. 119). Longman Scientific and Technical.

Kumar, S., & Upadhyay, J. P., (2013). Cultural morphological and pathogenic variability in isolates of *Fusarium udum* causing wilt of pigeon pea. *Journal of Mycology and Plant Pathology, 43*(1), 76–79.

Kumar, S., Upadhyay, J. P., & Rani, A., (2009). Evaluation of *Trichoderma* species against *Fusarium udum* Butler causing wilt of pigeon pea. *Journal of Biological Control, 23*(3), 329–332.

Mandhare, V. K., & Suryawanshi, A. V., (2004). Application of *Trichoderma* species against pigeon pea wilt. *JNKVV Res. J., 38*, 99–100.

Mayee, C. D., (2016). *Integrated Pest Management for Sustainable Crop Protection from Diseases: What are the Options?* (p. 70).6th International conference plant pathogens and people. New Delhi, India.

Sharma, M., (2016). *Climate Change and Emerging Diseases in Chick Pea and Pigeon Pea: Occurrence, Distribution and Epidemiology* (p. 53).6th International conference plant pathogens and people. New Delhi, India.

Shrivastava, V., Khare, N. K., Kinjulek, C. S., & Naberia, S., (2016). *Symposium on Physiological Approaches to Enhance Productivity in Pulses Under Changing Climate*. JNKVV, Jabalpur, MP.

Singh, J. P., & Singh, F., (2014). *Scenario of Pigeon Pea Research in India.* National conference of pulses: Challenges and opportunities under changing climate scenario. JNKVV Jabalpur, MP.

Vyas, S. C., (1993). In: *Hand Book of Systemic Fungicides, Disease Control* (Vol. III, p. 161). Tata McGraw Hill Publishing Company Limited, New Delhi.

CHAPTER 9

Important Diseases of Soybean Crop and Their Management

V. K. YADAV, [1] C. P. KHARE, [2] P. K. TIWARI, [2] and J. N. SRIVASTAVA[3]

[1]JNKVV, College of Agriculture, Ganj Basoda, Vidisha, Madhya Pradesh, India

[2]Division of Plant Pathology Indira Gandhi Agricultural University, Raipur, Chhattisgarh, India

[3]Department of Plant Pathology, Bihar Agricultural University, Sabour–813210, Bhagalpur, Bihar, India

Soybean, *Glycine max* (L.) Merr. genus name *Glycine* was originally introduced by Carl Linnaeus (1737) in his first edition of *Genera Plantarum*. The word glycine is derived from the Greek-*glykys* (sweet) refers to the sweetness of the pear-shaped edible tubers produced by the native north American twining or climbing herbaceous yam bean legume. The cultivated soybean first appeared in *Species Plantarum*, by Linnaeus, under the name *Phaseolus max* L. The combination *Glycine max* (L.) Merr. as proposed by Merrill in 1917 has become the valid name for this useful plant. Soybean is a species of legume native to East Asia.

The plant is classed as an oilseed rather than a pulse. Soybean is one of the most important crops for world food security, ranking first among field crops for protein production per hectare. Soybean has gained very high nutritional value with 20% oil and 40% high quality protein. Its protein has a good balance of amino acids except methionine and its oil is rich in unsaturated fatty acids. Soybean is commonly known as 'Wonder Crop.'

India is the fifth largest producer of soybean in the world. In India, soybean grows in mostly in Madhya Pradesh, Maharashtra, Karnataka, Andhra Pradesh, Tamil Nadu, Rajasthan, Gujarat, Uttar Pradesh, Punjab, and Haryana. Madhya Pradesh is the leading state in soybean production followed by Maharashtra and Rajasthan. About 80–85% acreage of soybean in India is concentrated in Madhya Pradesh. Soybean being the third in area and production of overall commercial oil seed crops of the world, contributes 33% of our commercial oil seeds and 21% of total pulse production. Soybean being a potentially high yielding crop can play a greater role in boosting oil seed production in the country.

The optimum growing conditions for soybean are mean temperatures of 20–30°C, temperatures of below 20°C and over 40°C retard growth significantly. It can be grown in a wide range of soils with optimum growth in moist alluvial soils with a good organic content.

Several factors account low productivity, among them climatic conditions, differences in rainfall pattern, outbreak of diseases and pests are important. Soybean plant is known to suffer from many fungal, bacterial, viral, and nematode diseases. Among them, some economically important diseases are charcoal rot, aerial blight, anthracnose, rust; cercospora leaf spot (CLS), Yellow mosaic and bacterial pustule are common in soybean growing areas. The important diseases of soybean are described in this chapter.

9.1 CHARCOAL ROT

9.1.1 ECONOMIC IMPORTANCE

The fungus causing charcoal rot can infect over 500 different kinds of plants. Charcoal rot is a soil borne root and stem disease of soybean that develops in the mid to late season when plants are under stress, especially heat and drought stress. Plants of any age can be affected by charcoal rot. However, symptoms remain latent unless stressful environmental conditions, especially hot, dry weather prevails. Usually irregularly sized patches of infected plants are seen in the field, but sometimes individual plants in the row infected. Infected plants may die prematurely and are often wilted and stunted. It is also called "summer wilt" or "dry-weather wilt." Yield loss from charcoal rot is highly variable.

9.1.2 SYMPTOMS

Often, infections occur early, when soil moisture is good, and symptoms become obvious late in the season as plants become stressed from drought. In soybean, Plants that have been infected early in the season may not display symptoms until midseason during or after flowering. Symptoms tend to occur first in small groups of plants, often on ridge slopes where soil is thin and along the edges of fields, especially where there is competition for soil moisture by bordering trees. Charcoal rot is hard to diagnose in dry years, since it is difficult to distinguish between the symptoms of the disease and those of general drought stress. However, plants with charcoal rot die more quickly during periods of drought stress than those without the disease. To accurately identify charcoal rot, pull symptomatic plants and split the lower stems and taproot to confirm the pathogen. The fungus causes a general root rot in soybean, infecting the roots and lower stems.

The name charcoal rot is descriptive of the small black fungal structures, called microsclerotia, on the lower stem of infected plants. In more mature plants, the fungus can cause reduced vigor, yellowing, and wilting. The charcoal rot fungus can infect seeds, seedlings, or mature plants. If infected at the seed or seedling stages, plants may not emerge or seedlings may become discolored and die. Infected seedlings show a reddish brown discoloration at the soil line extending up the stem that may turn dark brown to black. Foliage of infected seedlings can appear off-color or begin to dry out and turn brown and infected seedlings may die, particularly under hot, dry conditions. Under cool, wet conditions, young plants that are infected may survive but carry a latent infection that will express symptoms later in the season with hot, dry weather. The earliest symptoms of charcoal rot include leaf rolling and wilting during the heat of the day. During periods of hot, dry weather, leaves turn yellow and wilt, but remain attached to the plant takes on a dull greenish-yellow appearance prior to wilting. This is easy to mistake for normal maturity. Beginning at flowering, a light gray discoloration develops on the epidermal and sub-epidermal tissues of both tap and secondary roots and lower stems. Small black dots may form beneath the epidermis of the lower stem and in the tap root to give the stems and roots a charcoal-sprinkles appearance, a diagnostic symptom of charcoal rot. In addition, reddish brown discoloration and black streaks can form in the pith and vascular tissues of the root and stem. In some cases, the upper one third of the plant may have only flat pods without seed.

9.1.3 CAUSAL ORGANISM

Macrophomina phaseolina.

Macrophomina phaseolina (Tassi) Goid. (syns. *M. phaseolina* (Maubl.) Ashby, *Rhizoctonia bataticola* (Taub.) Briton-Jones, is highly variable, with isolates differing in microsclerotial size and presence or absence of pycnidia. The pycnidial stage is not common on soybean, but is on peanut.

9.1.3.1 PYCNIDIA

Initially immersed in host tissue, then erumpent at maturity. They are 100–200 μm in diameter; dark to grayish, becoming black with age; globose or flattened; membranous to subcarbonaceous with an inconspicuous or definite truncate ostiole. The pycnidia bear simple, rod-shaped conidiophores, 10–15 μm long.

9.1.3.2 CONIDIA

Single celled, hyaline, elliptic or oval, (14–33 x 6–12 μm).

9.1.3.3 MICROSCLEROTIA

Microsclerotia are formed from aggregates of hyphal cells joined by a melanin material with 50 to 200 individual cells composing individual microsclerotia. Microsclerotia of *M. phaseolina* are jet black in color and appear smooth and round to oblong or irregular. Across isolates, microsclerotia vary on size and shape and on different substrates.

9.1.4 DISEASE CYCLE AND EPIDEMIOLOGY

M. phaseolina survives as microsclerotia in the soil and on infected plant debris and it can be spread with contaminated seed. The fungus can survive on seeds in small cracks or in seed coat, on available nutrients in plant debris or in the soil, or by infecting alternate hosts. Infected seeds either do not germinate or produce seedlings that may die soon after emergence. The microsclerotia serve as the primary source of inoculum and have been found to persist within the soil up to three years. The microsclerotia are black, spherical to oblong structures that are produced in the host tissue and released in to the soil as the infected plant decays. These multi-celled structures allow the persistence of the fungus under adverse conditions such as low soil nutrient levels and temperature above 30°C. Germination of the microsclerotia occurs throughout the growing season when temperatures are between 28 and 35°C. Microsclerotia germinate on the root surface, germ tubes form appresoria that penetrate the host epidermal cell walls by mechanical pressure and enzymatic digestion or through natural openings. The hyphae grow first intercellularly in the cortex and then intracellularly through the xylem colonizing the vascular tissue and plug the vessels. The mechanical plugging of the xylem vessels by microsclerotia, toxin production, enzymatic action, and mechanical pressure during penetration lead to disease development. The rate of infection increases with higher soil temperatures and low soil moisture will further enhance disease severity. The population of *M. phaseolina* in soil will increase when susceptible hosts are cropped in successive years and can be redistributed by tillage practices. After pod set, the fungus colonizes the stem more aggressively, choking off water movement from the roots to the foliage. This is results in wilt and plant death.

Charcoal rot Infection is favored by abundant soil moisture early in the season, late plantings seem to be more susceptible to seedling infection, plants under stress from moisture or nutrients, excessive plant population and soil compaction increase the severity of the diseases. This disease is also worsened in plants weakened by such conditions as poor soil fertility, excessive seeding rates, soil compaction, and insect damage.

Disease is most severe where plants have been growing under conditions of stress or injury. Many environmental factors affect microsclerotia survival, root infection, and disease development.

9.1.5 MANAGEMENT

- Limit drought stress during the reproductive stages of growth.
- Avoid narrow row widths and high plant populations and use recommended plant population according to variety. High plant populations can contribute to increase plant stress and competition for water increasing charcoal rot potential.
- Crop rotation with non-host crops (cereals) for 2 to 3 years can reduce inoculum levels.
- Irrigate where possible, to keep soil moisture high, or flood fields for 3–4 weeks before planting, maintaining good soil moisture with irrigation from planting to pod fill may reduce disease potential.
- Plant high-quality disease-free seed.
- Use Resistant/tolerant varieties i.e., J. S.97-52, J. S.335, NRC 2, NRC 37, LSB 1, MACS 13.
- Treat the seeds with Thiram + Carbendazim (2:1) @ 2.5–3 g or *Trichoderma viride* or *Pseudomonas fluorescens* @ 5 g per kg seed.
- Mix 5 kg *Trichoderma viride* (commercial product) in 12–15 kg farm yard manure (FYM) and apply it in one ha. area before flowering.

9.2 SOYBEAN RUST

9.2.1 ECONOMIC IMPORTANCE

Soybean rust was first reported from Japan in 1903. In India, the disease was first noticed at Pantnagar of Uttaranchal during 1970 and subsequently it was observed at Kalyani in West Bengal and the foothills of Uttar Pradesh. The disease was severe in 1970, 1971, 1974 and mild in 1972 and 1973. It was disappeared then onwards and appeared suddenly in epiphytotic form and caused substantial yield losses up to 80% in Northern Karnataka and parts of Maharashtra during *Kharif* season 1994 and 1995. The disease also appeared in epiphytotic form during *Kharif* 1994 and 1995 in Madhya Pradesh and continued to be a major threat in soybean cultivation until 1999. Yield losses in soybean due to soybean rust have been reported to range from 30 to 100%. Before 1992, soybean rust was known to cause significant losses in Asia and Australasia, inclusive of the following countries: Australia, India, Indonesia, Japan, Korea, China, Philippines, Taiwan, Thailand, and Vietnam. There have been several early reports of soybean rust in equatorial Africa, but the first confirmed report of *Phakopsora pachyrhizi* on the African continent was in 1996 from Kenya, Rwanda, and Uganda.

The soybean rust pathogen is known to naturally infect 95 species from 42 genera of legumes, inclusive of important weed species like Kudzu vine (*Pueraria lobata*) and major crop species such as common bean (*Phaseolus vulgaris*). The significance of the numerous alternative host possibilities for the soybean rust pathogen is that these may serve as an inoculum reservoir or a 'green bridge' from one soybean planting season to the next.

It is now accepted that there are two different fungal species, *Phakopsora pachyrhizi,* and *Phakopsora meibomiae*, that cause soybean rust. *Phakopsora meibomiae*, referred to as the New World type, is a much weaker pathogen, and is the pathogen that has been found in limited areas in the Western Hemisphere (primarily the Caribbean). *P. pachyrhizi*, referred to as the Asian or Australasian soybean rust, is the more aggressive pathogen. It was initially limited to tropical and subtropical areas of Asia and Australia.

Soybean rust is a major disease of soybeans in many parts of the world. Affected plants are quickly defoliated, reducing pod set and fill, which results in reduced yields and seed quality. Soybean rust has devastated soybean crops in many parts of the world. The severity of losses varies depends on susceptibility of the soybean variety, time of the growing season in which the rust becomes established in the field and weather conditions during the growing season.

9.2.2 SYMPTOMS

Soybean rust is difficult to identify in the early stages of infection, as symptoms are very small, poorly defined and occur in the lower-middle canopy of the plant, where it is humid. As leaves show symptoms, infection may be mistaken for spider/mite damage or foliar diseases such as Septoria brown spot, bacterial blight or bacterial pustule. It is important to note that soybean rust pustules frequently lack the yellow halo associated with bacterial pustule.

Soybean rust tends to start in the lower part of the canopy and can move up through the plant quickly if weather conditions are favorable for disease development. Soybean plants are susceptible to soybean rust at any stage of development, but symptoms are most common during and after flowering. The disease usually starts within the low to mid canopy and moves up the plant.

Early symptoms appear as numerous yellow lesions followed by the appearance of a brown speck almost in the center of these yellow lesions that soon develop into light brown to dark pustules. The lesions tend to be angular to somewhat circular in shape and may be concentrated near leaf veins. Initially the lesions are small, barely larger than a pinpoint.

As the soybean plants mature, lesions may be found in the middle and upper canopy. When conditions are favorable for disease development, yellowing of the foliage may be evident and cause defoliation and premature death. Losses are due to a reduction in photosynthetic area of the plants and resulting reduction in pod and seed numbers and in seed weight. Once pod set begins on soybean, infection can spread rapidly to the middle and upper leaves of the plant. After infection, pustules can be seen after about 10–14 days. These pustules are most common on the underside of leaves but may also develop on petioles, pods, and stems especially at or near flowering. At first these pustules might appear to be small, raised blisters on the lower leaf surface. But as the rust pustules mature, they begin to produce large numbers of light colored, powdery spores (urediospores), which emerge through a distinct hole or pore in the cone-shaped pustule. High levels of infection in soybean fields result in a distinct yellowing and browning of fields and commonly, premature senescence in plants. Soybean plants are susceptible to soybean rust at any stage of development, but symptoms are most common during and after flowering.

9.2.3 CASUAL ORGANISM

Phakopsora pachyrhizi.

There are approximately 80 species of *Phakopsora* known worldwide, six of which occur on legumes. Scientists believe that the soybean rust pathogen can have as many as five stages in its life cycle, though only three have been observed. Spore-bearing bodies of those three stages (uredinia, telia, and basidia) grow on the underside of infected leaves, below the epidermal layer.

Spermogonia (pycnia) and aecia are unknown. Uredia are globose, subeidermal, erumpent, light cinnamon to reddish brown and opened through a central pore. They form abundantly on the lower surface of leaves, their size range from 100–200 µm in diameter. They are fever and smaller on the upper surface, within lesion new uredia develops continuously for few weeks. Paraphyses united at the base, form a dome like covering over the sporophores. Papraphyses hyaline to subhyaline and prominently capitates at the apex and measure about 7–15 µm toward the apex.

Uredospores are globose, subglobose, ovate, or ellipsoid and are hyaline to light yellow brown, minutely, and densely echinulate, and the walls are about 1 micron thick. The size of spores varied from 18–45 X 13–28 µm, depending on host and environmental conditions. Germ pores are not obvious in the finely echinulate, hyaline wall of the uredospore.

Telia form subepidermally mostly on the lower surface of the leaves, among the uredia and the edge of the lesion. Orange brown or light brown when young, they become dark brown to black with age. They are irregular to round, sparse to aggregated, 150–250 µm in diameter, with 3 to 5 irregular layers of teliospores. The telutospores are yellow to brown in color, one celled smooth walled and mostly clavate, oblong or angulate. These spores vary in size, in the range of 13–35 X 5–15 µm. Germination and production of basidia and basidiospores have been reported. However, their role in life cycle of pathogen is unknown.

9.2.4 DISEASE CYCLE AND EPIDEMIOLOGY

The soybean rust pathogen does not survive on residues left in the field and is not seed borne. Rust pathogens are considered to be obligate parasites in that they survive on living plant material. The soybean rust produces large numbers of spores on infected plants. Soybean rust spreads primarily by wind-borne spores across regions depending upon prevailing winds and other environmental conditions conducive to disease development. These spores of the soybean rust pathogen are transported by air currents can be carried hundreds of miles in a few days. Additional hosts can serve as overwintering reservoirs for the pathogen and allow for build-up of inoculum.

Numerous spores are released if infected plants are disturbed by wind or by individuals walking through rust-infected areas. If clothing is exposed to spores, care should be taken to prevent the spread of soybean rust to uninfected locations. Seed borne transmission of the disease has not been documented, but there is some concern that seed lots may contain small amounts of infected plant debris capable of spreading the pathogen. To date, seed lots have not proven to be a pathway for the disease.

Rust is a multi-cyclic disease. After the initial infection is established, the infection site can produce spores for 10 to 14 days. Abundant spore production occurs during wet leaf periods (in the form of rain or dew) of at least 8 hours and moderate temperatures of 15.6 to 26.7°C. Under these conditions, pustules form within 5–10 days and spores are produced within 10–21 days. The rapid development of epidemic involving more number of repeated cycles of uredospores, which, indicates short latent period of the pathogen. Physiological age (pod formation) of the soybean plant plays a significant role in rust development.

The infection process starts when urediniospores germinate to produce a single germ tube that grows across the leaf surface, until an appressorium forms. Urediniospores are able to penetrate the plant cells directly, rather than through natural openings or through wounds in the leaf tissue. Thus infection is relatively quick, about 9 to 10 days from initial infection to the next cycle of spore production. Penetration of epidermal cells is by direct penetration through the cuticle by an appressorial peg. When appressoria form over stomata, the hyphae penetrate one of the guard cells rather than entering the leaf through the stomatal opening. This rust and related species are unique in their ability to directly penetrate the epidermis; most rust pathogens enter the leaf through stomatal openings and penetrate cells once inside the leaf.

Uredinia may develop for up to 4 weeks after a single inoculation, and secondary uredinia will arise on the margins of the initial infections for an additional 8 weeks. Thus, from an initial infection, there could be first generation pustules that maintain sporulation for up to 15 weeks. Even under dry conditions this extended sporulation capacity allows the pathogen to persist and remain a threat. Successful infection is dependent on the availability of moisture on plant surfaces. At least 6 hours of free moisture is needed for infection with maximum infections occurring with 10 to 12 hours of free moisture.

9.2.5 FAVORABLE CONDITIONS

- Optimum temperature for urediniospore germination ranges between 12 to 27°C.
- The urediniospore germination is greater in darkness, light either inhibiting or delaying germination.
- Maximum infection at 20 to 23°C.
- No lesions >28°C.
- Twelve to 15 rain days per month.
- Late vegetative to late reproductive growth stages.
- The development of soybean rust is favored by prolonged periods of leaf wetness 6–12 hours.
- Relative Humidity (RH) between 75–80% required for spore germination and infection.
- Dense canopies may provide an ideal microclimate that encourages disease development.
- Cloudy weather during the growing season would favor soybean rust infection.

9.2.6 MANAGEMENT

- Wider row width and reduced plant populations could potentially decrease the severity of rust by decreasing the length of time leaves remain wet.
- Use resistant/tolerant varieties, PK 1024, PK 1029, JS 80-21, Indira Soybean 9, MAUS 61-2.
- Foliar spray of hexaconazole (0.05%), triadimefon (0.1%), propiconazole (0.1%), difenconazole (0.1%), or mancozeb (0.25%) at 15 days interval starting from the onset of disease were found effective in reducing soybean rust severity.

9.3 ANTHRACNOSE AND POD BLIGHT

9.3.1 ECONOMIC IMPORTANCE

Anthracnose is a fungal disease of soybean that occurs worldwide wherever soybean is grown and reduces plant stand, seed quality, and yield by 16–26%.

It is one of the major production constraints in all soybean growing areas of India. Now anthracnose has become a major threat for soybean cultivation during last few years due to favorable weather conditions during different crop growth stages.

9.3.2 SYMPTOMS

Soybeans are susceptible to infection at all stages of development. Plants and seed may be infected. In some cases, pods can be diseased, and the seed may be infected but they do not show any symptom. Infected seeds in diseased pods may be shriveled, dark brown and moldy or fail to form. Less severely infected seed may show no external sign of infection. Germinating seeds may be killed before or after emergence. If infected seed is planted, early disease development may result in damping-off (seed or seedling rot). Sometimes seedlings may also be infected, but may show no symptoms until the plants begin to mature. The anthracnose fungi may grow from infected cotyledons into young stems where small deep seated cankers may form that kill young plants. Dark, sunken cankers can occur on the cotyledons, epicotyls, and radical of seedlings to cause pre and post emergence damping-off. The symptoms on leaves are characterized by lesions appeared dull brown initially causing chlorosis. On such lesions, the pathogen produced plenty of acervuli. As the disease advanced, the entire infected lesions appeared dark brown and later turned necrotic and finally dried giving blighted appearance to seedlings.

Anthracnose is generally present in soybean fields to some degree every season. The pathogen primarily attacks on older, mature plants. Stem, pods, and leaves may be infected without showing external symptoms until the weather is warm and moist. Most commonly, however, plants become infected during bloom and pod fill stages. Symptoms appear on stems, pods, and leaf petioles as irregularly shaped brown blotches. Such dark brown areas turned necrotic and withered away giving shot hole type of symptoms on the leaves. Under high humidity, symptoms on leaves are veinal necrosis, leaf rolling, cankers on petioles premature defoliation, and stunted plants. Irregular brown spots may develop in a random pattern on stems and pods. Young pods may be attacked and killed. Infection of pods may result in few or small seeds per pod. Early season infection of pods or pedicles can result in shriveled pods, fewer, smaller, moldy seed or no seed development. However, the most significant damage occurs during the reproductive phase, including twisted and aborted pods, which direct impact on the yield. At advanced stages of disease development, near soybean maturity, black fungal fruiting bodies called acervuli that produce minute black spines (setae) are abundant and randomly distributed on infected tissue. Infected pods generally contribute more yield loss than infected stems or petioles.

9.3.3 CAUSAL ORGANISM (COLLETOTRICHUM TRUNCATUM)

The most common pathogen associated with anthracnose disease in soybeans, is *Colletotrichumi truncatum* (Schw.) Andrus and W. D. Moore. Although at least four other species have been associated to the disease, i.e., *C. coccodes, C. destructivum, C. gloeosporioides,* and *C. graminicola*. In addition, Yang et al., (2015) have described two other species with falcate conidia causing soybean anthracnose in the USA.

C. chlorohyti and *C. incanum*. *C. truncatum* is seed borne, produces hyaline, septate, and branched mycelium. The acervuli are oval to conical, hemispherical to truncate, erumpent, appearing singly, blackish brown in color and measuring 178.6 × 256.0 μm. The acervulus was embedded with honey colored conidial mucilaginous mass containing numerous conidia. Setae arose through this conidial mass were longer than conidial mass, with 1–3 septation, measuring 60–300 x 3–8 μm. They were broader at the base with dark brown to black color and tapering at apex with brown color. The conidia are single celled, smooth, hyaline, curved, or truncate containing oil globules and measured 17–31 x 3–4.5 μm in size. Conidia germinated terminally or sub-terminally with one to two short germ tubes. When germ tubes come in contact with the solid surface produces dark, sticky appressoria. Infection pegs were developed from appressoria that penetrated the epidermal cells.

9.3.4 DISEASE CYCLE AND EPIDEMIOLOGY

Pathogen over season as mycelium in diseased crop residue or in infected seeds. Spores from infected plant residue are distributed by wind and rain. Spores can infect soybean plants at any stage. Mycelium may also become established in infected seedlings without symptom developing until the plants begin to mature.

Anthracnose is favored by:

- Prolonged periods of wet, humid, warm weather, which occurs late in the season.
- Secondary stem and pod infections mostly occur in the reproductive stages from bloom to pod fill during warm, moist weather, temperature below 35°C and when free moisture is present for 12 hours or longer.
- Plants under shade showed more disease than under nonshade conditions.

- Early maturing varieties are frequently become infected by pathogen and fields that have soybeans in the previous year are more likely to have anthracnose problems.

9.3.5 MANAGEMENT

- Crop rotation with non-host.
- Use healthy or certified seeds.
- Completely remove plant residue from the field soon after harvest.
- Maintain well drained field.
- Avoid narrow row widths and high plant populations.
- Seed treatment with Thiram or Captan or Carbendazim 3 g per kg seed.
- Use Mancozeb @ 3 g or Carbendazim 1–1.5 g/l of water between early pod development and initial seed formation.
- Harvest soybeans promptly at maturity.

9.4 AERIAL BLIGHT

9.4.1 ECONOMIC IMPORTANCE

Aerial blight is an important disease of soybean. It was first reported from Philippines in 1980. Later it was reported from Malaysia, Mexico, Puerto Rico, China, Taiwan, Louisiana, North America, South America, Brazil, and Argentina. In India, it was first reported from Pantnagar (Uttarakhand) in 1967. Yield losses of 35% have been attributed to the disease in soybean when conditions favor disease development. Aerial blight, also called aerial web blight or Rhizoctonia foliar blight is a common disease on soybean. *Rhizoctonia solani* Kühn [teleomorph: *Thanatephorus cucumeris* (Frank) Donk] is reported to cause economic losses in soybean crops throughout the world. Symptoms observed on soybean plants and associated with *R. solani* infection include damping-off, roots, and hypocotyl rots, web blight, and aerial blight.

9.4.2 SYMPTOMS

The pathogen may infect leaves, pods, and stems at all stages of development through maturity. Disease symptoms first appear on the lower portion of the plant growing in closure canopy. Foliar symptoms often occur during late vegetative growth stages on the lower portion of the plant. Infected plants showing symptoms include leaf spots, leaf blighting, or defoliation. Initially leaf symptoms appear as small, circular, water-soaked, light greenish-brown spots appearing on the leaf lamina. The spots increase in size, finally becoming oblong or irregular that turn tan to brown at maturity with slight reddening of the veins on the lower surface. In older spots, a dark pinkish-brown margin around the light tan-brown central necrotic tissue is common. Diseased tissue in old lesions that are no longer expanding generally fall out during dry weather, creating ragged, shot hole effect. Later the spots may coalesce and cover major or entire portions of the leaf. Such leaves turn yellow, dry, and finally drop off prematurely. Infected leaves droop and may adhere to pods and stem, thus becoming source of infection for pods and seeds. Reddish-brown lesions can form on infected petioles, stems, pods, and petiole scars. Symptoms on petioles and stems are light to dark-brown, oblong, 1–2 mm wide and up to 3–4 cm long. Lesions on pods may be small, brownish spots or may blight the whole pod. Seed infection is associated with pod infection. Pods and stem tissues that are infected will be greasy brown and shriveled. Under field conditions, a web-like mycelium with micro- and/or macrosclerotia may form on plant surfaces.

9.4.3 CASUAL ORGANISM

Rhizoctonia solani.

9.4.3.1 MYCELIUM

Hyaline, Pale to dark-brown of relatively large diameter with branching near the distal septum of hyphal cells, septate, often at nearly right angles in older hyphae, constriction of branch of hyphae and formation of a septum in the branch at the point of origin.

9.4.3.2 SCLEROTIA

Micro and macrosclerotia are irregular or oval, nearly uniform texture and varying in size and shape, white when young, changing from light to dark brown at maturity. Sclerotia are composed mainly of monilioid cells. On sterile media, sclerotia are composed entirely of undifferentiated hyphae, brown in color and variable in shape. Sclerotia have been variously described as being irregular with smooth surfaces, globose, flat, and elongated, crusty or irregularly globose with a pitted surface. Sclerotia range in size from less than 1 mm to more than 2 cm in diameter.

R. solani has been variously separated into at least seven distinct anastomosis groups (AGs). Anastomosis will occur among isolates within a group but not between isolates from different groups. Within the groups the anastomosis varied from complete fusion to cell contact without fusion. Most strains of *R. solani* causing foliar blight of several agronomic plants belong to anastomosis group 1 and produces both macro and micro sclerotia.

9.4.4 DISEASE CYCLE AND EPIDEMIOLOGY

The fungus survives on plant residue or in soil as sclerotia. When soil warm, the fungus becomes active and infection may occur soon after seed is planted. The fungus grows better in aerated soils; thus, disease is more severe on light and sandy soils. Sclerotia require 24 to 48 hours of continuous free moisture at 25°C for germination. Mycelia of the fungus may grow up the stem of the plants or the sclerotia or mycelial fragments may be splashed by rain onto the foliage. During warm, wet weather, mycelium spreads extensively on the surface of plants, forming localized mats of "webbed" foliage.

Infection usually begins at flowering during prolonged periods of high humidity and warm temperatures. When plants are mature with a closed canopy of leaves across the field, narrow rows, lodged plants and when frequent rains occur, aerial blight will spread fast. Small fields bordered by trees or poorly drained fields are more suitable to have severe aerial blight. The pathogen spreads mainly through contact between infected and non-infected parts of the plant by the formation of mycelial bridges. Sclerotia which survive in the soil for extended period serve as primary inoculum.

9.4.5 MANAGEMENT

- Deep summer plowing.
- Use resistant/tolerant varieties: PK 472, PK 564, PK 1042, and SL 295.
- Seed treatment limits an early season disease development. Treat the seeds with Thiram + Carbendazim (2:1) @ 2.5–3 g/kg seed.
- Improved drainage help in reducing the disease.

- Wide row spacing or lower plant populations help in reduction of aerial blight.
- Foliar spray of carbendazim @ 1 g/l water, on 45–60 days after sowing (DAS).

9.5 FROG EYE LEAF SPOT

9.5.1 ECONOMIC IMPORTANCE

Frogeye leaf spot (FLS) is also known as CLS. It was first reported from Japan in 1915. Later, it was also reported from soybean production areas in the world including Brazil, Argentina, India, and China. FLS usually develops during reproductive growth stages (blooming to maturity) but may develop sooner in continuous soybean fields and/or under optimal environmental conditions (warm, humid conditions). In fields where soybean has been planted for consecutive seasons, FLS may develop in the lower canopy initially, but with prolonged adequate moisture, infection of new leaves that are fully expanded can develop resulting in symptoms throughout the soybean canopy. Yield loss from FLS is mainly the result of reduction in photosynthetic area and premature defoliation. Estimates of yield loss due to FLS have been reported to be 22%.

9.5.2 SYMPTOMS

Infection can occur at any stage of soybean development, but most often occurs after flowering. The most common initial symptoms are lesions of varying size are formed on leaves. They are circular to angular spots up to 5 mm in diameter on the upper surface of the leaf. These dark, water-soaked spots develop into lesions with dark brown centers surrounded by narrow, purple, or dark reddish brown margins. As lesions age, the center becomes light brown to light gray, and the border remains dark reddish purple in color. These leaf spots are diagnostic symptoms but are often mistaken for herbicide drift or other leaf diseases. Older lesions are translucent and have whitish centers containing black dots (stromata). In severely infected plants, several lesions may coalesce into larger irregular shaped spots. When lesions cover about 30% of the leaf area, a blighting phase often occurs and leaves wither quickly and fall prematurely. On the underside of the leaf, the lesion appears brown with tiny dark "hairs." These hairs are the long conidia of the fungus. Young leaves are extremely susceptible while older leaves are more resistant.

Although less frequent, lesions also develop on stems and pods if rainfall and humidity persist. But these lesions are less diagnostic than leaf lesions. Stem and pod symptoms are less common, but may appear late in the growing season. Stem lesions are oblong, they are two to four times as long as they are wide, somewhat red brown with narrow dark brown to black margins when young, and mature lesions are slightly sunken with light gray centers and brown boarders.

Lesion development on pods is similar to that of the leaves. Pod lesions are circular to elongated, slightly sunken, and reddish-brown in color. The center of stem and pod lesions become gray to brown as they mature. The fungus that causes the disease may grow through the pod walls and invade seed. Infected seeds may show a range of symptoms. They may have no symptoms, or they may appear dark, shriveled, and may have a cracked seed coat. These seeds may show cracking of the seed coat and discoloration ranging from small specks to large blotches. Infected seed may germinate poorly, and plants that do emerge from infected seed are often stunted and may have lesions on the cotyledons.

9.5.3 CAUSAL ORGANISM

Cercospora sojina

C. sojina produces long, narrow spores that arise in clusters from dark cushions at the centers of lesions during warm, humid weather. A related fungus, *C. kikuchii*, causes CLS and purple seed stain, but its spores are much longer than those of *C. sojina*. *C. sojina* conidiophores arise in fascicles of 2–25 from a thin stroma. They are light to dark brown and are 52–120 x 4–6 μm. They have one to several septations with prominent geniculations and spore scars. Conidia are borne on the tip of conidiophores and are pushed aside as the conidiophores continue their indeterminate growth. Generally, 1 to 3 but sometimes up to 11 conidia are formed on a single conidiophores. The conidia are zero to 10 septate, hyaline when young and elongate to fusiform and tapering toward the tip, the base of conidium is usually rounded and bears circular scars markings, the conidia are 24–108 x 3–9 μm.

9.5.4 DISEASE CYCLE AND EPIDEMIOLOGY

The *C. sojina* overwinters as mycelium on soybean residue and seed. There are reports of the fungus being transmitted by seed; it contributes to long distance spread of the pathogen. Infected seed germinates poorly and the resulting seedlings are often weak. Spores produced on the cotyledons of infected seedlings are the main source of inoculum for the leaf phase of the disease. Leaf infection can occur at any stage of soybean development but most often occurs after flowering, typically in the upper canopy. Leaves are most susceptible to infection when they are just emerging and become less susceptible as they mature and symptoms become visible 7 to 14 days after infection. Spores produced on infected plants can move to new plants in the same field, and wind can also disperse the spores to nearby fields. The fungus can also produce spores on the residue of a previous soybean crop. Wind and splashing water may disperse spores from one field to another.

The primary and secondary inoculum is conidia that form on tips of conidiophores and are pushed aside as the conidiophores continue to grow. Conidia can germinate on a leaf surface within an hour of deposition in the presence of water. Conidia can be carried short distances by air currents and rain splashes, whereas secondary infections can continue throughout the soybean-growing season under favorable conditions.

FLS development is favored by warm (25–30°C), humid (>90% RH) conditions coupled with cloudy days and extended periods of wet weather. These conditions promote sporulation of the pathogen that can create cycles of leaf infection throughout the season. Frequent rains following disease onset can lead to serious epidemics, dry weather severely limits disease development. Overhead irrigation may increase the risk of severe FLS compared to flood or furrow irrigation or dryland production systems. Fields will have higher risk for FLS if, there is a susceptible soybean variety grown in a field with a history of FLS, or fields have continuous soybean production with conservation tillage practice. This is a polycylic disease, in which infection, symptom development, and production of conidia are repeated throughout the growing season. If the first symptoms of this disease are detected late in the season, there is very little impact on the plant. However, if this cycle begins prior to or at flowering, then substantial amounts of disease can develop on plants and the greater the reduction in yield.

C. sojina, like other *Cercospora* plant pathogens produces a toxin, cercosporin, which has been shown to play an important role. Cercosporin is a photosensitizer (photoactivated and lacks toxicity in the dark), its toxicity is due to oxidative damage to lipids, proteins, and nucleic acids of cell membranes.

Exposure of plant cells and tissues to cercosporin results in peroxidation of the membrane lipids, leading to membrane breakdown and death of the cells. Membrane damage cause by cercosporin allows for leakage of nutrients into the leaf intercellular spaces, allowing for fungal growth and sporulation.

9.5.5 MANAGEMENT

- Removal of diseased plant debris.
- Apply crop rotation with nonhost crops, such as corn, small grains, or sorghum to break the disease cycle and reduce disease inoculum.
- Consider tillage to reduce infected residue (primary inoculum) left on soil surface.
- Use disease free healthy seeds.
- Grow resistant/tolerant varieties, i.e., Bragg, JS 80-21, KHSB 2, and VLS 21.
- Treat the seeds with Thiram + Carbendazim (2:1) @ 2.5–3 g per kg seed.
- Spray Mancozeb @ 2 g or Carbendazim 1.5 g or Thiophonate methyl 1.5 g/l water.

9.6 BACTERIAL PUSTULE

9.6.1 ECONOMIC IMPORTANCE

This disease has been reported in most parts of the world where soybeans are grown. It is most prevalent after 30–35 days of sowing to throughout the season. It is difficult to distinguish symptoms of this disease from bacterial leaf blight and leaf rust, especially in the early phase of disease progression. Bacterial leaf blight lesions appear water-soaked while the lesions of bacterial pustule do not. In case of soybean rust, pustules have a small, round opening at the top for spore release. Whereas, in bacterial pustule lesions normally lack an opening. If an opening is present with bacterial pustule, it is typically a linear crack across the surface of the pustule. This disease causes 15–20% yield losses.

9.6.2 SYMPTOMS

The symptoms are characterized by minute to large, irregular, reddish brown spots appear on the leaf that may develop on either surface of the leaf but are more common on the lower surface, which are surrounded by haloes. Small, raised, light red to brown colored pustules, later form in the centers, usually in lesions on the lower leaf surface. Spots with elevated centers may form on either or both leaf surface. Diseased leaves develop a ragged appearance when the necrotic areas are torn away by stormy or windy weather. Severe disease often results in yellowing of leaves and premature defoliation that may decrease yield by reducing seed numbers and size.

9.6.3 CAUSAL ORGANISM

Xanthomonas axonopodis pv. *glycines*

Xanthomonas axonopodis (syn. *campestris*) pv *glycines* is a motile gram negative rod, 0.5–0.9 x 1.4–2.3 µm, with a single polar flagellum.

9.6.4 DISEASE CYCLE AND EPIDEMIOLOGY

Pathogen overwinters in infested seed and soil on crop residue. The bacteria spread from crop residue or nearby diseased plants by splashing water, windblown rain and by farm machinery during cultivation when the foliage is wet. Disease outbreaks usually occur 5 to 7 days after wind-driven rains. The bacterium enters the plant through stomata and wounds. Warm weather with frequent showers promotes the development of this disease. Unlike bacterial leaf blight pathogen, *X. axonopodis* development is not hampered by high temperatures.

9.6.5 MANAGEMENT

- Deep summer plowing.
- Rotate soybeans with non-host crops.
- Use pathogen-free seed.
- Use resistant or tolerant varieties; PK 1029, PK 1042, Indira soya 9, J. S.95-60, J. S.97-52, J. S.335, J. S.93-05, Bragg, KHSB 2, MAUS 32, NRC 7, NRC 37 and VLS 2.
- Avoid dense cropping.
- Seed treatment with Streptocyclin @ 250 ppm (2.5 g per 10 kg seeds).
- Avoid field cultivation when the foliage is wet to reduce disease spread.

- Spray Copper oxychloride (1 kg) + Streptocycline (100 g) or Kasugamycin/Validamycin (1 kg) + Copper oxychloride (1 kg) in 500 liters of water per hetare.

9.7 YELLOW MOSAIC

9.7.1 ECONOMIC IMPORTANCE

About 50 viruses have been reported to occur in soybean in various parts of the world. Among all these viruses MYMV causes economic yield loss and are sometime a major limiting factor in soybean production. Yellow Mosaic Disease of soybean was first described by Nariani (1960) and since then it had spread at alarming proportions. The virus causes 15–75% reduction in yield of the soybean.

9.7.2 CAUSAL ORGANISM

Mungbean yellow mosaic virus (MYMV)

MYMV is a Gemini virus, belonging to the genus begomovirus of the family, *Geminiviridae*. The family geminiviridae have been classified into four genera, namely Begomovirus, Curtovirus, Topocuvirus, and Mastrevirus, depending on their genomes, mode of transmission and host range. Begomoviruses have characteristic icosahedral geminate particles (30 x 18 nm), *Nucleic acid:* Circular single-stranded (ss) DNA, about 20% of particle weight, M. Wt 0.8 x 10^6. They infect dicots and are transmitted by the whitefly *Bemisia tabaci*, Gennadius in the persistent (circulative) manner. They have monopartite or bipartite genome. The genus begomovirus generally comprises bipartite (two components, namely DNA-'A' and 'B') genome, both being 2.5–2.7 kb in size, which replicates via rolling circle (RCR) model with the help of few viral and several host factors. Both components share a common region (CR) of about 200 bp containing the important cis-elements for viral DNA transcription and rolling circle replication (RCR).

In bipartite begomoviruses, DNA A encodes proteins required for replication, transcription, and encapsidation whereas DNA B encodes proteins required for movement functions inside the host. According to the Baltimore classification, they are considered to be class II viruses. The genome can either be a single component between 2500 and 3100 nucleotides, or, in the case of some begomoviruses, two similar-sized components each between 2600, and 2800 nucleotides. They have elongated, geminate capsids, the capsids range in size from 18–20 nm in diameter with a length of about 30 nm.

MYMV has a thermal inactivation point between 40–50°C, dilution end point between 10^{-2}–10^{-3}, particle weight 4.0 x 10^{6} Daltons. Loose aggregate of particles can be observed in the nucleus of phloem cells.

9.7.3 OTHER HOSTS

The host range of MYMV is: mungbean, black gram, kidney bean, azuki bean, *Alternanthera sessilis, Sida rhombifolia,* etc.

9.7.4 TRANSMISSION

The virus reported from India has not been transmitted by mechanical inoculation but has been transmitted by the whitefly (*Bemisia tabaci*) in a persistent manner, not only to several species in the Leguminosae (Nariani, 1960) but also to *Brachiaria romosa* (Gramineae) and *Cosmos bipinnatus, Eclipta alba* and *Xanthium strumarium* (Compositae) (Nene et al., 1971; Nene, 1973).

9.7.5 SYMPTOMS

Disease may appear from early vegetative stage to physiological maturity stage of the plant. Disease first appears on young leaves with yellow mottling of the leaves with an intense contrast between the yellow and the green areas. Yellow areas are either scattered or produced in indefinite bands along the major veins. Rusty necrotic spots appear in the yellow areas as the leaves mature. Symptoms are more severe at lower temperature. Under severe condition plants produces shriveled and lightweight seeds or sometimes fail to form flowers and pods.

9.7.6 MANAGEMENT

- Destruction of weed hosts.
- Initially rogue out infected plants and burn them.

- Use resistant or tolerant varieties; JS 97-52, JS 20-29, JS 95-60, PK 472, PK 564, PK 1024, PK 1029, PK 416, PK 1042, Pusa 37, SL 295, SL 525 and SL 688.
- Seed treatment with Thiamethoxam (70WS) 3 g per kg seed.
- Spray insecticide to reduce insect population, i.e., Metasystox 25 E. C.1 ml or Imidachloprid 0.5 ml per liters of water or Thiamethoxam (25 WG) 100 g/ha, after 35 days of sowing.

KEYWORDS

- ***Brachiaria romosa***
- **common region**
- **frogeye leaf spot**
- **mungbean yellow mosaic virus**
- **rolling circle replication**
- **yellow mosaic**

REFERENCES

Anonymous, (2007). *Directors' Report and Summary Table of Experiments of AICRP on Soybean*. National Research Centre for Soybean, Indore.

Anonymous, (2012). *Directors Reports and Summary Tables of Experiment*. AICRP on soybean, directorate of soybean research, Indore.

Athow, K. L., & Probst, A. H., (1952). The inheritance of resistance to frogeye leaf spot of soybeans. *Phytopathology, 42*, 660–662.

Backman, P. A., Williams, J. C., & Crawford, M. A., (1982). Yield losses in soybean, from anthracnose caused by *Colletotrichum truncatum. Pl. Dis., 66*, 1032–1034.

Bartlett, D. W., Clow, J. M., Goodwin, J. R., Hall, A. A., Hamer, M., & Parr-Dobrzanski, B., (2002). The strobilurin fungicides. *Pest Management Science, 58*, 649–662.

Begum, M. M., Sariah, M., Puteh, A. B., & Abidin, M. A. Z., (2008). Pathogenicity of *Colletotrichum truncatum* and its influence on soybean seed quality. *Int. J. Agri. Biol., 10*, 393–398.

Bowen, C. R., & Schapaugh, J. W. T., (1989). Relationships among charcoal rot infection, yield, and stability estimates in soybean blends. *Crop Sci., 29*, 42–46.

Bowers, G. R., & Russin, J. S., (1999). Soybean disease management. In: Heatherly, L. G., & Hodges, H. F., (eds.), *Soybean Production in the Mid-South*. CRC Press.

Bowers, G. R., (1984). Resistance to anthracnose in soybean. *Soybean Genet. Newslett., 11*, 150, 151.

Bradbury, J. F., (1986). *Xanthomonas* dowson 1939, In: Krieg, N. R., & Holt, J. G., (eds.), *Bergey's Manual of Systematic Bacteriology* (Vol.1, pp. 199–210). Williams and Wilkins, Baltimore.

Bromfield, K. R., & Hartwig, E. E., (1980). Resistance to soybean rust and mode of inheritance. *Crop Sci.*, *20*, 254, 255.

Bromfield, K. R., (1980). Soybean rust: Some considerations relevant to threat analysis. *Prot. Ecol.*, *2*, 251–257.

Bromfield, K. R., (1984). *Soybean Rust*. American Phytopathological Society, St Paul, Minnesota.

Chacko, S., & Khare, M. N., (1978). Reaction of soybean varieties to *Colletotrichum truncatum*. *Jawaharlal Nehru Krishi Vishwa Vidyalaya Res. J.*, *12*, 138, 139.

Cloud, G. L., & Rupe, J. C., (1991). Comparison of three media for enumeration of sclerotia of *Macrophomina phaseolina*. *Plant Disease*, *75*, 771, 772.

Cockerham, C. C., (1963). Estimation of genetic variance components. In: Hanson, W. D., & Robinson, H. F., (eds.), *Statistical Genetics and Plant Breeding* (pp. 53–94). National academy of sciences-national research council publ., UK.

Dadke, M. S., (1996). *Studies on Rust of Soybean* [*Glycine max.* (L.) Merrill] caused by *Phakopsora pachyrhizi* Syd. *M. Sc. (Agri.) Thesis*. Uni. Agric. Sci., Dharwad (India).

Dashiell, K. E., Bello, L. L., & Root, W. R., (1987). Breeding soybeans for the tropics. In: Singh, S. R., Rachie, K. O., & Dashiell, K. E., (eds.), *Soybean for the Tropics* (pp. 3–16). Research, Production, and Utilization. Wiley, Chichester.

Daub, M. E., & Briggs, S. P., (1983). Changes in tobacco cell membrane composition and structure caused by the fungal toxin, cercosporin. *Plant Physiol.*, *71*, 763–766.

Daug, M. E., & Chung, K. R., (2007). *Cercosporin: A Phytoactivated Toxin in Plant Disease*. Online APS net Features. doi: 10.1094/APSnetFeature/2007–0207.

De Candolle, A. P., (1815). Uredo rouille des cereals In Forafran caise, famille de champigons. *Mem. Mus. D. Hist. Nat.*, *2*, 209–216.

Dhinga, O. D., & Sinclair, J. B., (1977). *An Annotated Bibliography of Macrophomina Phaseolina: 1905–1975*. Universidade Federal de Vicosa, Minas Gerais, Brazil.

Dhinga, O. D., & Sinclair, J. B., (1978). *Biology and Pathology of Macrophomina Phaseolina* (p. 166). Universidade Federal de Vicosa, Brazil.

Dow, J. M., & Daniels, M. J., (1994). Pathogenicity determinants and global regulation of pathogenicity in *Xanthomonas campestris* pv. *Campestris*. In: *Molecular and Cellular Mechanisms in Bacterial Pathogenesis of Plants and Animals, 192*, 29–41.

Embrapa Recomendações técnicas para a cultura da soja Na região central do Brasil 1999/2000, (1999). Londrina. *Embrapa Soja. Documentos*, 132.

Fauquet, C. M., & Stanley, J., (2003). Gemini virus classification and nomenclature; progress and problems. *Ann. Appl. Biol.*, *142*, 165–189.

Fett, W. F., Dunn, M. F., Maher, G. T., & Maleeff, B., (1987). Bacteriocins and temperate phage of *Xanthomonas campestris* pv. *Glycines Current Microbiology, 16*, 137–144.

Gawade, D. B., Suryawansh, A. P., Patil, V. B., Zagade, S. N., & Wadje, A. G., (2009). Screening of soybean cultivars against anthracnose caused by *Colletotrichum truncatum. J. Pl. Dis. Sci.*, *4*(1), 124–125.

Ghawde, R. S., Gaikwad, S. J., & Borkar, S. L., (1996). Evaluation of fungicides and screening of varieties against pod blight of soybean caused by *Colletotrichum truncatum* (Schw.) Andrus and Moore. *J. Soils and Crops, 6*, 97–99.

Girish, K. R., & Usha, R., (2005). Molecular characterization of two soybean infecting begomoviruses from India and evidence for recombination among legume-infecting begomoviruses from South-East Asia. *Virus Research*, *108*, 167–176.

Grau, C. R., Dorrance, A. E., Bond, J., & Russin, J. S., (2004). Fungal diseases. In: Boerma, H. R., & Specht, J. E., (eds.), *Soybeans: Improvement, Production, and Uses* (3rd edn., pp. 732, 734). Monogr.16. Am. Soc. of Agron., Madison, WI.

Hartman, G. L., (2007). Soybean rust: The first three years. *Proceedings of the 2007 Illinois Crop Protection Conference*.

Hartman, G. L., Manandhar, J. B., & Sinclair, J. B., (1986). Incidence of *Colletotrichum* spp. On soybeans and weeds in Illinois and pathogenicity of *Colletotrichum truncatum*. *Pl. Dis., 70*, 780–782.

Hepperly, F. R., Mignucci, J. S., Sinclair, J. B., & Smith, R. S., (1982). Rhizoctonia web blight of soybean in Fuerto Rico. *Plant Disease*, *66*, 256, 257.

Hernández, J. R., Levy, L., & De Vries-Paterson, R., (2005). First report of soybean rust caused by *Phakopsora pachyrhizi* in the continental United States. *Pl. Dis., 89*, 774.

Hokawat, S., & Rudolph, K., (1993). The hosts of *Xanthomonas*, 44–48. In: Swings, J. G., & Civerolo, E. L., (eds.), *Xanthomonas*. Cloning and characterization of pathogenicity genes from *Xanthomonas campestris* pv. *Glycines Journal of Bacteriology, 174*, 1923–1931.

Jagtap, G. P., & Sontakke, P. L., (2009). Taxonomy and morphology of *Colletotrichum truncatum* isolates pathogenic to Soybean. *Afric. J. Agri. Res*., *4*(12), 1483–1487.

Jones, S. B., & Fett, W. F., (1987). Bacterial pustule disease of soybean: microscopy of pustule development in susceptible cultivar. *Phytopathology*, *77*, 266–274.

Khare, M. N., & Chacko, S., (1983). Factors affecting seed infection and transmission of *Colletotrichum truncatum* in soybean. *Seed Science and Technol., 11*, 853–858.

Killgore, E. M., & Heu, R., (1994). First report of soybean rust in Hawaii. *Pl. Dis., 78*, 12–16.

Kochman, J. K., (1979). The effect of temperature on the development of soybean rust (*Phakopsora pachyrhizi*). *Aust. J. Agric. Res., 30*, 273–277.

Laviolette, F. A., Athow, K. L., Probst, A. H., & Wilcox, J. R., (1976). Effect of bacterial pustule on yield of soybeans. *Crop Sci*., *10*, 150, 151.

Laviolette, F. A., Athow, K. L., Probst, A. H., Wilcox, J. R., & Abney, T. S., (1970). Effect of bacterial pustule and frogeye leaf spot on yield of Clark soybean. *Crop Sci., 10*, 418, 419.

Manandhar, J. B., Hartman, G. L., & Sinclair, J. B., (1985). Anthracnose disease development, seed infection, and resistance in soybean. *Phytopath., 75*, 1317.

Mehan, V. K., & McDonald, D., (1997). Charcoal rot. In: Kokalis-Burelle, N., et al., (eds.), *Compendium of Peanut Diseases* (2ndedn.). APS Press. St. Paul MN.

Mengistu, A., Kurtzweil, N. C., & Grau, C. R., (2002). First report of frogeye leaf spot (*Cercospora sojina*) in Wisconsin. *Plant Disease, 86*, 1272.

Mian, M. A. R., Phillips, D. V., Boerma, H. R., Missaoui, A. M., & Walker, D. R., (2008). Frogeye leaf spot of soybean: A review and proposed race designations for isolates of *Cercospora sojina* Hara. *Crop Sci., 48*, 14–24. doi: 10.2135/cropsci2007.08.0432.

Miles, M. R., Hartman, G. L., Levy, C., & Morel, W., (2003). Current status of soybean rust control by fungicides. *Pesticide Outlook* (pp. 197–200).

Nariani, T. K., (1960). Yellow mosaic of mung (*Phaseolus aureus* L.). *Indian Phytopathol., 13*, 24–29.

Nene, Y. L., (1973). Viral diseases of some warm weather pulse crops in India. *Pl. Dis. Reptr., 57*, 463.

Nene, Y. L., Naresh, J. S., & Nair, N. G., (1971). Additional hosts of mungbean yellow mosaic virus. *Indian Phytopath., 24*, 415–417.

Olaya, G., & Abawi, G. S., (1996). Effect of water potential on mycelial growth and on production and germination of sclerotia of *Macrophomina phaseolina*. *Plant Disease, 80*, 1347–1350.

Parmeter, J. R. J., (1970). *Rhizoctonia solani Biology and Pathology* (p. 255). University of California Press, Berkeley.

Patil, P. V., (2008). Evaluation of botanical products against soybean rust caused by *Phakopsora pachyrhizi* Syd. *J. Ecofriendly Agric.*, *31*, 62–64.

Philips, D. V., (1999). Frogeye leaf spot. In: Hartman, G. L., Sinclair, J. B., & Rupe, J. C., (eds.), *Compendium of Soybean Diseases* (4th edn., pp. 20, 21). American phytopathological society, St Paul, MN.

Pretorius, Z. A., Kloppers, F. J., & Frederick, R. D., (2001). First report of soybean rust in South Africa. *Pl. Dis.*, *85*, 1288.

Ramteke, R., & Gupta, G. K., (2005). Field screening of soybean, *Glycine max* (L.) Merrill. Lines for resistance to yellow mosaic virus. *J. Oilseeds Res.*, *22*, 224, 225.

Roy, K. W., (1996). Falcate spored species of *Colletotrichum* on soybean. *Mycologia.*, *88*(6), 1003–1009.

Schneider, R. W., Dhingra, O. D., Nicholson, J. F., & Sinclair, J. B., (1974). *Colletotrichum truncatum* borne within the seed coat of soybean. *Phytopath.*, *64*, 154, 155.

Sharma, N. D., & Mehta, S. K., (1996). Soybean rust in Madhya Pradesh. *Acta Botanica Indica*, *24*, 115–116.

Sinclair, J. B., & Hartman, G. L., (1999). Soybean rust. In: Hartman, G. L., Sinclair, J. B., & Rupe, J. C., (eds.), *Compendium of Soybean Diseases* (4thedn., pp. 25–26). American phytopathological society, St Paul, Minnesota.

Sinclair, J. B., (1982). *Compendium of Soybean Diseases* (2nd edn., p. 104). The American Phytopathological Society. St. Paul, Minn.

Suteri, B. D., (1974). Occurrence of soybean yellow mosaic virus in Uttar Pradesh. *Current Science*, *43*, 689–690.

Sutton, B. C., (1962). *Colletotrichum dematium* (Pers. Ex Fr.) Grove and *C. trichellum* (Fr. Ex Fr.) *Duke Phytopath.*, *45*, 222–232.

Tschanz, A. T., Wang, T. C., & Tsai, B. Y., (1983). Recent advances in soybean rust research. *International Symposium on Soybean in Tropical and Sub-Tropical Cropping Systems.* Tsukuba, Japan.

Usharani, K. S., Surendranath, B., Haq, Q. M. R., & Malathi, V. G., (2004). Yellow mosaic virus infecting soybean in Northern India is distinct from the species infecting soybean in Southern and Western India. *Current Science*, *86*, 845–850.

Usharani, K. S., Surendranath, B., Haq, Q. M. R., & Malathi, V. G., (2005). Infectivity analysis of a soybean isolate of mungbean yellow mosaic India virus by agroinoculation. *J. Gen. Plant Pathol.*, *71*, 230–237.

Valenzeno, D. P., & Pooler, J. P., (1987). Photodynamic action. *Bio. Science.*, *37*, 270–275.

Van Regenmortel, M. H. V., (2000). *Virus Taxonomy: Classification and Nomenclature of Viruses* (p. 1162.). Academic Press, San Diego, USA.

Varma, A., & Malathi, V. G., (2003). Emerging Gemini virus problems: A serious threat to crop production. *Ann. Appl. Biol.*, *142*, 145–164.

Wang, T. C., & Hartman, G. L., (1992). Epidemiology of soybean rust and breeding for host resistance. *Plant Prot. Bull.*, *34*, 109–124.

White, D. G., (1999). Fungal stalk rots. In: White, D. G., (ed.), *Compendium of Corn Diseases* (3rd edn.) APS Press: St. Paul, MN.

Wong, C. F. J., Niik, W. Z., & Lim, J. K., (1983). Studies on *Colletotrichum dematium* f. sp. *truncatum* on soybean. *Pertanika.*, *6*, 33.

Wyllie, T. D., (1988). Charcoal rot of soybean-current status. In: Wyllie, T. D., & Scott, D. H., (eds.), *Soybean Diseases of the North Central Region*. APS Press, St. Paul, MN.

Wyllie, T. D., (1993). Charcoal rot. In: Sinclair, J. B., & Backman, P. A., (eds.), *Compendium of Soybean Diseases* (3rd edn.) APS Press: St. Paul, MN.

Yang, H. C., & Hartman, G. L., (2015). Methods and evaluation of soybean genotypes for resistance to *Colletotrichum truncatum*. *Pl. Dis.*, *99*, 143–148.

Yang, X. B., Uphoff, M. D., & Sanogo, S., (2001). Outbreaks of soybean frogeye leaf spot in Iowa. *Plant Disease*, *85*, 443.

Zhang, G., Newman, M. A., & Bradley, C. A., (2012). First report of the soybean frogeye leaf spot fungus (*Cercospora sojina*) resistant to quinone outside inhibitor fungicides in North America. *Plant Disease, 96*, 767.

CHAPTER 10

Current Status of Castor (*Ricinus communis* L.) Diseases and Their Management

N. M. GOHEL,[1] B. K. PRAJAPATI,[2] and HARSHIL V. PARMAR[1]

[1]*Department of Plant Pathology, B. A. College of Agriculture, Anand Agricultural University, Anand–388110, Gujarat, India, E-mail: nareshgohel@aau. in (N. M. Gohel)*

[2]*Agricultural Research Station, S. D. Agricultural University, Aseda, Ta. Deesa, Banaskantha–385535, Gujarat, India*

10.1 INTRODUCTION

Castor is a Latin word, also known as Palm of Christ and scientifically known as *Ricinus communis* L. The castor plant appears to have originated in eastern Africa, especially around Ethiopia and cultivated around the world. It grows throughout the warm-temperate and tropical regions and flourishes under a variety of climatic conditions. Asian country (India) is the main producer of castor within the world (Sudha et al., 2016). The most important castor manufacturing states in this country are Gujarat, Rajasthan, and Andhra Pradesh. Together, these States account for quite 90% of total domestic production with Gujarat being the most important physic seed manufacturing State (Anon, 2011a). Important diseases affecting this crop are described in detail along with their management practices.

DISEASES

1. **Seedling Blight:** *Phytophthora colocasiae, Phytophthora parasitica.*
2. **Alternaria Blight:** *Alternaria ricini.*
3. **Cercospora Leaf Spot:** *Cercospora ricinella.*
4. **Powdery Mildew:** *Leveillula taurica.*

5. **Wilt:** *Fusarium oxysporum* f. sp. *ricini.*
6. **Root Rot/Charcoal Rot:** *Macrophomina phaseolina.*
7. **Gray Mold/Gray Rot/Blossom Blight:** *Botrytis ricini.*
8. **Bacterial Leaf Spot:** *Xanthomonas campestris* pv. *ricinicola.*

10.2 SEEDLING BLIGHT

10.2.1 ECONOMIC IMPORTANCE

The disease was initially noted from Pusa within the year 1909. An average loss of 10% happens in crop stand because of this disease.

10.2.2 SYMPTOMS

1. **Dead Seedling:** The disease initially makes its look on each the surfaces of the cotyledonary leaves in the kind of round patch of dull green color that shortly spreads to the point of attachment inflicting the leaf to rot and droop down.
2. **Spot on Older Leaf:** The infection additionally spreads to the stem with the result that the seed plant is killed either because of the destruction of growing point or by the collapse of stem.

The true leaves of seedlings and therefore the very young leaves of older plants can also be affected; however usually not much injury is caused. The leaf spots flip yellow then brown and concentrical zones of lighter and darker brown color are formed. The disease spots coalesce at a later stage and cover the complete leaf. The affected leaves shed untimely beneath wet conditions, a very fine whitish haze is found on the under-surface of the leaf spots just in case of mature plants conjointly the disease might spread from young leaves to the stem through the stalk (Anon., 2014).

10.2.3 CAUSAL ORGANISM

Phytophthora colocasiae, Phytophthora parasitica

The parasite produces non-septate and hyaline mycelium. Sporangiophores rise through the stomata on the lower surface separately or in gatherings. They are unbranched and bear single celled, hyaline, round or oval sporangia at the tip separately. The sporangia develop to create inexhaustible zoospores. The parasite additionally creates oospores and chlamydospores in adverse seasons.

10.2.4 DISEASE CYCLE AND EPIDEMIOLOGY

The causal organism, *Phytophthora colocasiae* is soil borne in nature, comprises of inter and intra-cell mycelium which creates inside the host tissue. Following a couple of day's development various branches rise up out of the lower epidermis of the leaf by and large through stomata as sporophores either independently or in twos or threes. A solitary dry ovoid or roundish sporangium is borne at the tip of sporophore. A ready sporangium frees zoospores when put in water. The quantity of zoospores shifts from 5 to 45 in every sporangium. The zoospores sprout promptly by one or infrequently two germ tubes. These are formed freely during hot and dry months when sporangia are sparse and hold the intensity of germination for a long time. Secondary disease spreads quickly through sporangia gave the climate conditions are positive. The sporangia are effectively scattered by the breeze and develop promptly on the leaves delivering zoospores which enter by methods for germ tubes either through stomata or straightforwardly and create unhealthy spots inside 24 hours and the following product of sporangia show up in around two days (Anon., 2016).

10.2.5 FAVORABLE CONDITIONS

Rainy climatic conditions, low temperature (20–25°C), low lying, and not well depleted soils.

10.2.6 MANAGEMENT

- Ill depleted; clammy and low lying territories ought to be maintained a strategic distance from for sowing castor (Anon., 2015).
- Seed dressing with 4 g Trichoderma viride definition on 3 g Metalaxyl per kg seed can decrease infection frequency (Anon., 2015).
- Soil dousing with Copper oxychloride @3 g/lit or Metalaxyl 2 g/lit (Anon., 2015).
- Remove and decimate tainted plant deposits.

- Avoid low-lying and not well depleted fields for sowing.
- Treat the seeds with Metalaxyl at 3 g/kg or T. viride at 4 g/kg.
- Soil soaking with Metalaxyl @ 0.2% or COC@0.3%
- Give require-based shower of Copper oxychloride @ 0.3% to maintain a strategic distance from additionally spread of the infection.

10.3 ALTERNARIA BLIGHT

10.3.1 ECONOMIC IMPORTANCE

The disease has been accounted for from various parts of the nation now and again and is accepting genuine extents in the ongoing years especially in Bombay territories. In some different nations additionally, Alternaria leaf spot is thought to be one of the genuine infections of castor. In a few fields, around 70% of the plants are accounted for to be influenced with the infection causing genuine misfortunes in yield and oil content (Anon., 2016).

10.3.2 SYMPTOMS

All the elevated parts of the plant, i.e., stem, leaves, inflorescence, and containers are at risk to be assaulted. These may show up on any part of the leaf and are sporadic, scattered, and have concentric rings. These are dark colored and later wind up secured with pale blue green or dirty development. At the point when the assault is serious the spots combine and shape enormous patches bringing about untimely defoliation of the plant which step by step shrivels away. In one case the containers, when half develop, shrivel abruptly, turn dark colored and because of crumple of the pedicel, the cases fall or hand down. They are littler in size and have immature and wrinkled seeds with little oil content (Anon., 2018).

10.3.3 CAUSAL ORGANISM

Alternaria ricini.

The pathogen produces erect or somewhat bended, light dim to dark colored conidiophores, which are incidentally in gatherings. Conidia are delivered in long chains. Conidia are obclavate, light olive in shading with 5–16 cells having transverse and longitudinal septa with a bill at the tip (Anon., 2011b).

10.3.4 DISEASE CYCLE AND EPIDEMIOLOGY

The conidia are delivered plentifully on the ailing bit under sodden conditions and borne in chains on the conidiophores. The illness is helped over through the seed both remotely and inside. It is accounted for that the sickness causes pre and post-development damping off and a seedling and foliage scourge when the infected seeds are sown. A portion of the conditions which administer the seriousness of the ailment are accounted for to be the nearness of a defenseless assortment, high air moistness and low temperature (16–20°C) (Anon., 2016).

10.3.5 MANAGEMENT

- Seed treatment might be valuable in fighting the underlying period of the illness.
- Use of healthy seed and treat the seed with Captan or Thiram @ 4 g/kg seed or spray Mancozeb at 2.5 g/lit concentration at an interim of 15 days beginning from 90 long periods of crop development.

10.4 CERCOSPORA LEAF SPOT (CLS)

10.4.1 ECONOMIC IMPORTANCE

- The disease has been accounted for from Bihar, Uttar Pradesh and Hyderabad and is most likely present in numerous different parts of the nation.
- It makes significant damage the leaves and is the wellspring of loss of sustenance for the Eri-silkworm which is kept up on castor plant.

10.4.2 SYMPTOMS

- The disease appears as minute black or brown points surrounded by a pale green ring.
- These spots are visible on both the surfaces of the leaf.
- As the spots enlarge, the center turns pale brown and then greyish-white surrounded by a deep brown band which may be narrow and sharp or broad and diffused.

- The fructifications of the fungus appear as tiny black dots in the white center.
- The diseased spots often occur in great numbers scattered over the leaf and are roundish when young but may become irregularly angular when mature.
- When the spots are close together, the intervening leaf tissue withers and large brown patches of dried leaf may result (Anon., 2016).

10.4.3 CAUSAL ORGANISM

Cercospora ricinella.

- The pathogen hyphae gather underneath the epidermis and shape a hymenial layer. Groups of conidiophores develop through stomata or epidermis.
- They are septate and unbranched with profound darker base and light dark colored tip. The conidia are extended, dismal, straight or somewhat bended, truncate at the base and tight at the tip with 2–7 septa.

10.4.4 DISEASE CYCLE AND EPIDEMIOLOGY

The hyphae of the causal organism gather underneath the epidermis and form little stomata. Bunches of conidiophores as a rule in gatherings of 10–20 develop through any piece of the leaf tissue and form the fructification of the parasite (Anon., 2016).

10.4.5 FAVORABLE CONDITIONS

High moistness and warm climate conditions are in charge of the disease.

10.4.6 MANAGEMENT

- Treat the seed with Thiram or Captan 3 g/kg seed.
- Spraying twice with Mancozeb 2.5 g/lit or Carbendazim 0.5 g/lit at 10-multi day interim diminishes the disease power.

10.5 POWDERY MILDEW

10.5.1 ECONOMIC IMPORTANCE

Leveillula taurica causing powdery mildew of castor (*R. communis*) has been reported worldwide by Ramakrishnan and Narasimhalu (1941), Chiddarwar (1954), Amano (Hirata) (1986), Farr and Rossman (2010). Powdery mildew of castor caused by *Leveillula taurica* was first recorded by Mirzaee et al. (2011) from Iran. The disease in India is accounted for to be predominant amid November to March at Coimbatore (Anon., 2016).

10.5.2 SYMPTOMS

It is portrayed by average mold development which is for the most part restricted to the under-surface of the leaf. At the point when the disease is serious the upper-surface is likewise secured by the whitish development of the organism. Light green patches, relating to the diseased regions on the under surface, are noticeable on the upper side particularly when the leaves are held against light (Anon., 2016).

Broad abaxial front of more seasoned leaves by white patches of outside shallow mycelium and conidiophores, alongside chlorotic and necrotic locales on the upper leaf surface. Provinces were likewise present on the upper leaf surface as appeared in (Mirzaee et al., 2011).

10.5.3 CAUSAL ORGANISM

Leveillula taurica.

Conidiophores developed through leaf stomata, independently or fanned, as a rule in gatherings of a few, and formed dimorphic conidia. Essential conidia lanceolate with particular apical focuses, 12.5–19×37.5–70 μm; auxiliary conidia round and hollow, 12.5–20×37.5–77.5 μm; the two conidia hyaline with precise/reticulated wrinkling of the external dividers. These morphological highlights are run of the mill of the anamorphic stage depiction of *Leveillula taurica* (Lév.) Arnaud (Braun, 1987).

10.5.4 DISEASE CYCLE AND EPIDEMIOLOGY

The pathogen is endophytic and comprises of hyphae which are intercellular and involve the light parenchyma of the mesophyll. The haustoria infiltrate into a portion of the parenchymatous cells. The conidiophores of the parasite are stretched and more often than not rise through stomata in conglomeration. The conidia are hyaline, shifting fit as a fiddle, bear minute papilla-like projection at the expansive end, and are borne separately at the tip of each branch. These develop promptly in water delivering a germ tube from one end (Anon., 2016).

10.5.5 FAVORABLE CONDITIONS

Cool (10–20°C) and wet climate (90% RH) favors disease development (Anon., 2018).

10.5.6 MANAGEMENT

- Two spray of wettable sulfur 3 g/lit at 15 days interim, beginning from 3 months in the wake of sowing.
- Spray 1 ml Hexaconazole or 2 ml Dinocap/lit of water at fortnight interims.

10.6 WILT

10.6.1 ECONOMIC IMPORTANCE

It is the most imperative disease of castor at present in India. Castor shrink was first time recorded in Morocco (Reiuf, 1953). In India, wither was recorded out of the blue from Udaipur, Rajasthan by Nanda and Prasad (1974) and later from Gujarat amid 1980–81. Event of wither from USSR was accounted for by Andreeva (1979). The degree of disease occurrence was up to 80 for every penny in Russia (Moshkin, 1986). Misfortunes in yield were seen in all developed castor half and halves in Gujarat and as high as 85 for each penny shrink occurrence was accounted for under North Gujarat conditions (Dange, 1997; Dange, 2003).

10.6.2 SYMPTOMS

Castor plants are powerless to wither at all development arranges yet disease for the most part shows up at blooming and spike formation organize and turns out to be more conspicuous in later phase of the harvest. Youthful seedling at two-three leaves organizes display discoloration of hypocotyl and loss of bloat with or without change in shading. Debilitated plants either don't bear cases or give diminutive seeds (Moshkin, 1986). Youthful plants at sprouting stage are likewise seriously assaulted, which display progressive yellowing of apical leaves, shrinking with minor rot and dry totally. The mycelium enters the vascular arrangement of the roots, stems, and leaves causing corruption, which prompt withering lastly passing of the plant (Sviridov, 1989). Tainted stem indicates blackish sores over the neckline locale and these injuries additionally spread up to a separation of 15 to 20 cm over the ground level (Reiuf, 1953). Some of the time these dark stripes may cover the entire stem. At the season of blooming and spike formation organizes, the disease is portrayed by progressive yellowing, withering with peripheral and between-veinal putrefaction of clears out. At long last the leaves with petiole become scarce and hang down (Reiuf, 1953; Nanda and Prasad, 1974). The parts of contaminated plants likewise end up wilted and stained. Contaminated plants once in a while bear seeds and such seeds are deformed and light in weight. Underlying foundations of withered plant demonstrate darkening and putrefaction, while in the event of halfway shrinking just a single side of root framework is watched blackish and necrotic and the opposite side root framework stays sound. At the point when the stem of withered plant is splitted open, white cottony parasitic development is seen in the substance area and the essence end up blackish. Transverse and longitudinal segments of the influenced roots uncover the nearness of the organism in vascular tissue and the xylem parenchyma. Formation of tyloses is additionally seen in tainted roots (Nanda and Prasad, 1974). The contaminated stem tissue demonstrates between cell mycelium in vessels and hypertrophy of xylem parenchymatous cells.

10.6.3 CAUSAL ORGANISM

Fusarium oxysporum f. sp. *ricini* Nanda and Prasad

The causal organism of wilt was identified as *Fusarium oxysporum* f. sp. ricini Nanda and Prasad. The fungus produces abundant fluffy white mycelial growth on PDA, which turns pinkish when incubated under fluorescent light. Micro-conidia are hyaline, round to oval in shape, single celled but rarely septate. The size of single celled micro-conidia ranged from 5.25–14.00 x 3.50–7.00 μm. Macro-conidia are also hyaline, few in number having 2–6 septa (mostly 3 septa), straight, spindle as well as sickle shaped and measure 17.50–70.00 x 3.50–5.25 μm (Desai et al., 2003).

10.6.4 DISEASE CYCLE AND EPIDEMIOLOGY

The pathogen is fundamentally soil-borne and gets by as small scale conidia, large scale conidia and chlamydospores. Seed borne nature of the pathogen has additionally been accounted for. Seeds from wilted castor plants conveyed inoculum at the micropylar end in 2–19 for every penny seeds and seed contamination was bound to testa, tegmen, and endosperm (Naik, 1994). *Fusarium oxysporum* f. sp. ricini was discovered seed borne in 10.8 for each penny seeds of castor assortment Aruna gathered from wilted plants (Chattopadhyay, 2000). In this manner, seeds from contaminated region may likewise assume an essential job in the dispersal of the pathogen in the new zones. Tainted seeds assumed essential job in the propagation and spread of the pathogen (Dange, 2003).

10.6.5 FAVORABLE CONDITIONS

The disease appears at all growth stages of the crop but becomes more prominent and severe at the time of flowering and spike formation. Favorable temperature for infection is 13–15°C and for symptom expression is 22–25°C (Andreeva, 1979).

10.6.6 MANAGEMENT

- Select disease free seeds for planting.
- Rogue out and burn disease affected plants and crop debris regularly.
- Follow crop rotation for 2–3 years with non-host plants like pearl millet, finger millet or other cereals.
- Follow intercropping with red gram.
- Seed treatment with *Trichoderma viride* @ 4 g/kg and Thiram @ 3 g/kg seed or carbendazim @ 2 g/kg seed.

- Multiplication of 2 kg *T. viride* formulation by mixing in 50 kg FYM. Sprinkle water and cover with polythene sheet for 15 days and then apply between rows of the crop.
- Use wilt resistant varieties Jyothi, Jwala, and hybrids *viz.*, DCH 32, DCH 177, DCH 519, GCH 4, GCH 7 and GCH 8.

10.7 ROOT ROT/CHARCOAL ROT

10.7.1 ECONOMIC IMPORTANCE

Economic yield losses caused by the disease are 20 to 60% (Savalia et al., 2003).

10.7.2 SYMPTOMS

Sudden wilting of plants in patches under high soil dampness stretch combined with high soil temperatures is a typical side effect. The plants hint at water deficiency. Inside seven days, the leaves and petiole hang down and inside a fortnight, the tainted plants go away. Dim dark colored injuries are seen on the stem close to the ground level. The taproot hints at getting and root bark sheds dry effectively. Fruiting bodies (pycnidia) of the growth are viewed as minute dark specks on woody tissues and in essence district. In serious disease, whole branch or best of the branch shrinks away. Youthful leaves twist inwards with dark edges and drop off later. Such incredible diseased plants blossom rashly. Occurrence at development causes spike curse. Seed improvement is influenced.

10.7.3 CAUSAL ORGANISM

Macrophomina phaseolina.

10.7.4 DISEASE CYCLE AND EPIDEMIOLOGY

Pathogen survives in soil, plant debris and many cultivated and wild plants as sclerotia and pycnidia. Secondary spread is through sclerotial bodies.

10.7.5 FAVORABLE CONDITIONS

Disease is favored by soil temperature of 35°C and moisture stress conditions preceding crop maturity and application of more nitrogenous fertilizers.

10.7.6 MANAGEMENT

- Burn crop debris containing the sclerotia of the fungus.
- Seed treatment with *Trichoderma viride* @ 4 g/kg seed or carbendazim@1 g/kg seed.
- Seed treatment of carbendazim 50% WP @ 2 g/kg (17.13%) followed by the seed treatment of carboxin 37.5% + thiram 37.5% @ 3 g/kg (18.18%) (Parmar et al., 2017).
- Seed treatment of *Trichoderma harzianum* followed by soil application and drenching of same bio-agent (Parmar et al., 2017).
- Soil drenching with carbendazim @ 0.1%, 2–3 times at 15 days interval.

10.8 GRAY MOLD/GRAY ROT/BLOSSOM BLIGHT

10.8.1 ECONOMIC IMPORTANCE

This fungus had caused serious losses of castor crop in the summer of 1918 mainly in Florida and others southern States, where it was responsible for losses up to 100% of castor yield (Godfrey, 1923). Later, the disease was reported in almost all countries where castor has been cultivated (Kolte, 1995), having nowadays a worldwide distribution. In India, gray mold is found in few states and is regarded as troublesome only in Andhra Pradesh and Tamil Nadu, in the South, where the weather conditions are more favorable for disease development where in 1987, an epidemic outbreak of gray mold occurred (Dange et al., 2005).

10.8.2 SYMPTOMS

The essential focuses of the growth are the inflorescence and the containers, in any advancement organize (Araujo et al., 2007; Dange et al., 2005). A few creators (Drumond and Coelho, 1981; Batista et al., 1996) guarantee that the male blossoms are the first to be contaminated, yet it isn't generally the case in light of the fact that any piece of the inflorescence can be tainted, the female blooms being the special target. That case originated from the way that the male blooms are the first to be uncovered, at the prior phase of inflorescence formation; thusly such blossoms are presented longer to the contamination units of the organism. In any case, when the male blooms endure anthesis they are never again an objective and are not really contaminated, negating the announcement of Drumond and Coelho (1981) that "the growth assaults first the male blossoms on the grounds that the anthers, being doused with the rain water or dew, effectively hold the organism spores conveyed by the breeze."

The principal manifestations are noticeable as somewhat blue spots on the inflorescences, on both female and male (before anthesis) blooms, and on creating organic products. On organic products, the side effects can advance to roundabout or elliptic, indented, dim hued recognizes that can result in crack of the container (Araujo et al., 2007). These side effects are normally more incessant when a time of low relative moistness troublesome to contagious sporulation happens not long after the growth enters the host tissues. Contingent upon climate conditions (e.g., significant lots with high relative stickiness not long after the parasite enters the host), the event of yellow overflow at the purpose of contamination is visit (Batista et al., 1996; Dange et al., 2005) because of the fast enzymatic tissue corruption. The indications on the male blossoms, previously anthesis, are little, pale darker, necrotic spots, which can advance to bigger darker spots with a darker edge.

The tainted blooms and youthful cases wound up mollified due the parasitic colonization and mycelial development is, at first, pale dim and later dim olivaceous. An abundant sporulation is generally seen in such stage. At the point when the contamination begins on youthful cases, they end up spoiled; if the disease begins later, with completely created containers, the seeds generally ended up empty, with coat staining and weight reduction (Dange et al., 2005). The male blossoms can be tainted first, however, the growth has a reasonable inclination for the female blooms. Disease can prompt finish pulverization of the raceme, especially on the off chance that it achieves the primary stem and the climate conditions are positive for the disease. A few other plant parts, e.g., leaves, petioles, and stem can likewise be contaminated, fundamentally because of the affidavit or fall of tainted material from the inflorescence or racemes. On leaves, the sores are typically sporadic, yet can accept an elliptic or round example, the size is extremely factor, here, and there mixing and bringing about a foliar curse. On petioles and stems, necrotic, indented sores ordinarily are formed which can cause the strangulation and thus passing of the parts over the disease point (Batista et al., 1996; Dange et al., 2005).

10.8.3 CAUSAL ORGANISM

Botryotinia ricini.

Conidiophores long, slender, branched, septate, apical cells enlarged or rounded bearing clusters of conidia on short sterigmata. Entire structure resembles a grape bunch. Conidia are hyaline or ash colored, grey in mass, one-celled, globose to ovoid.

10.8.4 DISEASE CYCLE AND EPIDEMIOLOGY

There are few studies of the epidemiological aspects of gray mold on castor. Godfrey (1923) mentioned that temperatures around 25°C and high relative humidity (RH) are highly favorable to disease development. The minimum and maximum temperature for mycelial growth was established by Godfrey as 12 and 35°C, respectively. Sussel (2008) concluded that the disease shows a random distribution pattern, matching its airborne nature. However, under high rainfall, the disease assumes an aggregate pattern, typical of those dispersed by water splash.

10.8.5 MANAGEMENT

- Use healthy seeds, removal of plant debris, adequate choice of planting area and growing season, and use of less susceptible cultivars (Sussel, 2009).
- In India, Dange et al. (2005) recommended two prophylactic sprays with carbendazim (0.05%): the first at 50% of flowering; and the second, when the first disease symptoms appear.
- Adjust sowing time in such a way that crop maturation occurs during dry season.
- Seed treatment with carbendazim @ 3 g/kg
- Adopt wider spacing (90 x 60 cm).
- Remove diseased spikes and destroy them.
- Grow varieties like Jwala with non-spiny capsules and less compact inflorescence.
- Spray carbendazim/Thiophanate methyl @ 0.1% before the onset of cyclonic rains based on weather forecast followed by second spray soon after rains have receded.

- Application of 20 kg urea and 10 kg of muriate of potash after removal of diseased panicles may be useful for the growth of panicles that subsequently develop.

10.9 BACTERIAL LEAF SPOT

10.9.1 ECONOMIC IMPORTANCE

The disease is caused by *Xanthomonas ricinicola*, which is gram negative, melts gelatine, hydrolyses starch, digests casein, peptonises drain, lessens litmus, produces smelling salts and hydrogen sulfide and does not diminish nitrates.

10.9.2 SYMPTOMS

The pathogen assaults cotyledons, leaves, and veins and creates few to various little round, water-doused spots which later end up precise and dim dark colored to coal black in shading. The spots are by and large accumulated towards the tip. At a later stage the spots end up unpredictable fit as a fiddle especially when they mix and territories around such spots turn pale-dark colored and weak. Bacterial overflow is seen on both the sides of the leaf which is as little sparkling dabs or fine scales (Anon., 2014).

10.9.3 CAUSAL ORGANISM

Xanthomonas campestris pv. *ricinicola*

10.9.4 DISEASE CYCLE AND EPIDEMIOLOGY

The optimum temperature for the growth of the bacterium and its thermal death point are 31°C and 51°C respectively (Anon., 2016).

10.9.5 MANAGEMENT

- Field sanitation help in limiting the yield misfortune as pathogen makes due on seed and plant flotsam and jetsam.

- Remove and obliterate the tainted plant flotsam and jetsam.
- Hot water treatment of seeds at 50–60°C for 10 minutes.
- Spray streptocycline @ 500 ppm in blend with copper oxychloride @ 0.3%.

KEYWORDS

- **bacterial leaf spot**
- **blossom blight**
- **cercospora leaf spot**
- **charcoal rot**
- **powdery mildew**
- **wilt**

REFERENCES

Amano, H. K., (1986). *Host Range and Geographical Distribution of the Powdery Mildew Fungi* (p. 742). Japan Scientific Society Press, Tokyo.

Andreeva, L. T., (1979). *Zashchita Rastenii, 7*, 22–23.

Araujo, A. E., Suassuna, N. D., & Coutinho, W. M., (2007). Doenças e seu Manejo. In: Azevedo, D. M. P., & De Beltrão, N. E. M., (eds.), *O Agronegócio da Mamona no Brasil*, (pp. 283–303). Embrapa Informação Tecnológica, ISBN 978-85-7383-381-2, Brasília, Brazil.

Batista, F. A. S., Lima, E. F., Soares, J. J., & Azevedo, D. M. P., (1996). Doenças e pragas da mamoneira (*Ricinus communis* L.) e seu controle. Embrapa Algodão, [Circular Técnica 21], ISSN 0100-6460, Campina Grande, Brazil.

Braun, U., (1987). A monograph of the erysiphaceae (powdery mildews). Beiheftezur. *Nowa Hedwigia, 89*, 1–700.

Chattopadhyay, C., (2000). *J. Mycol. Pl. Pathol., 30*, 265.

Chiddarwar, P. P., (1954). Occurrence of *Oidiopsis taurica* (Lev.) Salmon on a new host, *Ricinus Communis* L. *Curr. Sci., 23*, 198.

Dange, S. R. S., (1997). In: *Proc. of International Conference on Integrated Plant Disease Management for Sustainable Agriculture* (p. 107). IARI, New Delhi, India.

Dange, S. R. S., (2003). *J. Mycol. Pl. Pathol., 33*, 333–339.

Dange, S. R. S., Desal, A. G., & Patel, S. I., (2005). Diseases of castor. In: Saharan, G. S., Mehta, N., & Sangwan, M. S., (eds.), *Diseases of Oilseed Crops* (pp. 211–234). Indus Publishing Co, ISBN 81-7387-176-0, New Delhi, India.

Desai, A. G., & Dange, S. R. S., (2003). *Agric. Sci. Digest, 23*, 20–22.

Drumond, O. A., & Coelho, S. J., (1981). Doenças da mamoneira. *Informe Agropecuário, 7*(82), 38–42, ISSN: 0100-3364.

Farr, D. F., & Rossman, A. Y., (2010). *Fungal Databases, Systematic Mycology and Microbiology Laboratory*. ARS, USDA. Retrieved from: http://nt. ars-grin. gov/fungaldatabases.
Godfrey, G. H., (1923). Gray mold of castor bean. *Journal of Agricultural Research*, *23*(9), 679–715 + 13 plates, ISSN 0095-9758.
Kirk, P. M., Cannon, P. F., Minter, D. W., & Stalpers, J. A., (2008). *Dictionary of the Fungi* (10th edn.). Wallingford: CABI.
Kolte, J. S., (1995). *Castor: Diseases and Crop Improvement*. Shipra Publications, ISBN 81-85402-54-X, Delhi, India.
Mirzaee, M. R., Khodaparast, S. A., Mohseni, M., Ramazani, S. H. R., & Najafabadi, M. S., (2011). First record of powdery mildew of castor- oil plant (*Ricinus communis*) caused by the anamorphic stage of *Leveillula taurica* in Iran. *Australasian Plant Dis. Notes*, *6*, 36–38.
Moshkin, V. A., (1986). *Castor*. Amerind Publishing Co. Pvt. Ltd., New Delhi.
Naik, M. K., (1994). *J. Mycol. Pl. Pathol.*, *24*, 62, 63.
Nanda, S., & Prasad, N., (1974). *J. Mycol. Pl. Pathol.*, *4*, 103–105.
Parmar, H., Kapadiya, H. J., & Bhaliya, C. M., (2017). Integrated management of root rot of castor (*Ricinus communis* L.) caused by *Macrophomina phaseolina* (Tassi) Goid. *International Journal of Chemical Studies, 6*(1), 849–851.
Ramakrishnan, T. S., & Narasimhalu, I. L., (1941). A new host, *Ricinus communis*, for *Leveillula taurica* (Lev.) Arn. (*Oidiopsis taurica* (Lev.) Salm.). *Curr. Sci., 10*, 211, 212.
Reiuf, P., (1953). *Rev. Path. Veg.*, *32*, 120–129.
Savalia, R. L., Khandhar, R. R., & Moradia, A. M., (2003). Screening of castor germplasm against root rot caused by *Macrophomina phaseolina* under sick plot. *ISOR National Seminar: Stress Management in Oilseeds.*
Soares, D. J., (2012). In: Christian, J. C., (ed.), *Gray Mold of Castor: A Review, Plant Pathology*. ISBN: 978-953-51-0489-6.
Sudha, S. P., Kumar, O. A., & Sujatha, M., (2016). Importance and production strategies of castor in India: A review. *International Research Journal of Natural and Applied Sciences*, *3*(10), 122–149.
Sussel, A. A. B., (2008). *Epidemiologia do mofo-cinzento (Amphobotrys ricini Buchw.) da Mamoneira*. PhD Thesis (Phytopathology). Universidade Federal de Lavras, Lavras, Brazil.
Sussel, A. A. B., (2009). *Epidemiologia e Manejo do Mofo-Cinzento-da-Mamoneira*. Embrapa Cerrados, [Documentos 241], ISSN 1517–5111, Brasília, Brazil.
Sviridov, A. A., (1989). *Mikologiyai Fitopatologiya*, *23*, 91–97.

CHAPTER 11

Current Status of Groundnut (*Arachis hypogae* L.) Diseases and Their Management

KOTRAMMA C. ADDANGADI[1] and RANGANATHSWAMY MATH[2]

[1]*Regional Research Station, S. D. Agricultural University, Bhachau–370140, Gujarat, India,*
E-mail: kotramma. addangadi@gmail. com

[2]*Department of Plant Pathology, College of Agriculture, Jabugam, Anand Agricultural University, Gujarat–391155, India*

Groundnut is a key oil seed crop among nine oilseeds crops grown in our country. South America was the place from where cultivation of groundnut originated. India is the largest producers of oilseeds in the world and occupies vital position in the Indian agricultural economy. In India, major groundnut production takes place in Gujarat, Tamil Nadu, Andhra Pradesh, Maharashtra, and Karnataka states. Groundnut is considered as the 'king' of oilseeds. It is one of the important cash and food crops of our country, being a valuable source of all the nutrients; it is a low-priced commodity. Groundnut is wonder nut and poor man's cashew nut. It contains 25.20% protein, 48.20% oil, and 11.50% starch.

This crop is prone to attack by various diseases and pest to a greater extent compare to other crops. Most important factors contributing to low yield are disease and pest attack. Around 55 pathogens have been reported to affect groundnut. Management of pest and diseases help farmers to reduce crop damage from pests and increase food production. They are very important as they improve the quality and yield of produce. The important diseases which cause major losses of groundnut yield/quality and their management have been discussed here.

11.1 EARLY AND LATE LEAF SPOT OF GROUNDNUT

11.1.1 INTRODUCTION/ECONOMIC IMPORTANCE

Early leaf spot (ELS) and late leaf spots (LLS) are considered as the most important diseases of groundnut. Both together are called as 'Tikka' disease. Berkeley (1875) was first time described groundnut leaf spot and named the causal organism as *Cladosporium personatum* Berk and Curt. Ten years later Ellis and Everhart (1885) transferred genus to *Cercospora.* Woodroof (1933) gave a clear account of the existence of two distinct species of *Cercospora,* i.e., *C. arachidicola* causing "ELS" and *C. personata* (Berk and Curt). Ell. and Eve. causing "late leaf spot." In 1983, Von Arx recognized the perfect stage of the genus is *Mycosphaerella* and given the new combination *Phaeoisariopsis personata* (Berk. and Curt.). The perfect stage of *C. arachidicola* was *Mycosphaerella arachidicola* Jenkins whereas that of late leafspot fungus is *Mycosphaerella berkeleyii* Jenkins (Jenkins, 1939). The perfect stages are noticed only in U. S. A. The disease is noticed in all groundnut growing areas of India. It becomes more severe during *Kharif* season and causes yield losses up to 50%. In India LLS (late leaf spot) is more severe than ELS (Bharat et al., 2013).

11.1.2 SYMPTOMS

Initially small chlorotic spots will be noticed on the leaf around 30–35 days after sowing (DAS). Symptoms can appear on any above ground parts of the plant including leaves, petioles, stems, and pegs in the advanced stages of disease. Later defoliation will occur. Both the leaf spots can be easily differentiated.

Early Leaf Spot	Late Leaf Spot
Leaf spot color is reddish brown and spots are not typically circular and larger that late leaf spot	Color of the spot is dark brown and spots are more circular and small
Yellow halo is conspicuous and spreading	Yellow halo is dull and limited to margin and sometimes it may absent
Lesions are less in number	Lesions are more in number
Development of the disease is rapid but less dangerous than the late leaf spot	Development of the disease is slower than the early leaf spot but causes more loss
It occurs early in the season compare to late leaf spot due to presence more RH during early season	Usually it occurs late in the season

FIGURE 11.1 Early Leaf Spot of Groundnut

FIGURE 11.2 Late Leaf Spot of Groundnut

11.1.3 CASUAL ORGANISM

Early Leaf Spot: *Cercospora arachidicola* Hori. Perfect stage: *Mycosphaerella arachidicola.*

Late Leaf Spot: *Cercosporidium personatum* Perfect stage: *Mycosphaerella berkeleyii.*

11.1.4 TAXONOMY

Kingdom: Fungi
Phylum: Ascomycota
Class: Dothideomycetes
Order: Capnodiales
Family: Mycosphaerellaceae
Genus: *Cercospora*

C. arachidicola grows initially with intercellular and then intracellular when the host cells starts to die. No haustoria are produced. Olivaceous brown conidiophores are produced, continuous or 1–2 septate. Hyaline or pale yellow conidia, obclavate with rounded to distinctly truncate base and sub acute tip. Septate mycelium is produced by *C. personatum* in the host and intercellular sending haustoria into the palisade and mesophyll cells. Conidiophores are uniformly olivaceous brown, continuous, sometimes 1–2 septa are present, without branches and geniculate. Conidia are cylindrical, light colored, 1–7 septate with bluntly rounded ends (Singh, 1998).

11.1.5 DISEASE CYCLE

Both causal agents are soil borne in nature. Perfect stages of the fungi are known but ascospores does play important role as primary source of inoculum. Mycelium directly produces the conidia in crop debris in the soil immediately after early rains and gets deposited on the young leaves by wind or rain splash and disease cycle will get initiated. The oldest leaves which are near the soil surface are first to show the symptoms with small spots and the later conidia will be carried by wind, rain droplets and insects to the later-formed leaves and to the near plants. During the off-season pathogens survive on volunteer groundnut plants and on crop debris. There are no reports on internally seed borne nature of the pathogen (McDonald et al., 1985).

11.1.6 EPIDEMIOLOGY

Temperature, relative humidity (RH) and rainfall are the most important epidemiological factors which have tremendous impact on the outbreak and spread of leaf spots and rust of groundnut. Crop sown early will reduce the disease severity (Hazarika et al., 2000). Sulaiman and Agashe (1965) reported that the temperature of 20°C to 30°Cwas found optimum for development of leaf spots. Disease severity will be increased when RH >90%.

11.1.7 INTEGRATED MANAGEMENT PRACTICES OF DISEASE

Crop rotation is the best management practices to avoid the soil borne pathogen necessary. Avoiding growing susceptible varieties and growing moderately resistant varieties like ICGV 89104, ICGV 91114 (EM), ICGV 920920 and ICGV 92093 (MM). Proper sanitation and removal of infected crop plants and debris. Foliar application of *Prosopis juliflora* extract at 45, 75 and 90 DAS and chlorothalonil at 60 DAS effectively reduced foliar diseases severity and increases the pod yields by 81–98% (Krishna and Suresh, 2005). Foliar spray with tebucanozole 25 EC (0.1%) at 15 days interval will manage the disease.

11.2 RUST

11.2.1 INTRODUCTION/ECONOMIC IMPORTANCE

In India, rust has been first time recorded from Ludhiana of Punjab state (Chahal and Chohan, 1971). Around 57% loss have been estimated and even greater loss in haulm yield have been estimated (Subrahmanyam and McDonald, 1987). Epidemic of rust disease was appeared in Maharashtra during 1976–77. Spegazzini (1884) named causal organism as *Puccinia arachidis.*

11.2.2 SYMPTOMS

Symptoms start as white flecks on abaxial surface and later turns to yellowish. Orange red brown (uredinia) colored pustules will appear on the lower surface of the leaves. The pustules are circular to elliptical and raised. Size ranges from 0.3 to 2.0 mm in diameter.

Two days after the formation the pustule rupture and releases the uredospores. Pustules may appear on the upper surface of the leaf with necrosis around the pustule, these necrotic areas coalesced together and results in defoliation. Uredospores with maturity turn to cinnamon brown color (Savary et al., 1989). Petioles and stems also get infected. Disease spreads in a radiating pattern from a single spot to the entire field.

11.2.3 CAUSAL ORGANISM

Puccinia arachidis Spegazzini.

11.2.4 TAXONOMY

Kingdom: Fungi
Phylum: Basidiomycota
Class: Pucciniomycetes
Order: Pucciniales
Family: Pucciniaceae
Genus: *Puccinia*
Species: *P. arachidis.*

Different stages of groundnut rust:

- **Stage 0:** Spermagonia not known.
- **Stage 1:** Aecia not known.
- **Stage 2:** The uredinial stage commonly observed and predominant stage. Uredospores are broadly ellipsoid or obovoid, 16–22 x 23–29 μ in size.
- **Stage 3:** Teliospores are rarely. Teliospores are scattered, prominent, chestnut brown or cinnamon brown becoming grey from germination, predominantly two celled, sometimes 3–4 cells.
- **Stage 4:** Metabasidia and basidiospores are not reported (Tashildar, 2011).

11.2.5 DISEASE CYCLE

Uredospores present on self sown/voluntary groundnut plants acts as primary source of inoculum and helps for survival carry over the groundnut rust from season to season (Gururaj and Srikant, 2007). During season, these uredospores infect the leaves and form uredosori and release numerous uredospores and complete many life cycles by infecting groundnut crop. During nutrient stress condition and low temperature, the dikaryotic hyphae from uredospores form telia, only few reports are there for existence of teliospores (Rodrigues et al., 2006; Tashildar et al., 2012). Hence, it needs further more research to confirm the role of teliospore for causing groundnut rust.

11.2.6 EPIDEMIOLOGY

Infection present on voluntary/self sown plants carry uredospores, which act as primary source of inoculums.25°C temperature is congenial for germination of uredospores and for infection, 78% RH is optimum (Gururaj and Srikant, 2007). In India, a continuous dry period characterized by high temperature (> 26°C) and low RH (< 70%) is reported to delay rust occurrence and severity, whereas intermittent rain, high RH and 20 to 26°C temperatures favor disease development (Siddaramaiah et al., 1980).

11.2.7 MANAGEMENT

Crop rotation with non-host crops and field sanitation and early sowing in the month of June will reduce the disease incidence. Removal of volunteer and self-sown groundnut plants reduce the primary sources of inoculum. Intercropping with bajra or sorghum reduces the disease intensity. Spraying with aqueous neem extract @3–5% is economical for the control of rust. Application of hexaconazole (0.1%) thrice on 30, 45, and 60 DAS, reduced leaf spot and rust disease incidence and increased the yield of groundnut significantly. Hexaconazole treatment showed 71% increase in pod yield and 87% increase in fodder yield (Jadeja et al., 1999). Spray application of Saaf/Companion (carbendazim + mancozeb) at 40 and 50 DAS manages both leaf spot and rust disease.

11.3 CROWN ROT/COLLAR ROT

11.3.1 *INTRODUCTION/ECONOMIC IMPORTANCE*

The pathogen is seed-borne in nature and having important role in seed viability, seed germination and spread of disease. *A. niger* induces crown rot on emerging seedlings in light sandy soils and collar rot on slightly older seedlings in clay soils (Mayee et al., 1995). Collar rot is known to cause 27.9–46.9% avoidable losses (Backetia, 1983). In India it was first reported by Jain and Nema (1952) as *Aspergillus* blight.

11.3.2 *SYMPTOMS*

Symptoms of collar rot disease are commonly manifested as a pre and post emergence damping-off of the affected seedlings. In case of pre-emergence damping off, seed rot is the most obvious symptom wherein, seeds are infected by the black spore mass of *A. niger*, thereby resulting in rotting, less seed germination and crop stand. Post-emergence damping-off or seedling blight occurs on germinated seedlings and manifests as circular brownish small spots on the cotyledons, followed by rotting and spreading of the disease to the collar and hypocotyl regions, resulting in sudden wilting of the plants. Occasionally, collar rot can continue up to crop harvesting stage resulting in damage to the seeds (Divya Rani, 2015).

11.3.3 *CAUSAL ORGANISM*

Aspergillus niger Tiegh.

11.3.4 TAXONOMY

Kingdom: Fungi
Phylum: Ascomycota
Class: Eurotiomycetes
Order: Eurotiales
Family: Trichocomaceae
Genus: *Aspergillus*
Species: *A. niger*

Reddy (1993) described that the mycelium of *Aspergillus niger*as scanty, hyaline to white or light yellow in color. Conidiophores arise from the seed coat and were 3 mm long, hyaline to light brown long, thin unbranched erect brittle and terminate in inflated apex upon which phialides are found. Conidia are produced in chains on the sterigmata, conidia are single celled, pale to dark brown, more or less globose with low to prominent echinulations.

11.3.5 DISEASE CYCLE

Infected plant debris and seeds play important role for start of the disease. Conidia in the soil plays important role to carry over of the disease from season to season.

11.3.6 EPIDEMIOLOGY

Conidia and mycelium present on the infected seeds, plant debris and infected cotyledons act as primary source of inoculum. During the maturation if stress conditions prevail, seeds gets infected particularly harvesting, Seeds become infected during the last days of maturation in the soil and during harvesting, handling, and shelling. The pathogen infection takes place when seeds are stored below 14°C, fungus tolerates low moisture levels and optimum temperature ranged from 30 to 35°C. Deep sowing increases the disease incidence. Bunch type varieties are less susceptible.

11.3.7 MANAGEMENT

Seed treatment with carbendazim + mancozeb or Vitavax power were found most effective in reducing the disease or treating the seed with

Trichoderma harzianum (10 g/kg of seeds) (Kumari, et al., 2016). Disease can be effectively managed when seeds are treated with carbendazim and soil application of *T. Harzianum* (Karthikeyan, 1996). Application of FYM enriched with *Trichoderma harzianum* will manage the disease.

11.4 AFLATOXIN

11.4.1 INTRODUCTION/ECONOMIC IMPORTANCE

The aflatoxin has been first discovered as causative agent of Turkey-x disease in 1960 in England (Blount, 1961). Then the fungus was isolated and named as *Aspergillus fluvus* Link and reported that disease was caused by toxins produced by strains of the fungus (Sargeant et al., 1961). Several outbreaks of aflatoxin in poultry have been reported from India. Mycotoxin is derived from the Greek word "mykes, " meaning fungus and the Latin word "toxicum" meaning poison. Mycotoxins are secondary fungal metabolites that cause pathological or undesirable physiological responses in human and other animals. Aflatoxins have received greater attention than other mycotoxins because of their established carcinogenic effect in various animals and their acute toxicological effects in humans (Mehan, 2002). European Food Safety Authority (EFSA), 2009 decided total the aflatoxin limits as 10 µg/kg for all tree nuts except almonds, hazelnuts, and pistachios and concluded that there is no ill effects occurred up to this level (see, https://www. efsa. europa. eu/en/topics/topic/aflatoxins-food), while total aflatoxin limit for all commodities is 30 µg/kg in India (Sharma and Parisi, 2017).

11.4.2 SYMPTOMS

Yellow-green spore masses are commonly observed. Initially small discolored areas are often seen. Growth is very rapid and yellow to light green downy or powdery colonies will appear. Pre-harvest aflatoxin contamination in groundnut will occur when groundnuts mechanically or biologically damaged in the soil are predisposed to invasion by fungi. The saprophytic fungus will live predominantly on dead or dying tissue, therefore it may infect the maturing kernel if the pod is damaged while still in the ground (Smartt, 1994). In a standing groundnut crop, *A. flavus* invasion can occur in soil during pod development and maturation. Sometimes the mycelium of the fungus remains viable when the seed is sown and may contribute to either seed rot or seedling disease generally called as aflarot.

11.4.3 CAUSAL ORGANISM

Aspergillus flavus Link

11.4.4 TAXONOMY

Kingdom: Fungi
Phylum: Ascomycota
Class: Eurotiomycetes
Order: Eurotiales
Family: Trichocomaceae
Genus: *Aspergillus*
Species: *A. flavus.*

Conidiophores are colorless, thick walled, usually less than 1 mm in length. Vesicles and phialides are produced on the conidiophore, vesicles are elongate and become subglobose to globose (Hedayati et al., 2007). *A. flavus* often produces aflatoxins BI and B2 as well as cyclopiazonic acid. In general AFBL was the toxin produced in the highest quantity by the aflatoxicogenic strains (Magnoli, 1998). *A. flavus* can be divided into two strains. S and L (Bayman and Cotty, 1993). Isolates of the S strain produce numerous small sclerotia (<400 pm in diameter) and fewer conidia than L strains. Strain S isolates produce, on an average, more aflatoxin than L strain isolates.

11.4.5 DISEASE CYCLE

Pathogen overwinters in the soil and on infected decaying matter, survives as sclerotia or mycelia. During favorable condition, sclerotia germinate and produces the hypha and conidia. These conidia plays important role in the disease cycle as primary inoculum. The conidia can infect either grains or legumes (Diener et al., 1987)

11.4.6 EPIDEMIOLOGY

Infected seeds and soil plays important role to start disease. Pathogen is a heat tolerant; it also causes the post harvest storage rots, especially when the plant material is stored at high moisture levels. *A. flavus* grows and thrives in hot and humid climates (Agrios, 2005). Some predisposing factors like repeated cultivation of host plants or susceptible crops species on the same piece of land supports rapid buildup of *A. flavus* populations leading eventually to preharvest contamination of crops in field. Late planting, drought as well as insect pest attacks, especially termites increasing chances of *Aspergillus*.

11.4.7 MANAGEMENT

1. **Pre-Harvest Management:** Use of resistant varieties, early planting, and avoid monocropping will reduces the infection. *T. harzianum* (2 kg) enriched with FYM (500 kg/ha) should be applied 15 days before sowing, seed treatment with *Trichoderma harzianum* (10 g/kg) or seed treatment with Tebucanozole (3 g/kg) followed by spray with Tebuconazole @ 1 ml/l at 30 and 45 DAS. Application of gypsum @ 200 kg/ha during pod formation, avoid drought stress from flowering to pod maturation and avoiding the insect damage will controls the infection of *A. flavus*.
2. **Post Harvest Management:** Harvesting the crop at proper maturity, drying the seeds up to 8% moisture level, removing the immature and shriveled seeds and fumigation of seeds with insecticides to avoid the insect damage during the storage will reduces the post harvest infection.

11.5 STEM ROT

11.5.1 INTRODUCTION/ECONOMIC IMPORTANCE

Stem rot disease caused by *Sclerotium rolfsii* Sacc. is one of the significant factors contributing to yield loss. It is one of the most economically important diseases of groundnut (*Arachis hypogaea* L.), which accounts for 15 to 25% loss in groundnut yields annually (Storgeon, 1986). In India, stem rot occurs in all groundnut-growing areas and it is most severe in Maharashtra, Gujarat, Madhya Pradesh, Karnataka, Andhra Pradesh, Orissa, and Tamil Nadu. The name *Sclerotium rolfsii* was given by Saccardo (1911) who characterized the fungus as an imperfect form i.e., without sexual spores.

11.5.2 SYMPTOMS

Symptoms appear as both pre-emergence and post-emergence rot. It causes the seed rot, rotten, and soften seeds appears. After the germination, young seedlings will get infected and causes wilt. The disease can be easily identified by the presence of white fungal mycelium on the infected plants. Symptoms are typified by the development of white fungal thread over the affected plant tissue, later turns to brown color and production of mustard seed size sclerotia. Infection at the stem portion is most common and death of the young plants (Atla, 2017).

FIGURE 11.3 Stem Rot: White Threads of *S. rolfsii*

11.5.3 CAUSAL ORGANISM

Sclerotium rolfsii Sacc.

11.5.4 TAXONOMY

Kingdom: Fungi
Phylum: Basidiomycota
Class: Agaricomycetes

Order: Atheliales
Family: Atheliaceae
Genus: *Sclerotium*
Species: *S. rolfsii*

Initially white sclerotia are produced and later turn to brown color. Superficially, sclerotia are produced on the infected plant parts. Sclerotial size varies from 0.5 mm to 2.5 mm and spherical to ellipsoidal shape.

11.5.5 DISEASE CYCLE

Sclerotium rolfsii can over winter as sclerotia or mycelia in the soil or in the infected plant debris. Sclerostia are usually disseminated by through irrigation, cultural practices, infected transplant seedling and wind. In small amount, sclerotia may survive passage through sheep and cattle and thus could be spread through fertilizer (Stephen et al., 2001).

11.5.6 EPIDEMIOLOGY

Seclorotia present in the soil germinate under favorable environment condition and starts disease. They can be viable in the soil for 2 to 3 years. Mycelium present on the seed may also acts as primary source of infection but it is short lived.25–30°C temperature with high soil moisture enhances the diseases. Diseases are more severe in the area where frequent rain occurs.

11.5.7 MANAGEMENT

Cultural practices like deep summer plowing, groundnut crop should be rotated with non-host crops and burning of crop residues will reduce the disease. *T. harzianum* mixed with well decomposed FYM or castor cake @ 1.5 kg in 300 kg can be applied in furrow just before sowing.

11.6 BLACK HULL

Black hull is not severe disease in India. It is severe disease in some parts of New Mexico, recently observed in some parts of Kutch district of Gujarat. Disease is caused by *Theilaviopsis basicola.* The fungus survives in the soil as resistant spores for longer period of time. Disease is more severe in alkaline soil pH (7.0 and above), with more soil moisture and less temperature. Infection is mainly on the pods, initially black spots appears on the pod and later coalesced. Usually kernel discoloration is less observed. Excessive irrigation should be avoided and crop rotations with cotton and cereals. Crop rotation with grain sorghum is used to reduce black hull in New Mexico (http://osufacts. okstate. edu).

FIGURE 11.4 Black Hull of Groundnut

11.7 GROUNDNUT BUD NECROSIS DISEASE (BND)

11.7.1 INTRODUCTION/ECONOMIC IMPORTANCE

Many viral diseases are reported on groundnut in India. Among them, bud necrosis disease (BND) is the most severe which causes the maximum yield losses. In some parts of the world it is called as ring spot, in India it is called as bud blight or bud necrosis. Reddy et al. (1992) reported the disease for the first time in India during 1968 on groundnut. It is a major constraint in the cultivation of several leguminous and solanaceous hosts (Bhat et al., 2002). Yield losses up to 90–100% have been reported due to GBNV (Dwivedi et al., 1993).

11.7.2 SYMPTOMS

Chlorotic spots will appear on young leaflets later turned tonecrotic rings and streaks. Occasionally, the leaflets showed a general chlorosis with green islands. Necrosis of the terminal bud will occur. The bud necrosis symptom was common on crops grown in the dry (summer) and rainy seasons in India, indicating that this symptom was probably associated with high temperatures. Secondary symptoms start as stunting of the plants and proliferation of axillary shoots. Rarely, the lamina was reduced to the midrib, giving the leaflet a "shoe string" appearance and giving them a stunted and bushy appearance. Only a few branches on late-infected plants show these symptoms. Seeds from early-infected plants were small with shriveled testa and shows purple, red, or brown mottling (Reddy et al., 1992).

FIGURE 11.5 GBNS: Terminal Bud Necrosis Symptoms

11.7.3 CAUSAL ORGANISM

Peanut bud necrosis virus (PBNV) is a member of the genus Tospovirus of the family Bunyaviridae. It is transmitted by melon Thrips, *Thrips palmi*, in a propagative manner (Reddy et al., 1992).

11.7.4 TRANSMISSION

Tospoviruses are mechanically less transmissible, but there is no evidence for seed-transmission. Some reports said that the virus was detected in the shell and seed coat of groundnut, but not in the embryo (Pappu et al., 1999). *Thrips palmi* acquire PBNV as larvae and transmit it as adults.100% transmission will be achieved when there are ten adults per plant and cowpea is a diagnostic host (Jasani and Kamdar, 2015).

11.7.5 EPIDEMIOLOGY

The primary source of inoculum will be from infected crop plants, ornamentals, and weed hosts and also support thrips during summer season, play a major role in the spread of the virus. Thrips are mainly carried by wind (Reddy and Wightman, 1988). Most migrations occur when an air temperature exceeds 40°C. Secondary spread from infected groundnut plants within agroundnut field is considered to be negligible.

11.7.6 MANAGEMENT

Growing resistant or tolerant varieties is the cheap and best management practice. Some of the cultivars like ICGV 87157, ICGS 44, ICGV 87187, ICGV 87121, ICGV 87141, ICGV 87160, ICGS 11, ICGV 87119 and ICGV 86590 are resistant to the disease. Higher plant density, removal of alternate hosts and early sowing in the month of June, reduces the disease intensity. Spraying systemic insecticides like Monocrotophas or Dimethonate after 30–35 days of sowing would significantly reduce thrips population. Intercropping of groundnut with cereals like pearl millet will restrict spread of the virus.

11.8 GROUNDNUT STEM NECROSIS DISEASE (PBND/PSND)

Groundnut stem necrosis earlier it was considered as major disease now its severity is came down. The distinguishing features of two viral diseases have been given below.

Characters	PBND	PSND
Causal virus	Peanut bud necrosis virus	Peanut stem necrosis virus
Group	*Tospovirus*	*Ilarvirus*
Symptoms	Chlorotic lesions mainly on terminal leaflets, leads to necrosis of terminal bud and ring spots also formed. Proliferation of the axillary shoots, and stunted growth of the plant. No necrotic spots on pods and testa sometime discolored and mottled.	Necrotic lesions on terminal leaflets, complete stem necrosis and often total necrosis of entire plant. Proliferation of the axillary shoots restricted to apical portion may occur. Pods with necrotic spots. Testa discolored and mottled.
Serological cross reaction	Distinct *Tospovirus* in Bunya-viridae and serological reactions only with peanut bud necrosis virus antiserum.	Family is Bromoviridae *with genus Ilarvirus* serological reactions with many tobacco streak virus antisera.
Transmission	No reports on seed-transmission	Seed-transmitted in many hosts.
Vectors	Transmitted by *Thrips palmi.*	Transmitted by several thrips species, *Scirtothips dorsalis* and *Frankliniella schultzei,* they carry infected pollens.

11.9 GROUNDNUT ROSETTE VIRUS (GRV) DISEASE

11.9.1 INTRODUCTION/ECONOMIC IMPORTANCE

Groundnut rosette virus (GRV) was first time reported and described in Africa (1907) and causes severe losses to groundnut. In the Belgian Congo (1939) it has caused 80 to 90% yield loss.

11.9.2 SYMPTOMS

Due to the presence of SAT-RNA two variant symptoms appear as green rosette and chlorotic rosette resulted in reduced intermodal length, leaf lamina, leaf size which contributed for the stunted growth of the plant and bushy appearance. Main difference between two rosette is the appearance of bright yellow colored leaves with few green island in chlorotic rosette, in green rosette leaves are curled, dark green with light mosaic pattern (Naidu et al., 1999).

FIGURE 11.6 Groundnut Rosette Disease

11.9.3 CAUSAL ORGANISM

Three causal agents are involved in groundnut rosette disease. These are as:

1. Groundnut rosette assistor virus (GRAV);
2. Groundnut rosette virus (GRV);
3. Satellite-RNA (Sat-RNA) (Taliansky et al., 2000).

Groundnut rosette is an umbravirus and GRAV, a luteovirus. GRV needs this assistor or helper virus for transmission. Sat-RNA totally depends on GRV for its replication, encapsulation, and movement, both within and between the plants. Sat-RNA is responsible for variant rosette symptoms as green rosette and chlorotic rosette and plays a critical role in helper virus dependent transmission of GRV (Murant and Kumar, 1990; Taliansky et al., 1997).

11.9.4 TRANSMISSION

Aphis craccivora (cowpea aphid) transmit the rosette virus in both persistent and circulative manner (Waliyar, 2007). GRV and Sat-RNA packaged within the GRAV coat protein to be aphid transmissible.

11.9.5 EPIDEMIOLOGY

These viruses are not seed borne in nature, infected plants act as primary source of infection, and vectors should be present for transmission. The epidemiology of GRD is more complex, involves interactions between and among two viruses and a Sat-RNA, the vector, and the host plant and environment (Naidu et al., 1998). Intermittent wet and dry spells in the early time of the season, without heavy rainfall, are probably responsible for the development and successful dispersal of late aphids.

11.9.6 MANAGEMENT

Manipulation of the crop by planting early and at close spacing reduces the chance of aphids invading and multiplying within the crop. Field sanitation reduces the sources of inoculum. *Arachis* accessions ICGs 8946, 8128, 8186, 8904, 8970 and 11557 have been shown to be resistant to GRAV (Subrahmanyam et al., 2002) and management of aphids with insecticides.

11.10 KALAHASTI MALADY

11.10.1 INTRODUCTION/ECONOMIC IMPORTANCE

Many pod diseases have been reported on groundnut from different groundnut growing parts of the country. One of such pod diseases with restricted distribution is the Kalahasti malady. It is restricted to only Nellore and Chittoor districts of Andhra Pradesh, it has caused epidemic during 1976 in these districts. The disease was first recorded around Kalahasti area hence, commonly known as 'Kalahasti malady.' Losses ranging from 25–65% in pod yield are common, depending upon the severity.

11.10.2 SYMPTOMS

Initially brownish yellow lesions appear on the pegs, later infection will move to pod stalks and developing pods. Stalks length reduced and entire pod and roots will be discolored. Affected plants shows stunted growth and greener than normal foliage.

FIGURE 11.7 Kalahasti Malady: Brown Lesions on Pods

11.10.3 CAUSAL ORGANISM

Tylenchorhynchus brevilineatus Williams has been established as the causal agent of the disease (Reddy et al., 1984).

Kingdom: Animalia
Phylum: Nematode
Class: Secernentea
Order: Tylenchida
Super family: Tylenchoidea
Family: Belonolaimidae
Genus: *Tylenchorhynchus*
Species: *brevilineatus*

It is ectoparasite on epidermal cells between root hairs.

11.10.4 MANAGEMENT

Tirupathi-2 and 3 groundnut varieties are found resistant for the nematode infection. Soil application of carbofuran 3G @ 1.0 kg a. i/ha 25–30 DAS along with irrigation water is effective. Soil application of gypsum @ 200 kg/ha at the time of earthing-up will help to manage the disease. Crop rotation with non-host crops reduced nematode populations in soil (Naidu et al., 2000b). Poultry manure (50 q ha^{-1}), farmyard manure (100 q ha^{-1}) and saw dust (25 q ha^{-1}) reduced population of *T. brevilineatus* and disease severity, and in increasing groundnut yield over non-amended control plots. Maximum nematode control can be achieved by the application of poultry manure amendment followed by neem cake at the rate of 10 q/ha (Naidu et al., 2000a).

KEYWORDS

- **bud necrosis disease**
- **early leaf spot**
- **groundnut rosette assistor virus**
- **groundnut rosette virus**
- **late leaf spots**
- **peanut bud necrosis virus**

REFERENCES

Agrios, G. N., (2005). *Plant Pathology* (5th edn., p. 922) Elsevier Academic Press.

Atla, R. R., (2017). Stem rot of groundnut incited by *sclerotium rolfsii* sacc. And it's management-A review. *Inter. J. Agril. Sci. Res., 7*(3), 327–338.

Backetia, D. R. C., (1983). Control of white grub *Holotrichia consanquinea* and collar rot *Aspergillus Niger* of groundnut sown on different dates in Punjab. *Indian J. Agril. Sci., 56*(9), 846–850.

Bayman, P., & Cotty, P. J., (1993). Genetic diversity in *Aspergillus flavus*: Association with aflatoxin production and morphology. *Canadian J. Biol., 71*, 23–31.

Berkeley, M. J., (1875). Notices of North American fungi. *Grev., 3*, 106.

Bharat, C. N., Singh, J. P., Seweta, S., & Singh, R. B., (2013). Management of late leaf spot of groundnut by different fungicides and their impact on yield. *Pl. Pathol. J., 12*(2), 85–91.

Bhat, A. I., Jain, R. K., Varma, A., & Lal, S. K., (2002). Nucleocapsid protein gene sequence studies suggest that soybean bud blight is caused by a strain of groundnut bud necrosis virus. *Current Sci., 82*, 1389–1392.

Blount, W. P., (1961). Turkey 'X' disease and the labeling of poultry pods (Lener). *Veterinary Record*, *73*(9), 227.

Chahal, D. S., & Chohan, J. S., (1971). Puccinia rust on groundnut. *FAO Plant Prot. Bull.*, *19*, 90.

Diener, U. L., Cole, R. J., Sanders, T. H., Payne, G. A., Lee, L. S., & Klich, M. A., (1987). Epidemiology of aflatoxin formation by *Aspergillus flavus*. *Annual Review Phytopatho.*, *25*, 249–270.

Divya, R. V., (2015). Studies on stem rot and collar rot diseases of groundnut and their integrated management. *PhD (Agri.) Thesis*. Professor Jayashankar Telangana State Agril. Univ., Rajendranagar, Hyderabad (India).

Dwivedi, S. L., Demski, J. W., McDonald, D., Smith, J. J. W., & Smith, D. H., (1991). Bud Necrosis: A disease of groundnut caused by tomato spotted wilt virus. *Information Bulletin No.31* (p. 5). International Crops Research Institute for the Semi-Arid Tropics, Patancheru, Andhra Pradesh, India.

Dwivedi, S. L., Reddy, D. V. R., Nigam, S. N., Ranga, R. G. V., Wightoran, J. A., Amin, P. W., Nagabhushanam, G. V. S., Reddy, A. S., Scholberg, F., & Ramraj, V. M., (1993). Registration of ICGS 86031 peanut germplasm. *Crop Sci.*, *33*, 220.

Ellis, J. B., & Everhart, B. M., (1885). North American *Cercosporae*. *J. Mycol.*, *1*, 63.

Gururaj, S., & Srikant, K., (2007). Studies on perpetuation and carryover of groundnut rust (*Puccinia arachidis* Speg.) in Northern Karnataka. *Karnataka J. Agril. Sci.*, *20*(2), 297–300.

Hazarika, D. K., Dubey, L. N., & Das, K. K., (2000). Effect of sowing dates and weather factors on development of leaf spots and rust of groundnut. *J. Mycol. Pl. Pathol.*, *30*, 27–30.

Hedayati, M. T., Pasqualotto, P. A., Warn, P., Bowyer, & Denning, D. W., (2007). *Aspergillus flavus:* human pathogen, allergen and mycotoxin producer. *Microbiology*, *153*, 1677–1692.

Jadeja, K. B., Nandolia, D. M., Dhruj, I. U., & Khandar, R. R., (1999). Efficacy of four triazole fungicides in the control of leaf spots and rust of groundnut. *Indian Phytopatho.*, *52*, 421, 422.

Jain, A. C., & Nema, K. G., (1952). *Aspergillus* blight of groundnut seedling. *Sci. Culture.*, *17*, 348, 349.

Jasani, M. D., & Kamdar, J. H., (2015). Peanut bud necrosis disease. *Indian Farmer*, *2*(2), 242–244.

Jenkins, W. A., (1939). The development of *Mycosphaerella berkeleyii*. *J. Agril. Res.*, *58*, 617–620.

Karthikeyan, A., (1996). Effect of organic amendments, antagonist *Trichoderma viride* and fungicides on seed and collar rot of groundnut. *Pl. Dis. Res.*, *11*, 72–74.

Krishna, K. G., & Suresh, P., (2005). Integrated management of late leaf spot and rust diseases of groundnut (*Arachis hypogaea* L.) with *Prosopis juliflora* leaf extract and chlorothalonil. *Inter. J. Pest Management*, *51*(4), 325–332.

Kumari, M., Singh, M., Shailesh, G., Sangeeta, C., & Jitendra, S., (2016). Effect of different fungicides, plant extracts on incidence and varietal screening against collar rot of groundnut (*Arachis hypogaea*l.) caused by *Aspergillus niger* van tiegham. *The Bioscan*, *11*(4), 2835–2839.

Magnoli, C., Dalcem, A. M., Chiacchiera, S. M., Miavo, R., & Macnz, M. A., (1998). Enumeration and identification of *Aspergillus* group and Penicillin species in poultry feeds from Argentina. *Mycopathologia*, *142*, 27–32.

Mayee, C. D., (1995). Current status and future approaches for management of groundnut diseases in India. *Indian Phytopath.*, *48*(4), 389–401.

McDonald, D., Subrahmanyam, P., Gibbons, R. W., & Smith, D. H., (1985). Early and late leaf spots of groundnut. *Information Bulletin No.21* (pp. 8–19). ICRISAT, Patancheru, A. P. (India).

Mehan, V. K., (2002). Multimycotoxins in maizepose a colossal challenge to the food and feed industry. *Mycotoxin News on Newsletter of the Working Group on Mycotoxins, 1*(2), 3–6.
Murant, A. F., & Kumar, I. K., (1990). Different variants of the satellite RNA of groundnut rosette virus are responsible for the chlorotic and green forms of groundnut rosette disease. *Annals Applied Biology, 117*, 85–92.
Naidu, P. H., Mosas, G. J., & Reddy, D. D. R., (2000b). Influence of intercropping on Kalahasti malady (*Tylenchorhynchus brevilineatus*) in groundnut. *J. Mycol. Pl. Pathol., 30*, 207–209.
Naidu, P. H., Mosas, G. J., & Sitaramaiah, K., (2000a). Control of groundnut Kalahasti malady (*Tylenchorhynchus brevilineatus*) through organic and inorganic soil amendments. *J. Mycol. Pl. Pathol., 30*, 180–183.
Naidu, R. A., Bottenberg, H., Subrahmanyam, P., Kimmins, F., Robinson, D. J., & Thresh, J. M., (1998). Epidemiology of groundnut rosette disease: Current status and future research needs. *Annals Applied Biology, 132*, 525–548.
Naidu, R. A., Kimmins, F. M., Deom, C. M., Subrahmanyam, P., Chiyembekeza, A. J., & Van der Merwe, P. J. A., (1999). Groundnut rosette: A virus disease affecting groundnut production in Sub- Saharan Africa. *Plant Disease, 83*, 700–709.
Padma, B., (2013). *Kumbhalgarh Wildlife Sanctuary: An Overview* (Vol.47, p. 3). Conservation area series, Zoology Survey of India.
Pappu, S. S., Pappu, H. R., Culbreath, A. K., & Todd, J. W., (1999). Localization of tomato spotted wilt tospovirus in peanut pod. *Peanut Sci*., *26*, 98–100.
Reddy, D. D. R., Subramanyam, P., Sankara, R. G. H., Raja, R. C., & Siva, R. D. V., (1994). A nematode disease of peanut caused by *Tylenchorhynchus brevilineatus. Pl. Disease, 68*(8), 526–529.
Reddy, D. V. R., Ratna, A. S., Sudarshana, M. R., Poul, F., & Kirankumar, I., (1992). Serological relationships and purification of bud necrosis virus, a tospovirus occurring in peanut (*Arachis hypogea* L.) in India. *Ann. Appl. Biol., 120*, 279–286.
Reddy, G. R., Reddy, A. G. R., & Rao, K. C., (1993). Seed mycoflora of groundnut varieties collected from different sources. *J. Res., 19*, 82, 83.
Reddy, M., Reddy, D. V. R., & Appa, R. A., (1968). A new record of virus disease on peanut. *Plant Dis. Reporter, 52*, 494, 495.
Rodrigues, A. A. C., Silva, G. S., Moraes, F. H. R., & Silva, C. L. P., (2006). *Arachis repens*: Novo Hospedeiro de *Puccinia arachidis*. *Fitopatol Bras., 31*(4), 410–411.
Saccardo, P. A., (1911). Notes mycologiceae. *Ann. Mycol*., *9*, 249–257.
Sargeant, K., Sheirdan, A., Kelly, J., & Carnaghan, R. B. A., (1961). Toxicity associated with certain samples of groundnuts. *Nature*, *192*, 1096, 1097.
Savary, S., Subbarao, P. V., & Zadoks, J. C., (1989). A scale of reaction types of groundnut to *Puccinia arachidis* Speg. *J. Phytopath., 124*, 259–266.
Sharma, R. K., & Parisi, S., (2017). Aflatoxins in Indian food products. In: Sharma, R. K., & Parisi, S., (eds.), *Contaminants in Indian Food Products* (pp. 13–15). Compiled by Springer International Publishing, Switzerland.
Siddaramaiah, A. L., Desai, S. A., Hegde, R. K., & Jayaramaiah, H., (1980). Effects of different dates of sowing of groundnut on rust development in Karnataka. *Proceedings of Indian National Science Academy, 46*, 380–395.
Singh, R. S., (1998). *Plant Diseases* (p. 686). Oxford & IBH Publishing Co., New Delhi.
Smartt, J., (1994). *The Groundnut Crop: A Scientific Basis for Improvement* (pp. 509–547). Chapmanand Hall, London.
Spegazzini, C. L., (1884). Fungi Guaranitici. *Anales De La Sociedad Ceintifica Agentia*, *17*, 69–96.

Stephen, A., Ferreira, R., & Soley, A., (2001). *Sclerotium rolfsii* (Southern blight; Southern wilt). *Crop Knowledge Master*, pp. 1–6.

Storgeon, R. V., (1986). Peanut disease loss estimated for major peanut producing states in the United States for 1984 & 1985. *Proc. American Peanut Res. Edu. Scociety*, *18*, 24, 25.

Subrahmanyam, P., & McDonald, D., (1987). Groundnut rust disease: Epidemiology and control. *Groundnut Rust Disease: Proc Discuss Group Meet* (pp. 27–39). ICRISAT, Patancheru, AP (India).

Subrahmanyam, P., Vander, M. P. J. A., Chiyembekeza, A. J., & Chandra, S., (2002). Integrated management of groundnut rosette disease. *African Crop Sci., 10*(1), 99–110.

Sulaiman, M., & Agashe, N. C., (1965). Influence of climate on the incidence of Tikka disease of groundnut. *Indian Oilseeds J.*, *9*, 176.

Suvendu, M., & Badigannavar, A. M., (2015). Peanut rust (*Puccinia arachidis* Speg.) disease: Its background and recent accomplishments towards disease resistance breeding. *Protoplasma, 10*, 783–796.

Taliansky, M. E., Robinson, D. J., & Murant, A. F., (1997). Complete nucleotide sequence and organization of the RNA genome of groundnut rosette umbravirus. *J. General Virology, 77*, 2335–2345.

Taliansky, M. E., Robinson, D. J., & Murant, A. F., (2000). Groundnut rosette disease virus complex: biology and molecular biology. *Advance Virus Res., 55*, 357–400.

Tashildar, C. B., (2011). Studies on molecular variations in *Puccinia arachidis* speg. Causing rust of groundnut. *M. Sc. (Agri.) Thesis.* Univ. Agril. Sci., Dharwad, Karnataka, (India).

Tashildar, C. B., Adiver, S. S., Chattannavar, S. N., Palakshappa, M. G., Kenchanagaudar, P. V., & Fakruddin, B., (2012). Morphological and isozyme variations in *Puccinia arachidis* Speg. causing rust of peanut. *Karnataka J. Agric. Sci., 25*(3), 340–345.

Von Arx, J. A., (1983). *Mycosphaerella* and its anamorphs. *Proceedings of the Konenklijke Nederlandse Academie van Weten Schappen,* Series-C., *86*, 32–43.

Waliyar, F., Kumar, P. L., Ntare, B. R., Monyo, E., Nigam, S. N., Reddy, A. S., Osiru, M., & Diallo, A. T., (2007). A century of research on groundnut rosette disease and its management. *Information Bulletin No.75.* Technical Report. International Crops Research Institute for the semi-arid tropics, Patancheru, Andhra Pradesh, India.

Woodroof, N. C., (1933). Two leaf spots of peanut (*Arachis hypogaea* L.). *Phytopathol., 23*, 627.

CHAPTER 12

Current Status of Linseed (*Linum usitatisimum* L.) Diseases and Their Management

RANGANATHSWAMY MATH[1] and DURGA PRASAD AWASTHI[2]

[1]*Department of Plant Pathology, College of Agriculture, Jabugam, Anand Agricultural University, Gujarat–391155, India, E-mail: rangu. math@gmail. com*

[2]*Department of Plant Pathology, College of Agriculture, Tripura–799210, India*

Linseed belongs to the Linaceae family. It is an important oilseed crop grown for its seed and fiber. Linseed oil is rich (>66%) in linolenic acid. In India, linseed is majorly cultivated in Madhya Pradesh, Rajasthan, Bihar, Gujarat, and Maharashtra as Rabi crop. India is one of the major producers of linseed. Among various diseases rust, blight, and wilt are major ones affecting linseed production.

12.1 RUST

12.1.1 INTRODUCTION/ECONOMIC IMPORTANCE

Rust is a major disease limiting the linseed production in most parts of the world. With susceptible host, virulent pathogen and under favorable climatic condition disease spreads very rapidly within no time it may take epidemic form. The frequent development of new races that attacked hitherto resistant varieties has been the major challenge for the breeding the resistant cultivars (Flor, 1953b, 1954). In India, this disease appears in almost all linseed growing areas. Around 16.0 to 100.0% seed yield loss has been reported due to this disease.

12.1.2 SYMPTOMS

The disease attacks all the above ground parts. Initial symptom appears on leaves as yellow to orange pustules. On leaves pustules small, rounded and covered by chlorotic zone. On stem pustules are distributed irregularly and look elongated. Late in the season teleutosori form in place of uredosori. The teleutosori produce the overwintering hardy teliospores. Entire field look burnt appearance and plants get killed before maturity.

12.1.3 CAUSAL ORGANISM

Melampsora lini (Pers.) Lev.

Kingdom: Fungi
Phylum: Basidiomycota
Class: Pucciniomycetes
Order: Pucciniales
Family: Melampsoraceae
Genus: *Melampsora*
Species: *Melampsora lini*

Studies by Harold Flor demonstrated that single pair of allelic genes determines the avirulence/virulence phenotype on host lines with particular resistance which led him to propose his famous 'Gene for Gene hypothesis. It is an autoecious and macrocyclic rust pathogen producing all the five types of spores on linseed only. Uredospores are hyaline, ovate echinulated with equatorial pores. Numerous hyaline, thin walled paraphyses are mingled with urdeospores. Telia are irregularly elongate, sub-epidermal brown to black in color and usually formed on the stem. Teliospores are sessile, cylindrical, single-celled arranged in column. Teliospores germinate by producing four haploid basidiospores. The sporidia can infect and produce pycnia and aecia on linseed plant. The pycnia are small flask shaped pale yellow in color. Pycniospores areminute, single celled, hyaline, and measure about 3–4 x 2–3 μ. The aecia are cup like, orange yellow colored. Aeciospores are roundish, hyaline thin walled produced in chains. The aeciospores germinate by forming germ tube which later infects linseed to produce uredia.

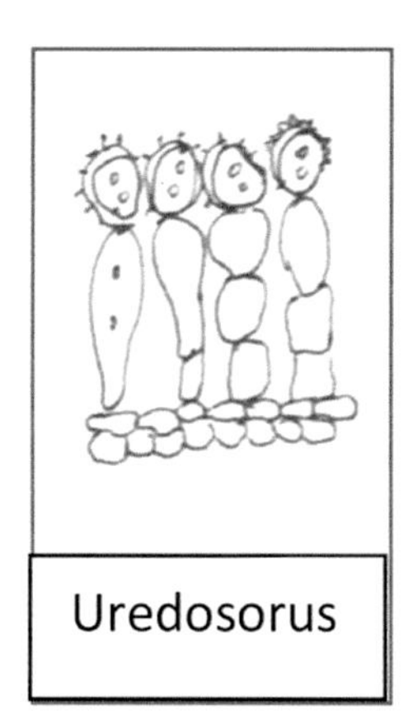

FIGURE 12.1 Uredosorus

12.1.4 DISEASE CYCLE AND EPIDEMIOLOGY

Pathogen is autoecious rust. It survives as teliospores during off season in the hills. The uredospores produced in the hills are carried by wind and initiate disease in the plains. Secondary spread of the disease takes place through uredospores produced as a result of primary infection. Disease is favored by high relative humidity (RH) and cool nights. Low temperature and moisture are necessary for the incidence of the disease. Temperature of 15–20°C with leaf wetness period of four hours will favor infection and spread of the disease. Uredospores are heat sensitive and get killed if temperature crosses 34°C.

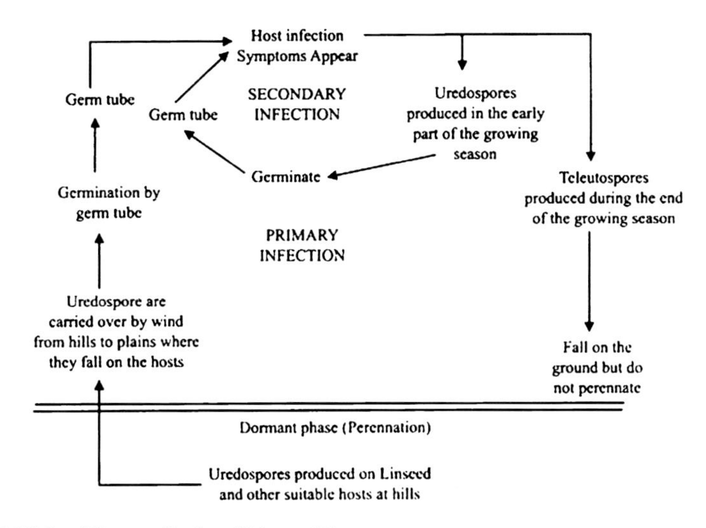

FIGURE 12.2 Disease Cycle of Linseed Rust

12.1.5 INTEGRATED DISEASE MANAGEMENT (IDM)

- Use resistant varieties like Himaline, Jawahar 7 and LCK 38.
- Use of healthy seeds for sowing.
- Rotate the crop with cereals and other non-hosts.
- Seed treatment with Carbendazim @ 2.0 gm per kg of seed.
- Avoid use of excessive nitrogenous fertilizers.
- Two sprays of Mancozeb @ 0.2% or Benomyl @ 0.1% will help to control disease.
- Destroy crop residues after harvesting of the crop.

12.2 *FUSARIUM* WILT

12.2.1 INTRODUCTION/ECONOMIC IMPORTANCE

The wilt of linseed is reported from most of the linseed growing countries. In India, it is most common in Northern India and Madhya Pradesh. The losses caused due to wilt vary based on infection time, cultivar, and environmental conditions. Continuous cropping of linseed in the same field will responsible for soil sickness problem. Sattar and Hafiz (1952) reported up to 80%loss, if infection occurs early in the stage.

12.2.2 SYMPTOMS

Plant may be get infected by the pathogen at any stage of growth. If infection occurs early in the season, then entire seedlings cease to grow, collapse, wilt, and killed. Infection on matured plant initially appears as small, dark brown spots on leaves. Infected leaves soon turn to yellow and droop down. The drooped plants die and remain standing in the field as dry and defoliated. Sometimes partial wilting of the plants is also observed. The vascular system of the infected plant is filled with mycelium. Brown to black streaks can be observed in the vascular bundle of the stem and root.

12.2.3 CAUSAL ORGANISM

Fusarium oxysporum f. sp. lini.

Mycelium is septate, branched, intercellular, and develops into stroma of various shades. They produce sporodochia directly on mycelium. Pathogen produces both micro as well as macro conidia. Micro conidia are hyaline, one-two celled and measure 4.8–14.4 X 2.2–4.8 µm. Macro conidia are fusiform to falcate, 3 spetate, hyaline to pinkish in mass and measure 21.0–53.0 X 2.4–5.6 µm. Conidiophores are short and branched and usually form in sporodochia.

12.2.4 DISEASE CYCLE AND EPIDEMIOLOGY

Pathogen is mainly soil borne and to some extent seed borne. It perennates in the soil as mycelia and spores for five to ten years in the debris. The pathogen perennating in the soil become active and cause primary infection. It mainly blocks the xylem vessels and result in wilting of the plants. Secondary spread of the disease takes place through conidia carried by irrigation water. Plants wilt rapidly under warm weather conditions. High temperature and low moisture are important factors in the development of the disease. Infection can occur at temperature between 25 and 28°C. Singh (1999) reported Nitrogen application up to 60 kg/ha significantly reduces the disease while increased application of phosphorus favored the wilt incidence.

12.2.5 INTEGRATED DISEASE MANAGEMENT (IDM)

- Rotate the crop with non-hosts for 3–5 years.
- Treat the seeds with bio-agents like *Trichoderma viride* @ 10 gm per kg/seed or with fungicides like Carbendazim @ 2.0 g/kg of seed.
- Use of resistant cultivars like NP 1, NP124, NP (RR) 9, 10, 80, 236, 439, 440 K1, and K2.
- Soil solarization will help to reduce the soil borne inoculum.
- Soil application of neem cake or mustard cake.
- Soil application of bio-control agent like *Trichoderma viride* @ 2.5 kg/ha mixed with 100 kg of well rotten farmyard manure (FYM) or neem seed kernel cake (NSKC).

12.3 ALTERNARIA BLIGHT

12.3.1 INTRODUCTION/ECONOMIC IMPORTANCE

This disease is quite severe in Northern in India. It was first reported in 1933 by Deyfrom Kanpur (UP). Now it is reported from most of the linseed cultivating states of India. The loss due to this disease was estimated to vary from 28 to 60% in Uttar Pradesh (Chauhan and Srivastav, 1975). This disease is most serious in ill drained areas. The losses may reach up to 74% on the susceptible variety in favorable conditions (Malik, 2000).

12.3.2 SYMPTOMS

Symptoms can observe on all the above ground parts of the plant. On leaves dark brown irregular patches are observed under severe condition leaves gets curled and dry up. Infected floral bud fails to open. On the calyx also small dark lesions appear. Infection spreads and reaches up to the pedicel. The infected floral parts shrink and rot. Infected capsules shows burnt appearance. Infected plant dries up before producing the seeds. Infected seeds discolored and contain inoculum to cause seedling blight.

12.3.3 CAUSAL ORGANISM

Alternaria lini Dey.

Mycelium is septate, branched, brown, and intercellular. Conidiophores are simple septate bearing conidia. Conidia are brown and muriform.

12.3.4 DISEASE CYCLE AND EPIDEMIOLOGY

The pathogen perennates in the diseased crop debries and seed. It is internally and externally associated with the linseed seeds (Shukla and Bhargav, 1979). Seed act as a primary inoculum. Secondary spread of the disease takes place through wind borne conidia. The temperature between 26 and 33°C and humid conditions are most favorable for disease initiation and spread.

12.3.5 INTEGRATED DISEASE MANAGEMENT (IDM)

- Treat the seeds with captan @ 2 gm/kg of seed.
- Cultivate linseed in well drained, high lying fields.
- Adjust sowing time to avoid heavy disease incidence.
- Under high disease pressure take spray of propiconazole 25 EC @ 0.1%.

12.4 POWDERY MILDEW

12.4.1 INTRODUCTION/ECONOMIC IMPORTANCE

This disease is reported from Europe, Australia Asia and America. Infection in the starting stage of crop growth results in higher yield loss.

12.4.2 SYMPTOMS

Symptoms can be observed on the above ground parts of the plant. The common symptom of the disease is appearance of powdery spots on the leaves. Entire leaf soon covered by powdery patches. This white powder contains the conidia and conidiophores of the pathogen. Severely infected leaves dry up, wither, and die. Early infections may cause complete defoliation and reduce the yield.

12.4.3 PATHOGEN

Oidium lini Skoric. *Leveillula taurica* (Lev.) Arnaud.

Kingdom: Fungi
Phylum: Ascomycota
Subphylum: Pezizomycotina
Class: Leotiomycetes
Order: Erysiphales
Family: Erysiphaceae
Genus: *Oidium/Leveillula.*

Mycelium is septate bearing conidiophores and conidia. Conidia measure about 12.19–39.99 x 12.19 μm. Perithecia are scattered and reddish brown and develop simple appendages. Asci are stalked, clavate hyaline and usually contain two oblong to ovate ascospores. Mycelium of *L. taurica* is septate, endophytic. Cleistothecia are 135–250 μm diameter, globose. Appendage numerous, like hyphae, simple, densely interwoven, short indistinctly branched. Ascospores are cylindrical to pyriform form.

12.4.4 DISEASE CYCLE AND EPIDEMIOLOGY

Pathogen perennatesas cleistothecia in the infected crop debries. Ascospores produced in an asci cause primary infection. Disease is favored by dry weather and moderate temperatures (20–25°C). In India there is report of cleistothecia formation. The fungus probably survives as hyphae in susceptible tissues of perennial secondary hosts.

12.4.5 INTEGRATED DISEASE MANAGEMENT (IDM)

- Remove and destroy infected plant debris after harvesting.
- Cultivation of resistant varieties like EC 9832, KD 5835, EC 9232, EC 9835-11, NP6, NP 9, Chambal, Garima, Sweta, and Shubra.
- Remove any unwanted plants or weeds.
- Foliar spray with Zineb @ 2.5 g/l of water before flowering will help to control the disease.
- Take two sprays of hexaconazole 5 SC @ 1 ml/l or wettable sulfur @ 4 g/l at 45 and 60 days after sowing (DAS) (Anjum Arshiya et al., 2017).

12.5 SEEDLING BLIGHT AND ROOT ROT

12.5.1 SYMPTOMS

Infected seedlings appear yellow, wither, and collapse. Infection in the field appears in patches. Root shows red to brown lesions which may later turn dark and shrivel.

12.5.2 CAUSAL ORGANISM

Several soil borne fungi like *Fusarium, Pythium,* and *Rhizoctonia* cause seedling blight and root rot. However, *Rhizoctonia solani* is the main causal agent and can be particularly devastating in soils that are loose, warm, and moist.

12.5.3 DISEASE CYCLE AND EPIDEMIOLOGY

Primary infection is caused by sclerotia perennating in the soil.

12.5.4 INTEGRATED DISEASE MANAGEMENT (IDM)

- Select healthy seeds for sowing.
- Seed treatment with *Trichoderma viride* @ 10 gm per kg of seed or Carbendazim @ 2.0 gm per kg of seed.

- Soil solarization along with soil amendments with neem cake or mustard cake help in reduction of inoculums.
- Soil application of bio-control agent like *Trichoderma viride* @ 2.5 kg/ha mixed with 100 kg of well FYM or NSKC.
- Rotate the crop with other non-hosts.

12.6 PASMO

12.6.1 INTRODUCTION/ECONOMIC IMPORTANCE

It was first reported from Argentina in 1911. Now this disease is present in most of linseed growing countries of the world.

12.6.2 SYMPTOMS

Disease may appear at any stage from the seedling to maturity. Symptoms appear on all the above ground plant parts. On leaves, spots are small, circular, and brown while they are elongated on stem. Later in the season, pathogen produces pycnidia in the infected plant parts. Severely infected plant defoliate and heavy boll-drop occurs under rain accompanied with wind.

12.6.3 PATHOGEN

Septoria linicola.

12.6.4 DISEASE CYCLE AND EPIDEMIOLOGY

Pathogen survives in the infected crop debries as pycnidium. Primary infection in the season is caused by pycnidiospores. High RH and moderate temperature favor the infection and spread of the disease.

12.6.5 INTEGRATED DISEASE MANAGEMENT (IDM)

- Deep plowing to bury the infected crop debries.
- Foliar spray with fungicides like mancozeb@ 0.2% or benomyl @ 0.1%.

12.7 SCLEROTINIA STEM ROT

12.7.1 SYMPTOMS

The pathogen mainly damages the stem near to the soil line. Initially water soaked lesions appear on the infected stems. Lesions enlarge and girdle the stem resulting in bleaching, shredding, breakage, and lodging of plant. The affected parts being killed quickly and completely decayed. Small black sclerotia form in the infected stem.

12.7.2 PATHOGEN

Sclerotinia sclerotiorum.

Taxonomy:
Kingdom: Fungi
Phylum: Ascomycota
Subphylum: Pezizomycotina
Class: Leotiomycetes
Order: Helotiales
Family: Sclerotiniaceae
Genus: *Sclerotinia*
Species: *S. sclerotiorum.*

Pathogen perrenates as sclerotia. Sclerotia on germination forms trumpet shaped cinnamon-brown apothecia, which contain asci and ascospores.

12.7.3 DISEASE CYCLE AND EPIDEMIOLOGY

Pathogen perennates as sclerotia in the previously infected crop debries. Sclerotia present in the soil remain viable for several years. They germinate under favorable conditions and produce small stalked, trumpet shaped cinnamon-brown apothecia. On the apothecia asci arranged side by side. Each ascus contains eight ellipsoidal ascospores. Ascospores expelled forcibly land on the linseed plant where they germinate and cause primary infection.

12.7.4 INTEGRATED DISEASE MANAGEMENT (IDM)

- Avoid linseed cultivation under low-lying fields.
- Long crop rotation with non-hosts
- Soil application of neem cake or mustard cake
- Soil application of biocontrol agent like *Trichoderma viride* mixed with FYM.

KEYWORDS

- **alternaria blight**
- **farm yard manure**
- ***Fusarium* wilt**
- **neem seed kernel cake**
- **Pasmo**
- **powdery mildew**

REFERENCES

Anjum, A., Aswathanarayana, D. S., & Ajithkumar, K., (2017). Management of linseed powdery mildew caused by *Leveillula taurica* (Lev.) Arn. *International J. Agric. Sci., 9*(32), 4479–4481.

Chauhan, L. S., & Srivastav, K. N., (1975). Estimation of loss of yield caused by blight disease of linseed. *Indian J. Farm. Sci., 3*, 107–109.

Dey, P. K., (1933). An alternaria blight of Linseed plant. *Indian J. Agric. Sci.3,* 881–896.

Flor, H. H., (1942). Inheritance of pathogenicity in *Melamposra lini. Phytopathology., 32*, 653–669.

Flor, H. H., (1953b). Wilt, rust, and plasma of flax: Plant diseases In: *The Year Book of Agriculture* (pp. 869–873). USDA, Washington.

Flor, H. H., (1954). Identification of races of flax rust by lines with rust conditioning genes. *US Dept. Agr. Tech. Bull., 1087*, 1–25.

Goel, L. B., & Swarup, G., (1964). Some observations on linseed wilt. *Indian Phytopath., 17*, 133–137.

Gorey, S. C., Khosla, H. K., Upadhyaya, Y. M., & Mandoli, S. C., (1989). Inheritance of resistance to *Melamposra lini. Indian Phytopath., 40*, 374–378.

Ji, T., (1975). Reaction of linseed varieties to alternaria leaf spot. *Indian J. Mycol. Pl. Pathol., 5*, 103.

Kar, A. K., & Lenka, D., (1998). Occurrence of alternaria blight and powdery mildew on linseed genotypes in Orissa. *Plant. Dis. Res., 13*, 160, 161.

Khalid, Y. R., (2003). Principal diseases of flax. In: Alister, D. M., & Neil, D. W., (eds.), *Flax the Genus Linum* (pp. 92–123). Published by CRC Taylor and Francis, Canada.

Kroes, G. M. L. W., Loffler, H. J. M., Parlevliet, J. E., Keizer, L. C. P., & Lange, W., (1999). Interactions of *Fusarium oxysporum* f. sp. *lini* the flax wilt pathogen with flax and linseed. *Pl. Pathol., 48*, 491–498.

Lawrence, G. J., (1988). *Melamposra lini* RUST of flax and linseed. *Advances Plant Path, 6*, 313–331.

Malik, Y. P., (2000). Assessment of yield loss due to host pest interaction in linseed (*Linum usitatissimum*) *Indian J. Agric. Sci., 70*, 53, 54.

Mercer, P. C., (1986). Linseed diseases. *Ann. Rep. Rei. Tehc. Work Dept. Agric.* (p. 201). N. Ireland UK.

Prasada, R., (1948). Studies in linseed rust *Melamposra lini* (Pers.) Lev. in India. *Indian Phytopath., 1*, 1–18.

Rangaswami, G., & Mahadevan, A., (2016). *Diseases of Crop Plants in India* (4th edn., pp. 270–273). PHI Learning Private Limited, New Delhi–110092.

Raut, B. T., & Somani, R. B., (1990). Efficacy of different fungicides to linseed rust. *P. K. V. Res. J., 12*, 168–170.

Reethi, S., Singh, U. C., Khare, R. K., & Sharma, B. L., (2005). Diseases of linseed and sesame and their management In: Thind, T. S., (ed.), *Diseases of Field Crops & Their Management* (pp. 135–154). Daya Publishing House, New Delhi–110035.

Saharan, G. S., & Saharan, M. S., (1994). Progression of powdery mildew on different varieties of linseed in relation to environmental conditions. *Indian J. Mycol. Pl. Pathol., 24*, 88–92.

Saharan, G. S., (1988). Plant disease management in linseed. *Rev. Trop. Plant Path., 5*, 119–140.

Sangwan, M. S., Naresh, M., & Saharan, G. S., (2002). Fungal diseases of linseed In: Saharan, G. S., Naresh, M., & Sangwan, M. S., (eds.), *Diseases of Oilseed Crops* (pp. 176–202). Indus Publishing Company, New Delhi.

Sattar, & Hafiz, A., (1952). Research on plant diseases of Punjab. *Mag. Pak. Assoc. Adv. Sci., 1*, 55.

Sharma, H. C., & Khosla, H. K., (1979). Fungicidal control of powdery mildew of linseed in relation to losses. *JNKVV Res. J., 10*, 61, 62.

Shukla, D. M., & Bhargawa, S. N., (1979). Leaf blight of *Linum usitatissimum. Acta Bot. Indian., 7*, 184.

Singh, B. P., Shukla, B. N., & Sharma, Y. K., (1981). Effect of rust infection on the oil and protein content of linseed. *JNKVV. Res. J., 12*, 101.

Singh, Jyothi, & Singh, J., (2002). Effect of fungicidal sprays against foliar diseases of linseed. *Ann. Plant. Prot. Sci., 10*, 169, 170.

Singh, S. N., (1999). Effect of different doses of N and P on the incidence of linseed wilt caused by *Fusarium oxysporum* f. sp. *Lini.* (Bolley) Synder and Hansen. *Crop Res., 17*, 112, 113.

Susan, J. M., Jellis, G. J., & Cox, T. W., (1986). *Sclerotinia sclerotiorum* on linseed. *Plant Patholo., 35*(3), 403–405.

Thakore, B. B. L., Sinene, M. S., & Singh, R. B., (1989). Management of rust and alternaria blight of linseed by fungicides in Rajasthan. *Indian J. Mycol. Plant Pathol., 17*, 328, 329.

CHAPTER 13

Present Scenario of Rapeseed-Mustard Diseases and Their Integrated Management

P. D. MEENA[1] and ASHISH SHEERA[2]

[1]ICAR-Directorate of Rapeseed-Mustard Research, Bharatpur–321303, Rajasthan, India

[2]Sam Higginbottom University of Agriculture, Technology, and Sciences, Allahabad–211007, Uttar Pradesh, India

13.1 INTRODUCTION

The family *Brassicaceae* includes 338 genera, and 3709 species consists of crops, weeds, and decorative vegetation (Warwick et al., 2006; Love et al., 2005). In *Brassicaceae* family, remains an enormous diversity of economically important crops forms, the main source of edible and non-edible oil, vegetable, forage, and fodder, condiments. In the genus *Brassica* of family *Brassicaceae*, rapeseed-mustard crops play tremendous role in human and animal diet. Crushed to yield oil (vegetable oil for cooking) and residual protein-rich meal is commonly used for animal feed. Oilseed Brassica crops in India consist of conventionally grown native species, namely three ecotypes of Brassica (2n = 20; AA), i.e., Toria (*Brassica rapa* L. var. Toria), Brown Sarson (*B. rapa* L. var. Brown Sarson), Yellow Sarson (*B. rapa* L. var. Yellow Sarson), Indian mustard [*B. juncea* (L.) Czern & Coss.]; 2n = 36; AABB), black mustard (*B. nigra*; 2n = 16; BB) and Taramira (*Eruca sativa/vesicaria*; 2n = 22), which have been cultivated since about 3, 500 BC besides non-traditional genus including Gobhi Sarson (*B. napus*; AACC) and Ethiopian mustard or Karan Rai (*B. carinata*; BBCC) (Kumar et al., 2015). Among these, Indian mustard (*B. juncea*) occupying nearly 80% acreage under Oilseed Brassica crops in India. Brassicas and other closely related crucifers are cultivated under extensive cropping system in the country and be principally grown as a winter (*Rabi*) season crop on preserved soil moisture from the preceding wet season with low input management.

The Indian vegetable oil economy is the world's fourth largest next only to USA, China, and Brazil. India's share in the global vegetable oil imports is 14%, and 10% of edible oils during 2016–17 with a total market size of about Rs 600 billion (US$ 13.4 billion). In the context of national agricultural system, oilseeds occupy 13% of the country's gross cropped area and 3% to gross national products and 10% value of all agricultural products. The main oilseeds crops are Soybean Groundnut, and oilseed Brassica contributed around 80% of production in India. Rapeseed-mustard alone contributed 24.2% to the total oilseed production in India, with average productivity of 1181 kg/ha during 2016–17. *Brassica juncea, B. rapa, B. napus,* and *B. carinata* are commonly grown in different agro-climatic situations varying from north-eastern/north-western hills to down south under irrigated/rainfed, timely/late sown and sole/mixed cropping in India. Indian mustard (*B. juncea*) alone covered about 80% of the total area under rapeseed-mustard crops.

The present productivity level is achieved mainly through varieties developed by pure line selection from indigenous germplasm; therefore, inter-varietal hybridization in brassicas is needed to widen the genetic base of germplasm. Reduction in maturity period of the genotypes provided greater scope for expansion of this crop in non-traditional areas to reduce the edible oil import. A wide gap exists between the potential yield and the yield realized at the farmers' field, which is mostly due to numerous biotic and abiotic stresses. Therefore, the additional efforts should be on higher productivity per unit area and per unit time, rather than area expansion. Production may be increased vertically considering the exploitable yield reservoir. The biotic stresses reduced yield about 19.9%, whereas, diseases alone causes yield reduction at different growth stages.

Among 30 diseases recognized to distress rapeseed mustard crops, about 18 are studied as economically important in diverse condition all over the world (Saharan et al., 2016). To overcome such losses, it is vital to be acquainted with the causal agents, their ecosystem and aggression to the vulnerable phase. The disease scenario is shifting through the emerging pathogens. In climate changing situations, the biology, behavior, and epidemiology of existing pathogens are affected unfavorably responsible for distressing the production.

Diseases like black spot (*Alternaria brassicae* (Berk.) Sacc. and *A. brassicicola* (Schwein.) Wiltshire), white rust (WR) (*Albugo candida* (Pers. Ex

Lev.) Kuntze), downy mildew (*Hyaloperonospora brassicae* (Gaum) Goker), powdery mildew (*Erysiphe cruciferarum* Opiz ex L. Junell), Sclerotinia rot (SR) (*Sclerotinia sclerotiorum* (Lib.) de Bary) and club root (*Plasmodiophora brassicae* Woronin) are of the major significance due to their worldwide dissemination and severe yield losses.

Recently, there are some other new pathogens has been reported responsible for minor disease so far namely stem blight (*Nigrospora oryzae* (Berk. & Broome) Petch), root rot (*Sclerotium rolfsii* Sacc.), bacterial stalk rot (*Erwinia carotovora* (Jones) Bergey) (Kolte, 1985; Meena et al., 2014 Saharan et al., 2016).

Low solar radiation and short-day periodicity are accountable for privileged infections by *Fusarium, Sclerotinia, and Verticillium*. In India, SR considered the number one problem among the rapeseed-mustard diseases while it emerged severely after the year 2000. Recently, the root rot disease caused by *Sclerotium rolfsii* is threatening rapeseed-mustard production system in some areas. Earlier it was reported as bacterial in combinations fungal incidence (*Erwinia carotovora* pv. *carotovora, Fusarium, Rhizoctonia solani* and *Sclerotium rolfsii*). Powdery mildew (*Erysiphe cruciferarum*) in oilseed Brassicas was mostly occurring in Gujarat state barring stray incidences elsewhere, the emergence of the disease use to occur from late January. However, now a days the disease is occurring in other oilseed *Brassica* growing states *viz.,* Haryana, Central UP, MP, parts of Rajasthan and Bihar with its emergence even in December probably due to shortening of cold spell during the crop period (Meena et al., 2014).

The plant diseases cause severe loss, both in physical and financial terms. The physical losses arise from the loss of production arising from the incidence of disease. The financial loss is on account of both reduced production and lower value of the produce due to reduction in quality of the produce. There are different techniques to evaluate the economic losses due to plant diseases. A proper prioritization of plant diseases about the magnitude of the economic loss is crucial to make economically feasible mitigation actions like disease control programs over large areas. The quantification of the economic loss is not an end in itself. It shows the prioritization to be given to tackle the production constraints arising from diseases and indicate the level of benefits that may be gained through implementation of mitigation strategies. The loss arising from pest and some other abiotic stresses to a large extent than the losses caused by diseases, but as a considerable production constraint, we have to develop suitable mitigation strategies for preventing the losses from diseases in rapeseed-mustard.

13.2 FUNGAL DISEASES

13.2.1 ALTERNARIA BLIGHT

Alternaria is a ubiquitous fungal genus which includes saprobic, endophytic, and pathogenic species. Species of *Alternaria* are known as serious plant pathogens, responsible for serious yield losses on a large array of crops (Saharan et al., 2016). More than 4, 000 *Alternaria*-host interaction evidenced in the USDA Fungal Host Index Ranking the genus 10th among nearly 2000 fungal genera. As a consequence of extensive harmful health effects of *Alternaria* on plants and their surroundings, an accurate and speedy detection of *Alternaria* species may be of immense significance to scientists, mycologists, and the public alike (Woudenberg et al., 2013). The widespread and destructive diseases of Brassicaceae crops worldwide are those incited by four *Alternaria* species viz., *A. brassicae* (Berk.) Sacc. ; *A. brassicicola* (Schwein.) Wiltsh. ; *A. raphani* Groves and Skolko, and *A. alternata* (Fr.) Keissl. However, *A. brassicae* is the most frequent and virulent on rapeseed-mustard.

13.2.1.1 DISTRIBUTION AND SCOPE OF DAMAGE

The disease is reported from all areas where the oilseed brassica crops are grown with no proven resistance cultivars against the pathogen (Meena et al., 2010). Yield losses have been reported in the range of 35–45% in Yellow Sarson, 25–45% in Brown Sarson and 17–48% in Indian mustard. In addition, the pathogen severely affects the size, color, and germination ability of the seed in addition to reduce oil content from 1–10% in infected seed (Saharan et al., 2016).

13.2.1.2 SYMPTOMS

The disease appears on leaves after 45 days of sowing as identified the critical stage for initiation of disease and progress from lower leaves with pinhead size light brown to black round spots reached its peak on leaves after 75 days of sowing (Meena et al., 2004). Later on, these spots enlarge and develop into black round spots. The characteristic concentric rings are formed in these lesions or spots. A zone of yellow hallow around the spots may be formed. These spots on leaves coalesce to cause leaf blight and ultimately leads to defoliation. Dark brown, circular to linear lesions also develop on stem and siliquae. Lesions on stem usually remain elongated and pointed at later stages. Infection on pods results in formation of small, discolored, and shriveled seeds (Meena et al., 2010).

13.2.1.3 SURVIVAL OF PATHOGEN

The pathogen survives by developing mycelium, conidia, chlamydospores, and microsclerotia on diseased plant debris in soil or as mycelium and conidia on weed hosts (Verma and Saharan, 1994). The leaves initially infected few plants to serve as sources of secondary infection. The pathogen survives and perpetuates through infected seeds, diseased plant debris, and pathogen spores in the soil and allied crucifers/weed hosts in a appropriate agro-ecosystem (Chupp and Sherf, 1960; Verma and Saharan, 1994; Meena et al., 2016).

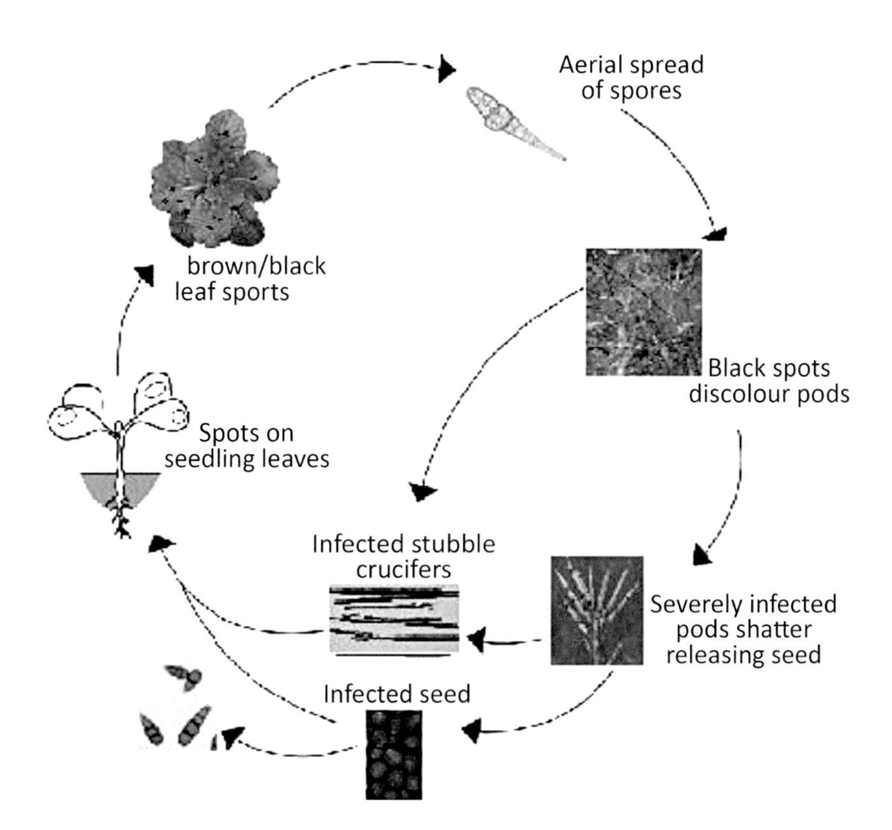

FIGURE 13.1 Disease Cycle

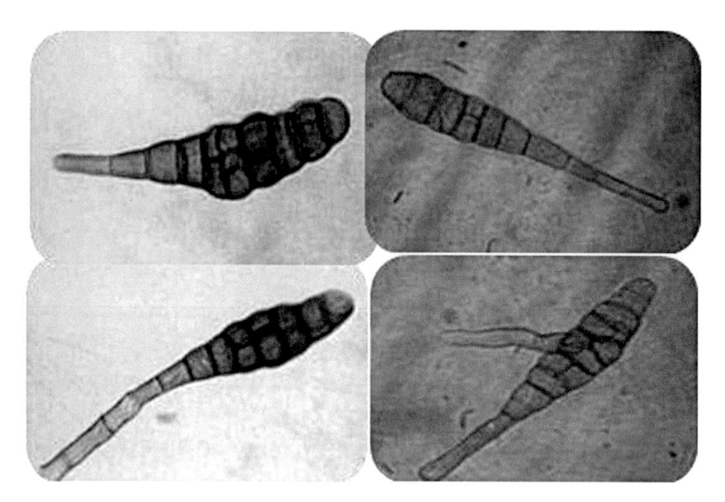

FIGURE 13.2 Conidia of *Alternaria Brassicae*

13.2.1.4 FAVORABLE WEATHER CONDITIONS

Alternaria blight disease severity on leaves are positively correlated to a maximum daily temperature of 18–27°C, minimum daily temperature of 8–12°C, daily mean temperature >10°C, >92% morning relative humidity (RH), >40% afternoon RH and mean RH of >70% in the preceding week. Severity of disease on siliquae is favored by a maximum daily temperature of 20–30°C, daily mean temperature of >14°C, morning RH of >90%, daily mean RH of >70%, >9 h of sunshine and >10 h of leaf wetness (Chattopadhya et al., 2005).

13.2.1.5 MANAGEMENT

Cultural practices including use of clean, bold, healthy, and treated seed of suggested varieties, extensive crop rotation (3–4 years), sanitation, weed control, shallow (2 cm depth) planting at optional time, use of balanced fertilization, proper plant density (45 x 30 cm), proper drainage in the field, plant debris management, use of tolerant/resistant cultivars, application of chemicals/bioagents at proper time with adequate foliage coverage, and education of farmers about the importance of appropriate practices have been advocated to control Alternaria blight of *Brassica* crops (Verma and Saharan, 1994; Kolte, 1985; Saharan and Chand, 1988; Peruch and Michereff, 2007; Saharan et al., 2016).

13.2.1.5.1 DISEASE RESISTANCE

Numerous tolerant sources against *Alternaria* species have been recognized from various *Brassica* species, their near and distantly related coeno species but some have been used to develop resistant cultivars (Verma and Saharan, 1994; Saharan et al., 2016). A short stature *Brassica juncea* cultivar Divya has been found tolerant to Alternaria blight (Kolte et al., 2000). Tolerant genotypes found in *B. juncea* including PHR-2, PAB-9511, PAB-9534, JMM-915, EC-399296, EC-399301 and RN-490, *B. carinata* are HC-1 and Kiran and *B. napus* are PBN-9501, PBN-9502, PBN-2001 and PBN-2002 can be exploited for further resistance development program (Kolte, 2005). Resistance sources against Alternaria blight have been spotted in wild Brassicas including *B. alba*, *Camelina sativa, Capsella bursa-pastoris, Eruca sativa, Neslia paniculata*, *Brassica desnottesii, Coincya pseuderucastrum, Diplotaxis berthautii, D. catholica, D. cretacea, D. erucoides,* and *Erucastrum gallicum* (Saharan et al., 2016).

13.2.1.5.2 CHEMICAL MANAGEMENT

Fungicides like iprodione, procymidone, and fludioxonil have shown resistance against isolates of *A. brassicicola,* which may affect their effectiveness under field situation to manage the disease (Huang and Levy, 1995; Iacomi-Vasilescu et al., 2004). Foliar application of mancozeb @ 0.25% after 40 and 60 days sowing is found reasonably effective for Alternaria blight management. Apply fungicides on the crop 2–3 times at 15 days interval, if needed (Meena et al., 2013).

Incorporation of all plant disease management approaches viz., cultural, chemical, biological, nutritional manipulation, host resistance, biotechnological, and genetic engineering together with other pest management is the best strategy to deal with Alternaria blight disease of rapeseed-mustard (Verma and Saharan, 1994; Kolte, 2005; Saharan et al., 2016).

13.2.2 WHITE RUST (WR) OR WHITE BLISTER

WR, white blister or white blister rust (liberate white powder or form galls) and stag heads (abnormality of inflorescence) are the familiar names of the disease caused by *Albugo* spp. associated with more than 400 species of plants worldwide (Saharan and Verma, 1992; Dick, 2002; Choi and Priest,

1995). The genus *Albugo* is a broadly scattered obligate fungus, having over 50 species infecting about 400 host plants belonging 31 families of 12 orders in Dicotyledoneae, and 1 family in Monocotyledoneae crops and common weeds (Biga, 1955; Choi and Priest, 1995; Walker and Priest, 2007). *Albugo candida* (Pers. Ex Lev.) Kuntze is recognized which is infecting oilseed Brassicas and crucifer vegetables.

13.2.2.1 DISTRIBUTION AND EXTENT OF DAMAGE

The WR disease is common and destructive in all oilseed brassica cultivated regions around the world. Leaf and floral phases of WR can cause yield losses ranging from 23–55% in rapeseed-mustard depending upon host genotype, planting time, plant population, nutrition, and environmental situations (Saharan and Verma, 1992; Saharan et al., 2014). The disease is responsible to cause up to 47% reduction in seed yield (Kolte, 1985), even though each percent severity of disease and staghead formation lead to reduce seed yield about 82 kg/ha and 22 kg/ha, respectively (Meena et al., 2002).

13.2.2.2 SYMPTOMS

In general, disease symptoms initially appear on lower leaves and spread gradually on stem, inflowerence, and siliqua of the infected plants. Shiny white to creamy yellow raised pustules develop on abaxial surface of the leaves, while the adaxial surface of the leaves becomes tan yellow. However, the initial symptom varies with the cultivar grown. Under favorable environment, both sides of leaves may have white pustules, which remain smooth initially but later rupture to release white to creamy colored mass of sporangia. Under severe cases, these white pustules can be seen on stem, inflorescence, and siliquae. The pathogen stimulates deformities like swelling and deformation of the stem and floral parts. The distortion of floral parts due to hyperplasia and/or hypertrophy of tissues causes deformed heads commonly called as stagheads mainly responsible for seed yield loss. Mixed infection of WR and downy mildew rarely develops under rainfed conditions, however, it is commonly observed in regions, where cool and wet conditions prevail.

13.2.2.3 SURVIVAL OF PATHOGEN

The pathogen perpetuates through oospores present in infected crop residue, soil, and seed, which serves as a primary source of infection. The *oospore germination* is possible throughout most of the *growing season* and subsequently releases zoospores, which infects lower leaves (Saharan and Verma, 1992; Saharan et al., 2005). Secondary spread is carried out by formation of sporangia and zoospores, which first make some loci of infection on isolated plants or on boarders of mustard field and then further spread in the whole field.

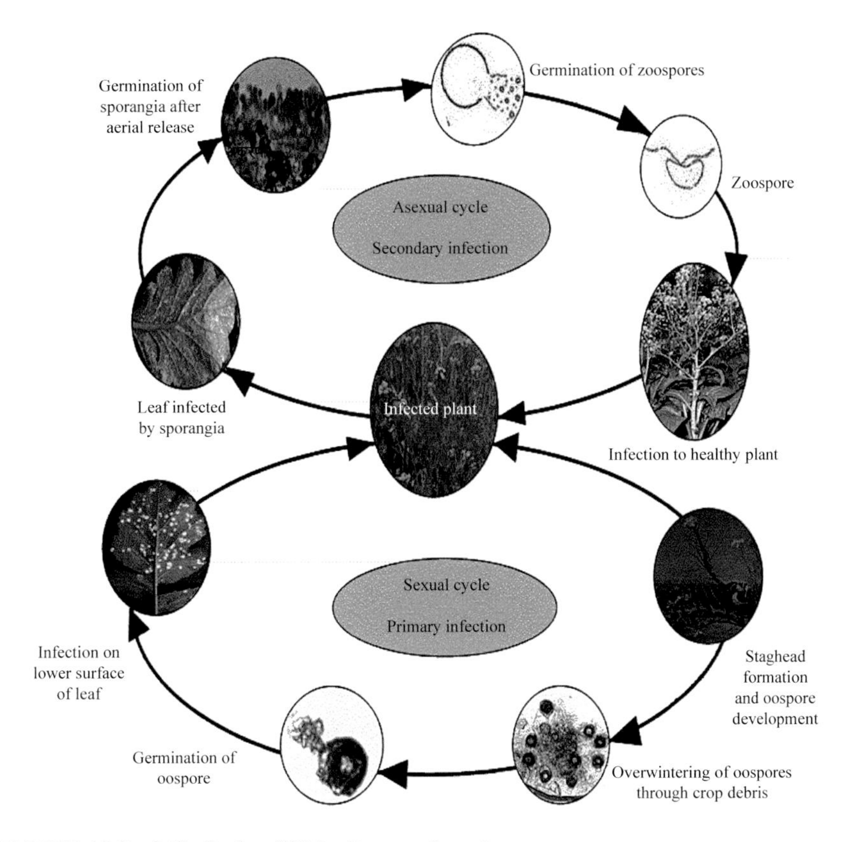

FIGURE 13.3 Life Cycle of White Rust on *Brassica*

13.2.2.4 FAVORABLE WEATHER CONDITIONS

WR disease severity on leaves is preferred by >40% afternoon (minimum) RH, > 97% morning (maximum) RH and 16–24°C maximum daily temperature. Staghead formation is influenced by 20–29°C maximum daily temperature and further aided by >12°C minimum daily temperature and >97% morning (maximum) RH (Chattopadhyay et al., 2011). About 117 races/pathotypes of *A. candida* have been reported so far, although their status still needs confirmation on a set of internationally accepted host differentials/isogenic lines (Saharan, 2010). However, the several genotypes of *Brassica* species demonstrate high degree of resistance from various countries, but very few have been exploited to develop resistant cultivars (Saharan and Verma, 1992; Saharan, 2010).

13.2.2.5 MANAGEMENT

Under field conditions not a single technique or approach is feasible, viable, cost-effective, and environmentally safe to manage WR disease. Therefore, there is a need to integrate all management approaches including, cultural, chemical, biological, and host resistance (Saharan and Verma, 1992; Mukerji et al., 1999; Saharan and Mehta, 2002).

- Crop rotation by non-host resistant crops, as well as burning of diseased debris helps in minimizing inoculum buildup.
- Balanced dosage of suggested fertilizers improves crop development. Excessive application of nitrogen makes the crop more vulnerable to pathogen attack.
- Timely sown crop (10–25th October) is likely to reduce disease incidence.
- Removal of weeds, rouging, and eradication of infected plants particularly stagheads and avoidance of over irrigation may reduce the disease incidence.
- Foliar spray of mancozeb @ 0.25% at initiation of symptoms is effective to minimize WR. Spray the crop 2–3 times at 15 days interval, if needed. Foliar spray of mixture of metalaxyl 64% and mancozeb 8% @ 0.25% at the time of symptoms initiation is also effective to manage the disease. However, many fungicides viz., captafol, benomyl (Gupta and Sharma, 1978), chlorothalonil (Verma and Petrie, 1979) are effectively reduce both leaf and staghead phase infection, mancozeb

(Dueck and Stone, 1979; Verma and Petrie, 1979; Mehta et al., 1996), combination of metalaxyl 35 ES 6 ml/kg seed treatment + 0.2 g/l spray of mixture of metalaxyl + mancozeb at 50, 65 DAS (Parui and Bandopadhyay, 1973; Verma and Petrie, 1979; Berkenkamp, 1980; Fan et al., 1983; Meena et al., 2003) are able to manage the disease.

Though, some resistance genotypes including [(*B. juncea*: PWR-9541, RLM-198, JMMWR 941-1-2, EC-399299, EC-399301, EC-399300, EC-399296, BIO YSR, DRMR2019, DRMR-2035) (*B. rapa*: PT-303, Tobin) (*B. carinata*: NRCDR-515, DLSC-1) (*B. napus*: PBN-2001, PBN-2002) (*Eruca sativa*: RTM-1471) (*B. alba*: Exotic-1, -2] have been recognized recently (Kolte, 1985; Pal et al., 1999; Mukherjee et al., 2001).

13.2.3 DOWNY MILDEW DISEASE

13.2.3.1 DISTRIBUTION AND EXTENT OF DAMAGE

The disease is scattered all over the rapeseed-mustard growing regions of the globe caused by *Hyaloperonospora brassicae* (Gaum) Goker. Seed yield reduction by downy mildew can be as high as 58%. Losses in seed yield caused by mixed infection of WR and downy mildew diseases have been recorded in range of 37–47%.

13.2.3.2 SYMPTOMS

The disease appears on seedlings as small creamy to light purple brown lesions on the lower surface of the lower leaves which enlarges in size, while upper surface of the leaves show water-soaked yellow patches. The foliage dry up ultimately and tear off easily. Downy mildew growth of fungus also appears on stems and stagheads formed by *Albugo* pathogen. Mixed infection of WR and downy mildew may results in malformation of stem and floral parts under wet and cool conditions (Meena et al., 2014).

13.2.3.3 FAVORABLE WEATHER CONDITIONS

The disease development is favored by wet (> 90% RH) and cool (10–20°C temperature) weather.

13.2.3.4 SURVIVAL OF PATHOGEN

The pathogen survives through oospores in infected crop residue, soil, and also as contaminant of seed. The oospores serve as primary source of inoculum and infect cotyledonary leaves and primary leaves. The secondary spread of the infection is through sporangia or zoospores.

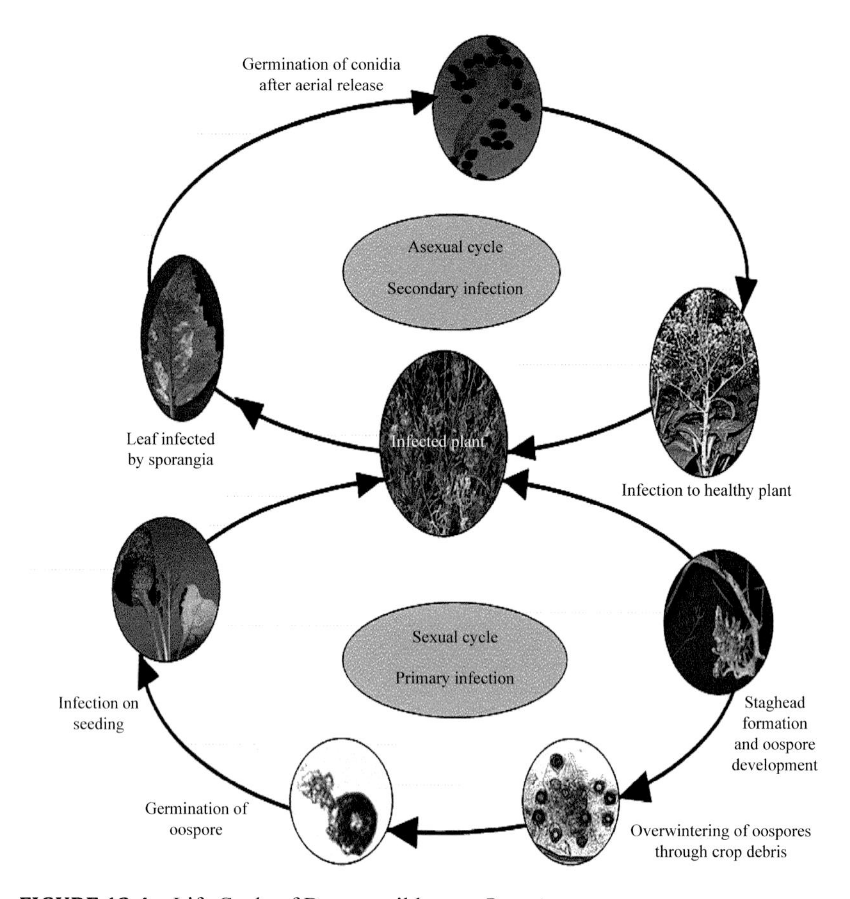

FIGURE 13.4 Life Cycle of Downy mildew on *Brassica*

13.2.3.5 MANAGEMENT

- Crop rotation with non-host crops, burning of diseased debris, helps in reducing inoculum buildup.

- Balanced dose of recommended fertilizers improves crop development. A higher dose of nitrogen makes the crop susceptible.
- Timely sown crop (10–25th October) is likely to have less disease incidence.
- Removal of weeds, roguing, and destruction of infected host and avoidance of over irrigation etc. is able to reduce disease severity.
- Foliar spray of mancozeb @ 0.25% at initiation of symptoms is able to reduce WR. Spray the crop 2–3 times at an interval of 15 days, if needed. Foliar spray of mixture of metalaxyl 64% and mancozeb 8% @ 0.2% at the time of initial symptoms is also found effective in controlling disease.

13.2.4 POWDERY MILDEW

Powdery mildew disease is incited by *Erysiphe cruciferarum* Opiz. ex. Junell usually appears at maturity stage and escapes infection, hence considered as less damaging in mustard, except during epidemic outbreak in late sown mustard crop.

13.2.4.1 DISTRIBUTION AND EXTENT OF DAMAGE

Recently, the disease is disseminated wherever the rapeseed-mustard grown in India. Seed yield losses due to powdery mildew is about 17.4–25.0%, besides 6.7% reduction in oil content. Considering the variation in disease severity over the period, it appears the loss is proportional to the severity of disease, which varies considerably depending on the occurrence at growth stage.

13.2.4.2 SYMPTOMS

Dirty white floury patches appear on all above ground parts of the host including leaves, stem, and pods. As the disease advances, entire parts of the host are enclosed by white floury patches and the plants looks to be dusted with white granular chalky powder. Under severe infection at siliqua development stage, the pods remain under sized and produce few small-size shriveled seeds. At maturity stage, minute, circular, black, spherical, and scattered cleistothecia develop on leaves, stems, and pods.

FIGURE 13.5 Typical symptoms of powdery mildew in Rapeseed - Mustard crop

13.2.4.3 FAVORABLE WEATHER CONDITIONS

Late sowing responsible for the coincidence of vulnerable growth stage of plants with warm (maximum temperature: 24–30°C; minimum temperature: > 5°C; ≥ 9.1 h sunshine) and lower (morning: < 90%; afternoon: 24–50%) RH conditions and less or no rainfall are congenial weather factors for disease development.

13.2.4.4 SURVIVAL OF PATHOGEN

The pathogen survives through cleistothecia (fruiting body of fungus) in diseased plant debris in soil and serves as source of primary inoculum. When environment becomes favorable, the ascospores formed in the cleistothecia are released which initiate infection on lower leaves. Secondary infection occurs through the conidia produced abundantly on these infected leaves.

13.2.4.5 MANAGEMENT

Balanced dose of recommended fertilizers improves crop development. A higher dose of nitrogen makes the crop susceptible. Choice of suitable planting dates appears to offer a promising technique of the disease management. There is a range of resistance levels in Brassica, although immunity is not apparent. Once the disease become serious, it is difficult to control the disease by dusting plants with sulfur or by spraying with wettable sulfur fungicides. Karathane @ 0.1% or Sulfex @ 0.2% sprayed thrice after appearance of initial symptoms at 10-days interval also gives good control over the disease (Singh and Solanki, 1974).

13.2.5 SCLEROTINIA ROT (SR)

Sclerotinia sclerotiorum (Lib.) de Bary causes SR is infecting above 500 host plants all over the world. The disease is threat to rapeseed-mustard cultivators in all continents of the globe. Disease start to occurs on leaves, stem, and inflorescence and siliquae at pod formation growth stages of the crop, reduces up to 40% seed yield, and considerable decline in oil content and quality.

13.2.5.1 DISTRIBUTION AND EXTENT OF DAMAGE

The SR disease is disseminated all rapeseed-mustard growing regions universally. Yield losses differ with the incidence of the disease and crop growth stage at the time of infection. Once the infection occurs at the early flowering stage produce little or no seeds, and when infection occurs at the late flowering stage will set seed and caused small yield reduction. The first record of its occurrence on oilseed brassica appears to have been made from India (Shaw and Ajrekar, 1915). Higher disease incidence up to 66% may cause severe yield losses up to 39.9% (Chattopadhyay et al., 2003). Premature ripening of the crop by disease incidence may cause additional reduction in yield due to the shattering of siliquae. The disease usually emerges at the time of pod formation, responsible for poor seed formation. Yield loss estimation caused by SR in rapeseed varied in Germany (Horning, 1983), 11.4–14.9% in Saskatchewan, Canada (Morrall et al., 1976), 5–13% in North Dakota, and 11.2–13.2% in Minnesota, USA during 1991–1997 (Lamey et al., 1998).

13.2.5.2 SYMPTOMS

On the basis of symptoms, the disease has been named white blight (Roy and Saikia, 1976), white rot (Rai and Dhawan, 1976), stem blight, stalk break, stem canker, or rape canker, Sclerotina stem rot, and Sclerotina rot (AICRP-RM, 2009). Disease starts as elongated, water soaked lesions on stem especially at base or at internodes. Soon after, white mycelial growth covers these lesions and affected plants look whitish from distance. The disease becomes air borne and spread through contaminated petals of flower which descend and stuck between the main shoot and secondary branches. Large oval to round shaped holes are also formed on leaves due to ascosporic infection. Under severe infection, defoliation, shredding of stem, wilting, and drying of plants occurs. Infected plants forced to premature and stand straight among immature plants. When the fungus completes its life cycle, sclerotia are formed in pith of the stalk, stems, and roots depending upon the mode of infection.

13.2.5.3 FAVORABLE WEATHER CONDITIONS

Development of the disease is preferred by average temperature (18–25°C) and high humidity (90–95% RH) along with winter rains and wind currents particularly at flowering to pod initiation stage.

13.2.5.4 SURVIVAL OF PATHOGEN

Sclerotinia is a necrotrophic fungus which preferred to lives on dead host tissue. *Sclerotinia* formed sclerotia as hard, black bodies resembling rat droppings to survive. These sclerotia spread in soil during harvesting or infected seed that serve as primary source of inoculum for the next crop season. Sclerotia require prolong period of wet soil to germinate myceliogenically and/or carpogenically. They germinate carpogenically to form golf tee shaped fruiting bodies that discharge ascospores. The ascospores are dispersed by airstream and are the main source of secondary infection. Initial mycelial infection at the base of the stem is an appearance of elongated saturated lesions that enlarge speedily. Carpogenic infection is fairly specific and occurs on the leaves or leaf axil.

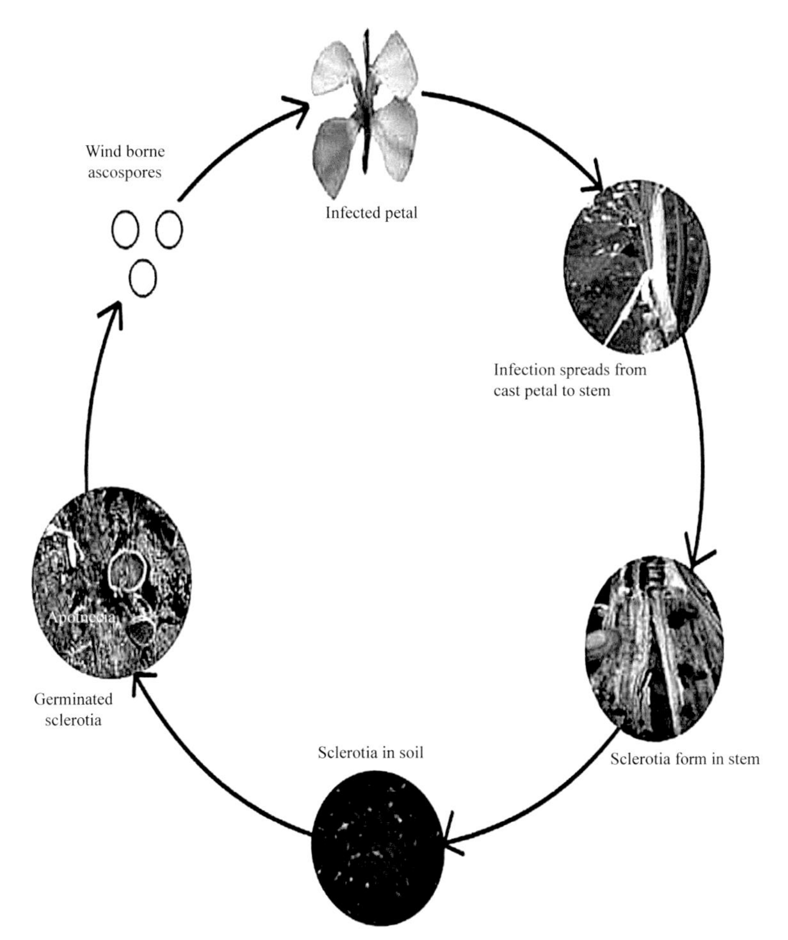

FIGURE 13.6 Life Cycle of Sclerotinia rot on *Brassica*

13.2.5.5 MANAGEMENT

Management of the disease is complex, unpredictable, and uneconomical because of wide range of hosts and long-term survival ability of sclerotia. In view of the fact that any single technique can't control *S. sclerotiorum* effectively. Therefore, the combination of different measures may be the best approach to control the pathogen. Destroy by fire all the infected stubbles to eradicate the sclerotia (Vasudeva, 1958). Deep summer plowing and crop rotation with the non-susceptible hosts (rice, maize), use of only suggested dosage of nitrogenous fertilizer, irrigation, and keeping plant population within limits of recommendation, flooding of soil, if possible, appear to minimize the sclerotial load in the soil, which subsequently might prove useful in disease management resulting from the soil-borne inoculum (Williams and Stelfox, 1980). Integration of the seed treatment followed by foliar sprays of carbendazim 0.1%, *Trichoderma viride* and *Allium sativum* aqueous bulb extracts at 45–50 and 65–70 days after sowing (DAS) provide better reduction of the disease (Chattopadhyay et al., 2002).

13.2.6 CLUBROOT DISEASE

The Russian biologist M. S. Woronin studied the disease in 1873 at St. Petersburg, and recognized *Plasmodipophora brassicae* Woronin as the causal agent of clubroot disease five years later (Woronin, 1878). Clubroot is recognized as a serious soil-borne disease, was reported for the first time in India from the South Indian Nilgiri Hills in Tamil Nadu by McRae (1928), and also from the Darjeeling Hills in the Eastern Himalayan region of West Bengal by Chattopadhyay and Sengupta (1952). The distribution of disease is primarily in cool, oilseed brassica growing regions of the world, recognized to occur in over 60 countries and responsible to cause up to 10–15% reduction in seed yields (Dixon, 2009).

13.2.6.1 SYMPTOMS

Seedlings infection may cause plant death, however, attack in later growth stages rarely kills the host plants. Diseased plants usually show reversible, foliar drooping under slight soil moisture stress. With the progress of disease, leaves turn into discolored, chlorotic, necrotic, and abscise. Both number of seeds per silquae and oil quality decreased (Tewari et al., 2005). The earliest above ground symptoms are stunting of plants, flagging, and yellowing of the leaves. When such plants are uprooted, the hypertrophied root system with typical galls on main or lateral roots having spindle shape can be easily seen. Numerous branch roots frequently grow from these galls to give them a hairy appearance. Symptoms depend on the host produces a fibrous root structure or the "root" is composed of generally a swollen hypocotyl as in swede and turnip. In both systems, the root tissue develop into deformed and composed of massively disordered cells overflowing with secondary plasmodia and ultimately liberate huge numbers of dormant spores.

13.2.6.2 FAVORABLE WEATHER CONDITIONS

The disease development is privileged by high soil moisture and cool weather; however, the disease can emerge at any soil temperature between 9°C and 30°C. There are requirements for favorable adaphic environments consist of acidic pH values (<pH 6.8), minimal calcium contents, nitrogen in an ammoniacal (N-NH4) state, poor soil makeup and impeded drainage being that the soil becomes waterlogged, and temperatures above 15°C (Dixon, 2002). Heat is the third constituent of the soil environment which affects the success of primary zoospore movement and infection into root hairs and epidermal cells. Around 20°C temperature is necessary to initiate the movement of *P. brassicae* and subsequent infection processes.

13.2.6.3 SURVIVAL OF PATHOGEN

The fungus carries on in soil as potential durable resting spores and is apparently capable of residual viable and dormant for partially 20 years (Dixon, 2009). Under favorable weather conditions, these resting spores sometimes swell from its original size and liberate zoospores, to initiate primary infection. The pathogen generates amoeboid thallus that grows to zoosporangia from where zoospores are moved out of diseased roots. Primary infection occurs on root hairs by zoospores produced from the resting spores lying free or in diseased debris in soil leads to the deformation and curling of the root hairs (Samuel and Garrett, 1945) and probably some root "epidermal" cells. Subsequently, secondary zoospores are the means of transportation for cortical infection which results in typical galling and clubbing of the main root systems. Local dissemination of secondary zoospores is by drainage water, implements, windblown soil, etc.

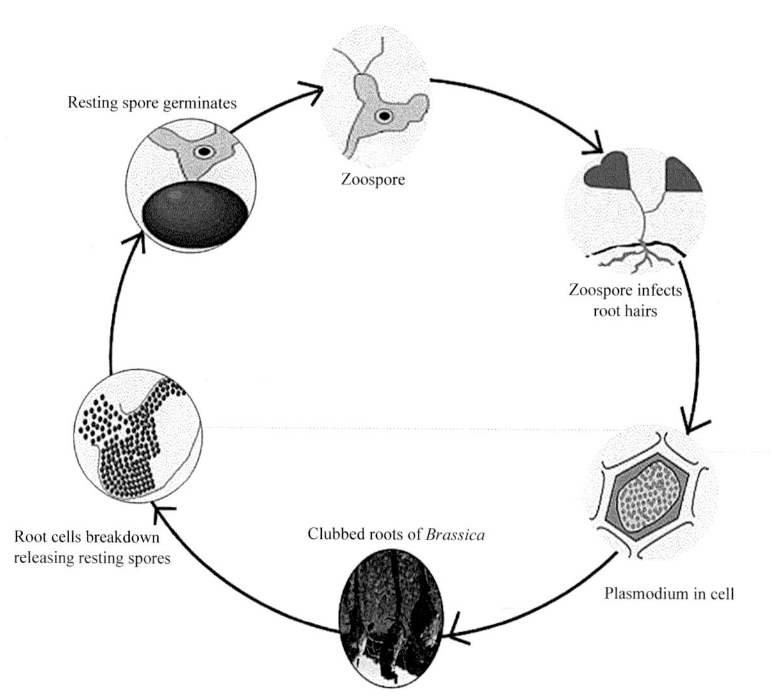

FIGURE 13.7 Life cycle of club root on *Brassica*

13.2.6.4 MANAGEMENT

Crop rotation with non-host crops, burning of diseased debris, deep plowing in summer months helps in minimizing inoculum buildup. Balanced dose of recommended fertilizers improves crop development. A higher dose of nitrogen makes the crop susceptible. Timely sown crop (10–25th October) is expected to hold less disease incidence. Removal of weeds, rouging, and devastation of infected plants and avoidance of over irrigation etc. are effective to reduce the disease incidence. There is a direct relationship between increased soil moisture content and increases the level of primary infection (Thuma et al., 1983). Application of lime @ 3 tonnes/ha and mixing of compost before sowing in infested fields reduces disease incidence. Soils with high calcium content and pH > 7.0 are antagonistic to the fungus and disease. Alkaline soils were thought to reduce symptom development, and it was recommended for good management of clubroot that the pH of the soil should be kept between 7.3 and 7.5. More recently, *Bacillus subtilis* has shown promise in commercial preparations as a bio-control agent (Lahlali et al., 2011; Li et al., 2012).

13.2.7 STEM BLIGHT

13.2.7.1 CAUSAL ORGANISM

Nigrospora oryzae (Berk. & Broome) Petch.

A new disease, first time reported on Indian mustard from Bharatpur district of Rajasthan in India during 2012. This minor problem may become serious problem as the disease incidence recorded more than 70% Indian mustard fields (Meena et al., 2014).

FIGURE 13.8 Typical symptoms of Stem Blight on Rapeseed - Mustard plants

13.2.7.2 SYMPTOMS

Initially, small (2–7 mm), circular to irregular, dark grey to black lesions with a slight bluish cast are formed on stem. They soon become discolored as numerous separate, irregular blotches are formed. In the advanced stages of disease, lesions reached a length up to 120 cm on stems and spread onto the petioles and midribs also (Meena et al., 2014).

13.2.7.3 SURVIVAL OF PATHOGEN

The fungus perpetuate on diseased plant debris in soil.

13.3 BACTERIAL DISEASE

13.3.1 BACTERIAL STALK ROT

13.3.1.1 CAUSAL ORGANISM

Erwinia carotovora pv. c*arotovora* (Jones) Bergy.

13.3.1.2 DISTRIBUTION AND EXTENT OF DAMAGE

The disease has turn into a threat to successful cultivation in some parts of Haryana, Madhya Pradesh and Rajasthan states of the country. The incidence of disease has been reported up to 60–80% at farmer's field.

13.3.1.3 SYMPTOMS

Water soaked lesions are formed at collar region of affected plants which further advances to cover larger area. The affected stalk and branches become soft, pulpy, and produce white dirty ooze of bacterial masses with foul smell. Within no time, infected collar region shows rotting and becomes sunken and turn yellow brown in color. The leaves wilted after infection and plants topple down within a week.

13.3.1.4 SURVIVAL OF PATHOGEN

The pathogen survives on diseased plant debris in soil.

13.3.1.5 FAVORABLE WEATHER CONDITIONS

The disease usually appears when first irrigation applied in Indian mustard. Warm and humid weather favors the disease development.

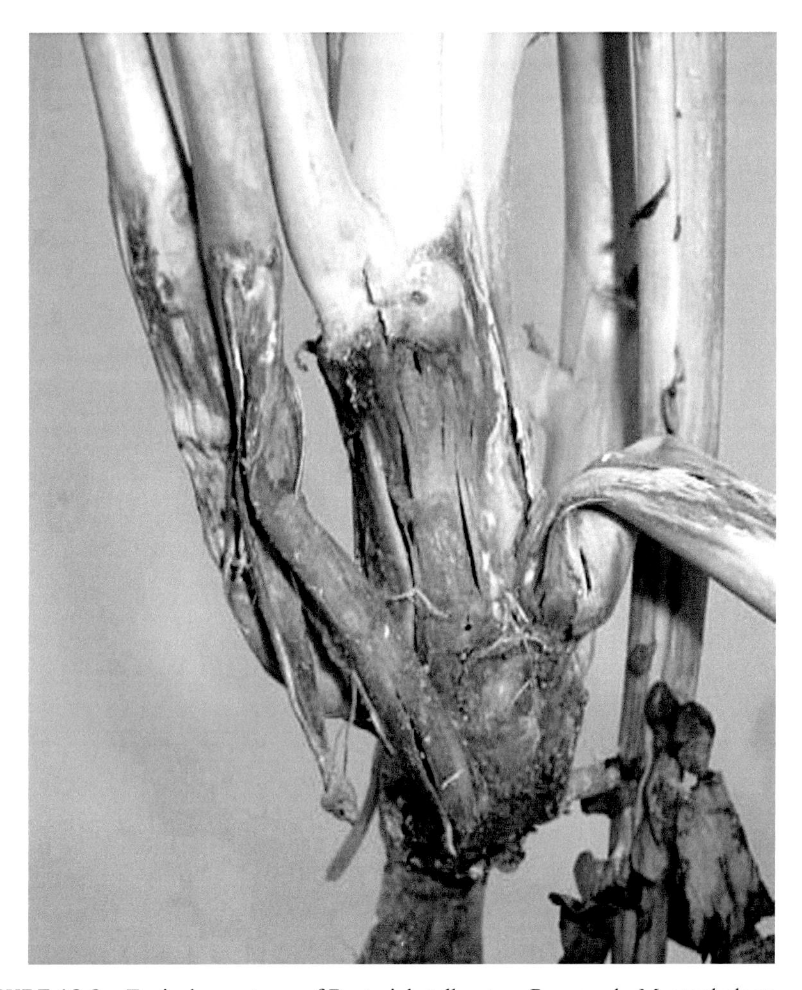

FIGURE 13.9 Typical symptoms of Bacterial stalk rot on Rapeseed - Mustard plants

13.3.1.6 MANAGEMENT

Crop rotation with non-host crops, burning of diseased debris, deep plowing in summer months helps in minimizing inoculum buildup. Timely sown crop (10–25th October) is expected to contain less disease incidence. Removal of weeds, roguing, and destruction of infected plants and avoidance of over irrigation are able to reduce the incidence of disease. Spray with Strepto-cycline 100 ppm + Copper oxychloride @ 0.2% are suitable to manage the disease.

13.4 PHYTOPLASMAL DISEASE

13.4.1 PHYLLODY

13.4.1.1 CAUSAL ORGANISM

Phytoplasma.

13.4.1.2 DISTRIBUTION AND EXTENT OF DAMAGE

The disease is common in *Brassica rapa* ssp. Toria and *B. rapa* ssp. Yellow Sarson growing states of the country. It may cause up to 70–90% yield losses in susceptible *B. rapa* ssp. Toria cultivars, if the disease occurs at early growth stage of the plant.

13.4.1.3 SYMPTOMS

The disease can be seen only at flowering stage and is describe by the alteration of floral parts into leafy structures. The corolla becomes green and saploid, while stamens are green and become indehiscent. The affected parts of the raceme do not form silquae. If the infection occurs at early stage, infected plants stay undersized and produce more branches, giving bushy appearance.

FIGURE 13.10 Typical symptoms of **Phyllody** in Rapeseed - Mustard plants

13.4.1.4 SURVIVAL OF PATHOGEN

The fungus survives on alternate host like sesame and some other plants which serves as primary source of infection. The disease is transmitted through leafhopper (*Orosius albicinctus* = *Deltocephalus* sp.). The disease spreads by repeated cycles of secondary infection through the process of transmission.

13.4.1.5 FAVORABLE WEATHER CONDITIONS

Prolong dry and warm weather is favorable for disease development through insect vector.

13.4.1.6 MANAGEMENT

- Toria sown around mid of September is likely to escape infection.
- Rouging and destruction of phyllody affected plants helps to reduce further spread of disease.
- Spray twice with Rogor or Metasystox @ 0.1% at an interval of 15 days starting from the initiation of symptoms to control the insect vector.

13.5 PHANEROGAMIC PARASITE

13.5.1 BROOMRAPE

Broom rape (*Orobanche aegyptiaca* Pers.) is commonly known orobanche, margoja, rukhri, sarson ka maama and khumbi. It is chlorophyll lacking root holoparasite completely reliant on roots of the host plant. The *Orobanche* genus includes six species (*O. aegyptiaca*, *O. ramosa*, *O. crenata*, *O. cernua*, *O. cumana,* and *O. minor*) that are of agricultural importance and are responsible for seed yield and quality losses to several crops including rapeseed-mustard, cabbage, cauliflower, turnip, brinjal, tomato, tobacco, and many Cruciferous and Solanaceous plants all around the world. However, *O. aegyptiaca* is more dominating one infesting mustard crop in India.

13.5.1.1 DISTRIBUTION AND EXTENT OF DAMAGE

Broomrape is most prevalent in Haryana, Rajasthan, Bihar, and Assam states of India. In case of severe infestation, broomrape can cause total collapse and death of mustard crop plants.

13.5.1.2 SYMPTOMS

The parasite emerges in the soil near to the mustard plant and appears as clumps of whitish, yellowish or purplish broom. It consists of a stout flashy stem, 25–35 cm tall and enclosed by small thin and brown scaly leaves. Sessile flowers appear in the axils of leaves that are bright bluish to violet in color and tubular in shape. There are around 100–200 capsules/plant which contain numerous tiny seed (800–1000 seeds/capsule). The seed is usually dark brown, oval shaped, dust sized weighing 3–6 μg and very difficult to observe by naked eyes. Affected mustard plants become weak, stunted in growth and in severe infestation, plants wilt and ultimately die. When the affected host plants are uprooted, the parasitic roots attached with the host root system are visible.

FIGURE 13.11 Typical symptoms of Broom rape (*Orobanche aegyptiaca* Pers.) or orobanche in Rapeseed - Mustard crops

13.5.1.3 SURVIVAL OF PARASITE

The plant parasite survives in soil as seed or as a seed contaminant along with the rapeseed mustard seed and serves as source of primary inoculum. Its seed can remain viable in soil over the years.

13.5.1.4 LIFE CYCLE OF BROOMRAPE

Broomrape is dicotyledonous annual plant that reproduces only by seeds. Its seed reach to the field through seed contamination, infested soil, manure, farm animal/implements, etc. Broomrape seeds germinate in the soil due to release of chemical exudates by mustard roots within 7–10 days of sowing. Only seeds within the close proximity of rhizosphere (3 mm) will germinate. The parasite seedling radicle recognized as germ tube can grow only a few millimeters and make contact with a host root after a few days of germination. A root like structure is formed after germination called "germ-tube" or "tube like organ" which affects root hairs and establishes association with the vascular tissues of the root through specialized structure called haustorium. Broomrape strongly competes with other parts of the host plant for absorption and translocation of water, mineral, and assimilates. It grows to develop flowers and seeds at maturity time of mustard crop. These seeds get mixed in soil at the time of harvesting of mustard crop through various farm operations and serve as primary source of inoculum for next year crop.

13.5.1.5 FAVORABLE WEATHER CONDITIONS

Favorable temperature for broomrape seed germination is 20–25°C.

13.5.1.6 MANAGEMENT

- Crop rotation with non-host crops particularly wheat, barley, gram, etc.
- Deep plowing in summer months helps in reduction of inoculum.
- Rouging and destruction of broomrape before seed setting.
- Sanitation practices like preventing movement of infested soil by vehicles, farm machinery and cleaning of farm machinery and equipment should be adopted.

- Use clean and certified seed free from contamination of parasitic plant.
- Do not use broomrape-contaminated water from ponds or reservoirs, since they may be a source of seed infestation.
- Post emergence application of glyphosate at 25 g/ha at 30 DAS and 50 g/ha at 55–60 DAS, helps to minimize *Orobanche* population but should be applied under irrigated conditions so as to avoid stress on crop.

KEYWORDS

- **bacterial stalk rot**
- **broomrape**
- **phyllody**
- **relative humidity**
- **sclerotinia rot**
- **stem blight**
- **white rust**

REFERENCES

Berkenkamp, B., (1980). Effects of fungicides and herbicides on stag head of rape. *Can J. Plant Sci., 60*, 1039–1040.

Biga, M. L. B., (1955). Review of the species of the genus *Albugo* based on the morphology of the conidia. *Sydowia, 9*, 339–358.

Chattopadhyay, C., Agrawal, R., Kumar, A., Bhar, L. M., Meena, P. D., Meena, R. L., et al., (2005). Epidemiology and forecasting of alternaria blight of oilseed *Brassica* in India: A case study. *Zeitschrift für Pflanzenkrankheiten und Pflanzenschutz. Journal of Plant Diseases and Protection, 112*, 351–365.

Chattopadhyay, C., Agrawal, R., Kumar, A., Meena, R. L., Faujdar, K., Chakravarthy, N. V. K., Kumar, A., Goyal, P., Meena, P. D., & Shekhar, C., (2011). Epidemiology and development of forecasting models for White rust of *Brassica juncea* in India. *Archives Phytopathology and Plant Protection, 44*(8), 751–763.

Chattopadhyay, C., Meena, P. D., & Kumar, S., (2002). Management of sclerotinia rots of Indian mustard using ecofriendly strategies. *J. Mycol. Pl. Pathol., 32*, 194–200.

Chattopadhyay, C., Meena, P. D., Kalpana, S. R., & Meena, R. L., (2003). Relationship among pathological and agronomic attributes for soilborne diseases of three oilseed crops. *Ind. J. Plant Prot., 31*, 127, 128.

Chattopadhyay, S. B., & Sengupta, S. K., (1952). Addition to the fungi of Bengal. *Bull. Soc. Bengal (India)*, *16*, 2–6.

Choi, D., & Priest, M. J., (1995). A key to the genus *Albugo*. *Mycotaxon, 53*, 261–272.

Chupp, C., & Sherf, A. F., (1960). Crucifer diseases. In: *Vegetable Diseases and Their Control* (Chapter 8, pp. 237–288). Ronald Press Company, New York.

Dick, M. W., (2002). Binomials in the peronosporales, sclerosporales and pythiales. In: Spencer-Phillips, P. T. N., Gisi, U., & Lebeda, A., (eds.), *Advances in Downy Mildew Research* (Vol.1, pp. 225–265). Kluwer academic publishers Dordrecht, Germany.

Dixon, G. R., (2002). Interactions of soil nutrient environment, pathogenesis, and host resistance. *Plant Prot. Sci.*, *38*, 87–94.

Dixon, G. R., (2009). The occurrence and economic impact of *Plasmodiophora brassicae* and clubroot disease. *J. Plant Growth Regul.*, *28*, 194–202.

Fan, Z., Rimmer, S. R., & Stefansson, B. R., (1983). Inheritance of resistance to *Albugo candida* in rape (*Brassica napus* L.). *Canadian Journal of Genetics and Cytology, 25*, 420–424.

Gupta, I. J., & Sharma, B. S., (1978). Chemical control of white rust of mustard. *Pesticides, 12*, 45–46.

Horning, H., (1983). Zur epidemiology und Bekafungder Weibstengelikeit (*Sclerotinia sclerotiorum*). *Raps, 1*, 32–34.

Huang, R., & Levy, Y., (1995). Characterization of iprodione-resistant isolates of *Alternaria brassicicola. Plant Disease, 79*, 828–833.

Iacomi-Vasilescu, B., Avenot, H., Bataille-Simoneau, N., Laurent, E., Guenard, M., & Simoneau, P., (2004). *In vitro* fungicide sensitivity of *Alternaria* species pathogenic to crucifers and identification of *Alternaria brassicicola* field isolates highly resistant to both dicarboximides and phenylpyrroles. *Crop Protection, 23*, 481–488.

Kolte, S. J., (1985). *Diseases of Annual Edible Oilseed Crops* (Vol. II, p. 135). CRC Press, Inc.

Kolte, S. J., (2005). Tackling fungal diseases of oilseed Brassicas in India. *Brassica, 7*, 7–13.

Kolte, S. J., Awasthi, R. P., & Vishwanath, (2000). Divya mustard: A useful source to create Alternaria black spot tolerant dwarf varieties of oilseed brassicas. *Pl. Varieties Seeds, 13*, 107–111.

Kumar, A., Banga, S. S., Meena, P. D., & Kumar, P. R., (2015). *Brassica Oilseeds Breeding and Management* (p. 280). CABI International, UK.

Lahlali, R., Peng, G., McGregor, L., Gossen, B. D., Hwang, S. F., & McDonald, M., (2011). Mechanisms of the biofungicide Serenade (*Bacillus subtilis* QST713) in suppressing clubroot. *Biocontrol. Sci. Tech.*, *21*, 1351–1362.

Lamey, H. A., Nelson, B. D., & Gulya, T. J., (1998). Incidence of *Sclerotinia* stem rot on canola in North Dakota and Minnesota, 1991–1997. *Proc.1998. Int. Sclerotinia Workshop* (pp. 7–9). Fargo, ND.

Li, X., Mao, Z., Wang, Y., Wu, Y., He, Y., & Long, C., (2012). ESI LC-MS and MS/MS characterization of antifungal cyclic lipopeptides produced by *Bacillus subtilis* XF-1. *J. Mol. Microbiol. Biotechnol., 22*, 83–93.

McRae, W., (1928). Report of the Imperial Mycologist Scientific Reports of the Agricultural Research Institute (pp. 56–70). Pusa, 1927–1928.

Meena, P. D., Awasthi, R. P., Chattopadhyay, C., Kolte, S. J., & Arvind, K., (2010). Alternaria blight: A chronic disease in rapeseed-mustard. *J. Oilseed Brassica, 1*, 1–11.

Meena, P. D., Chattopadhyay, C., Singh, F., Singh, B., & Gupta, A., (2002). Yield loss in Indian mustard due to white rust and effect of some cultural practices on alternaria blight and white rust severity. *Brassica, 4*, 18–24.

Meena, P. D., Gour, R. B., Gupta, J. C., Singh, H. K., Awasthi, R. P., Netam, R. S., et al., (2013). Non-chemical agents provide tenable, eco-friendly alternatives for the management of the major diseases devastating Indian mustard (*Brassica juncea*) in India. *Crop Protection, 53*, 169–174.

Meena, P. D., Meena, R. L., Chattopadhyay, C., & Arvind, K., (2004). Identification of critical stage for disease development and bio-control of alternaria blight of Indian mustard (*Brassica juncea*). *Journal of Phytopathology, 152*, 204–209.

Meena, P. D., Vinod, K., Rathi, A. S., & Singh, D., (2014). *Compendium of Rapeseed-Mustard Diseases: Identification and Management* (p. 30). ICAR-DRMR, Bharatpur.

Meena, R. L., Meena, P. D., & Chattopadhyay, C., (2003). Potential for biocontrol of white rust of Indian mustard (*Brassica juncea*). *Indian J. Plant Prot., 31*, 120–123.

Mehta, N., Saharan, G. S., Kaushik, C. D., & Mehta, N., (1996). Efficacy and economics of fungicidal management of white rust and downy mildew complex in mustard. *Indian Journal of Mycology and Plant Pathology, 26*(3), 243–247.

Morrall, R. A. A., Dueck, J., McKenzie, D. L., & McGee, D. C., (1976). Some aspects of *Sclerotinia sclerotiorum* in Saskatchewan, 1970–1975. *Canadian Plant Dis. Sur., 56*, 56–62.

Mukerji, K. G., Upadhyay, R. K., Saharan, G. S., Sokhi, S. S., & Khangura, R. K., (1999). In: Thind, T. S., (eds.), *Diseases of Rapeseed-Mustard and Their Management* (Vol.6, pp. 95–113). National Agriculture Technology Information Centre, Ludhiana, Chap.

Mukherjee, A. K., Mohapatra, T., Varshney, A., Sharma, R., & Sharma, R. P., (2001). Molecular mapping of a locus controlling resistance to *Albugo candida* in Indian mustard. *Plant Breed, 120*, 483–487.

Pal, S. S., Gupta, T. R., Kumar, V., & Dhaliwal, H. S., (1999). Transfer of white rust resistance from *B napus* to *B juncea* CV RLM 198. *Crop Improv., 26*, 249–251.

Parui, N. R., & Bandyopadhyay, D. C., (1973). A note on the screening of rai (*Brassica juncea*) against white rust (*Albugo candida*). (Pers) *Kuntz Curr. Sci., 42*, 798–799.

Peruch, L. A. M., & Michereff, S. J., (2007). Saprophytic survival of *Alternaria brassicicola* and management of broccoli leaf debris. *Ciencia Rural, 37*, 13–18.

Rai, J. N., & Dhawan, S., (1976a). Production of polymethyl galacturonase and cellulase and its relationship with virulence in isolates of *Sclerotinia sclerotiorum* (Lib.) de Bary. *Indian J. Exp. Biol., 14*, 197, 198.

Roy, A. K., & Saikia, U. N., (1976). White blight of mustard and its control. *Indian J. Agril. Sci., 46*, 274–277.

Saharan, G. S., & Chand, J. N., (1988). Diseases of rapeseed and mustard. In: *Diseases of Oilseed Crops* (in Hindi) (Vol.3, pp. 84–91). Directorate of Publication, Haryana Agric. Univ. Press, Hisar, India.

Saharan, G. S., & Mehta, N., (2002). Fungal diseases of rapeseed-mustard. In: Gupta, V. K., & Paul, Y. S., (eds.), *Diseases of Field Crops* (pp. 193–228). Indus Pub Co., New Delhi.

Saharan, G. S., & Verma, P. R., (1992). *White Rusts: A Review of Economically Important Species*. International Development Research Centre, Ottawa, Ontario, Canada. IDRC-MR315e: IV+65p.

Saharan, G. S., (2010). Analysis of genetic diversity in *albugo*-crucifer system. *Journal of Mycology Plant Pathology, 40*(1), 1–13.

Saharan, G. S., Mehta, N., & Sangwan, M. S., (2005). *Diseases of Oilseed Crops*. Indus Publishing Co, New Delhi.

Saharan, G. S., Naresh, M., & Meena, P. D., (2016). *Alternaria Blight of Crucifers: Biology, Ecology and Management* (p. 326). Springer Verlag, Singapore.

Samuel, G., & Garrett, S. D., (1945). The infected root hair count for estimating the activity of *Plasmodiophora brassicae* in the soil. *Annals of Applied Biology, 32*, 96–101.

Shaw, F. J. W., & Agrekar, S. L., (1915). The genus *Rhizoctonia* in India. *Department of Agriculture, Indian Botanical Survey*, *7*, 177–194.

Singh, R. R., & Solanki, J. S., (1974). Fungicidal control of powdery mildew of *Brassica juncea*. *Indian J. Mycol. PI. Pathol*., *4*, 210–211.

Tewari, J. P., Strelkov, S. E., Orchard, D., Hartman, M., Lange, R., & Turkington, T. K., (2005). Identification of clubroot of crucifers on canola (*Brassica napus*) in Alberta. *Can. J. Plant Pathol., 27*, 143–144.

Thuma, B. A., Rowe, R. C., & Madden, L. V., (1983). Relationships of soil temperature and moisture to clubroot (*Plasmodiophra brassicae*) severity on radish in organic soil. *Plant Dis*., *67*, 758–762.

Verma, P. R., & Petrie, G. A., (1979). Effect of fungicides on germination of *Albugo candida* oospores *in vitro* and on the foliar phase of the white rust disease. *Canadian Plant Disease Survey, 59*, 53–59.

Verma, P. R., & Saharan, G. S., (1994). *Monograph on Alternaria Diseases of Crucifers* (p. 162). Agriculture and Agri-Food Canada, Saskatoon Research Centre, Technical Bulletin 1994–6E.

Walker, J., & Priest, M. J., (2007). A new species of *Albugo* on *Pterostylis* (Orchidaceae) from Australia: confirmation of the genus *Albugo* on a monocotyledonous host. *Australasian Pl. Pathol., 36*, 181–185.

Warwick, S. I., Francis, A., & Al-Shehbaz, I. A., (2006). Brassicaceae: Species checklist and database on CD-Rom. *Pl. Syst. Evol., 259*, 249–258.

Williams, J. R., & Stelfox, D., (1980a). Occurrence of ascospores of *Sclerotinia sclerotiorum* in areas of Central Alberta. *Canadian Plant Dis. Surv., 60*, 51–53.

Williams, J. R., & Stelfox, D., (1980b). Influence of farming practices in Alberta on germination and apothecium production of sclerotia of *Sclerotinia sclerotiorum. Canadian J. Plant Pathol., 2*, 169–172.

Woronin, M., (1878). *Plasmodiophora brassicae*, Urheber der kohlpflanzen – hernie. *Jahrb Wiss Bot., 11*, 548–574. [translated by Chupp, C., (1934). *Phytopathological Classics, No 4*. American Phytopathological Society, St. Paul].

Woudenberg, J. H. C., Groenewald, J. Z., Binder, M., & Crous, P. W., (2013). *Alternaria* redefined. *Studies in Mycology, 75*, 171–212.

CHAPTER 14

Important Diseases of Sesamum (*Sesamum indicum* L.) and Their Management

V. B. SINGH, [1] A. K. SINGH, [2] J. N. SRIVASTAVA, [3] and S. K. SINGH[1]

[1]*Rainfed Research Sub-Station for Sub-Tropical Fruits, Raya, Technology (SKUAST-J), Jammu–180009, Jammu & Kashmir, India*

[2]*Division of Plant Pathology, Sher-E-Kashmir University of Agricultural Sciences and Technology (SKUAST-J), Chatha, Jammu–180009, Jammu & Kashmir, India*

[3]*Department of Plant Pathology, Bihar Agricultural University, Sabour–813210, Bhagalpur, Bihar, India*

Sesamum (*Sesamum indicum* L.) originated from Africa and probably the most ancient oil seed crop, it is cultivated in many parts of the world. Presently, India, China, and Myanmar are the world's largest producers of sesame, followed by Sudan, Nigeria, Pakistan, Bangladesh, Ethiopia, Thailand, Turkey, and Mexico (FAO, 2004). Sesame seed is a rich source of protein (20%) and edible oil (50%), and contains about 47% oleic acid and 39% linolenic acid. Sesame oil has excellent stability due to the presence of the natural antioxidants, i.e., sesamoline, sesamin, and sesamol. Its oil is used in cooking, salad preparation, margarine, and raw materials for the production of some industrial materials, including paints, varnishes, soaps, perfumes, pharmaceuticals, and insecticides, while sesame seeds are used in baking, candy, and in other food industries. Seeds with hulls are rich in calcium (1.3%) and provide a valuable source of minerals. The addition of sesame to the high-lysine meal of soybean makes a well-balanced animal food (Jin et al., 2001). Although sesame is widely used for different purposes, the crop has low yield capacity compared to other plants due to susceptibility to diseases, seed shattering, low harvest index, and indeterminate growth habit. Among the major constraints, several diseases *viz.* Phyllody, Alternaria blight, Phytophthora blight, Charcoal rot, Bacterial blight and Cercospora/ White leaf spot is the most common and occurred frequently in most of the Sesame growing areas of the world.

14.1 PHYTOPHTHORA BLIGHT

14.1.1 INTRODUCTION

Phytophthora blight of sesame was first reported from India by Butler (1918). The disease has also been reported from other parts of the world, i.e., India, Dominican Republic, Sri Lanka, Egypt, Iran, Argentina, and Venezuela (Verma et al., 2005). In India, the disease has been frequently occurred in most of the sesame growing states, i.e., Assam, Gujarat, Madhya Pradesh, and Rajasthan in India and in Sri Lanka (Kolte, 1985; Kalita et al., 2002). Due to infection of this disease the affected plants generally kills. It is also observed that the net loss is directly proportional to the incidence of the disease. Under favorable condition mortality of the plants due to the Phytophthora blight may be 72–79%. The disease is getting more important in Assam in recent years where the losses in yield in sesame crop range between 51–53% (Kalita et al., 2002). Besides causing blight, the pathogen is observed to be associated with vivipary in immature seeds of sesame contained in green pods of plants raised from naturally infected seeds. It is an unusual phenomenon that besides increasing the seed infection also renders poor-quality seeds. The host-pathogen interaction results in abnormal seedling emergence, which lacks vigor and further survival (Dubey et al., 2011).

14.1.2 SYMPTOMS

The disease can infect plants at all stages after they attain 10 days of age. Symptoms appear on all above ground parts of the plants. The disease is characterize by the appearance of water-soaked brown spots on leaves and stems near ground level. The spots gradually extend in size. Under favorable environmental conditions, the brownish discolored spots spread rapidly both upward and downward and also around the stem. The brownish area later turns deep brown and becomes black with the spread of the infection. In severe condition, the infection can also be seen on capsules also. In humid weather, the white woolly growth of the fungus can be seen on the surface of affected capsules. Capsules on affected branches are poorly formed. In the case of severe attack, seeds remain shriveled.

14.1.3 CAUSAL ORGANISM

Phytophthora parasitica

The pathogen is *Phytophthora parasitica* (Dastur) var. *sesami* Prasad (*P. nicotianae* B. de Haan var. *parasitica* [Dastur] Waterh). Mycelium of the fungus in young culture is coenocytic and profusely branched, but septa can be observed in 2-month-old cultures. The hyphae are hyaline. Generally the fungus does not form sporangia on culture media, but abundant sporangia can be observed in nature on woolly mycelium growing on infected capsules. The sporangiophores are branched sympodially and bear ovate-to-spherical sporangia terminally. They have a prominent apical papilla and measure 25–50 x 20–35 μ min size. The mycelium, when floated in tap water, forms zoosporangia readily in 48 h. The zoospores are formed inside, and they clearly get separated within the sporangium. The zoospores are liberated in water if the mycelium is flooded with water. The antheridium can be observed at the base, and attachment is typically amphigynous. The oospores are spherical, smooth, double walled, and hyaline.

14.1.4 DISEASE CYCLE

The fungus perpetuates in soil during the unfavorable condition in the form of dormant mycelium and/or in the form of chlamydospores. In addition to soil, seed also appears to play an important role in the recurrence and spread of the disease. In seed, the mycelium has been located in the embryo. However, there are reports that the fungus reduces seed viability but it is not seed borne (Maiti et al., 1988). The mycelium in the host tissue is inter- or intracellular, but it does not form haustoria. The sporangiophores emerge in groups by rupturing the epidermis, but sometimes they emerge through stomata (Verma et al., 2005). The zoosporangia are formed abundantly if humid weather prevails for 2–3 days but soon stop formation if a dry spell appears. The secondary infection occurs through zoospores. *P. parasitica* var. *sesami* is restricted in its pathogenicity to sesame plants only.

14.1.5 EPIDEMIOLOGY/FAVORABLE CONDITIONS

The pathogen can survive in mycelial form up to 50°C temperature, and culture having chlamydospores may survive up to 52°C. Viability of the culture can be kept in a refrigerator for 1 year at 5°C. The fungus can survive in soil during the summer and winter where temperature never rises beyond 50°C or drops below 5°C. Sehgal and Prasad (1966) have shown variation in virulence among various isolates of *P. parasitica* var. *sesami*. Single zoospore isolates show great variations, under similar conditions of infection, in virulence, which may range from nonpathogenic to highly pathogenic. A few isolates of *P. parasitica* var. *sesami* can lose virulence, but the loss in virulence is not permanent, since a few cultures can regain the loss of virulence after passage through the host. On repeated host passages, the culture can even become more virulent than the original ones (Sehgal and Prasad, 1971). Heavy rains for at least 2 weeks and high humidity (above 90%) for 3 weeks or more favor the development of the disease. When such favorable conditions persist for a longer time, the infection may be severe.

It is observed that the initial development of the disease is much earlier when the soil temperature is 28°C, while the initial appearance of the disease is delayed with an increase in the soil temperature up to 37°C. The pathogen is favored by 30°C, can tolerate 35°C, but fails to grow at 37°C. Hence, soil temperature of 28°C–30°C is necessary for disease development. It is further reported that incidence of the Phytophthora blight of sesame shows a close parallelism to the growth of the fungus. Bright sunshine hours for 2–3 days are not favorable for disease development since zoospore formation is stopped under such conditions. The disease appears to become more severe in heavy soils (Verma, 2002). The moderate nitrogenous fertilizer application leads to more incidence of Phytophthora blight of sesamum (Verma and Bajpai, 2001).

14.1.6 MANAGEMENT

14.1.6.1 HOST PLANT RESISTANCE

Out of several strains and varieties of sesame (*Sesamum orientate* L.) and five other closely related species *viz. S. occidentalis*, *S. indicum* L., *S. laciniatum* Willd., *S. prostratum* Betz., and *S. radiatum* Schum. and Thonn., tested for resistance to the disease, none is identified to be resistant against the Phytophthora blight (Kolte, 1985; Choi et al., 1987). However, under

All India Coordinated Research Project on Oilseeds, a number of sesame lines *viz.* TC-25, JLSC-8, TKG-21, AT-60, AT-64, B-14, Chopra-1, Durga (TKG-6), JLT-3, JLT-7, Lakhora-I, Phule till-1, RT-46, T-12, and T-13 over several years of crop season testing have been found to be tolerant to Phytophthora blight. These lines/strains that have shown tolerance over longer duration can be grown to manage the adverse effect of the disease on yield (Verma et al., 2005). In Venezuela, three lines, 71-184-1, 79-129-2, and 71-145-3 (selected from B5 of Ajinio Atar 55), are reported to be disease resistant. The National Institute of Crop Science in Korea has developed a new black-seeded variety Kangheuk, which is a high-yielding, high-lodging, and Phytophthora blight resistant variety (Shim et al., 2012). Epiphytotic conditions and nonavailability of resistant germplasm had prompted the use of gamma ray induced (450–600 Gy) mutation breeding for the development of Phytophthora blight.

14.1.6.2 CULTURAL CONTROL

Sanitation and clean cultivation should be followed as additional measures to control the disease. Use of sowing date depending upon the prevailing local conditions and crop fields with light soil with proper drainage should be preferred to avoid heavy losses due to disease. The intercropping of sesame with non-host crops, i.e., soybean, castor, maize, sorghum, or pearl millet in the ratio of 1:3 or 3:1 shows a low incidence of the disease with higher yield. Application of farmyard manure (FYM) or neem cake with inorganic fertilizers N60, P60, and K70 reduces the disease incidence (Verma et al., 2005). Planting of sesame in 0.2 mm wide ridge in plots mulched with vinyl reduces the spread of the disease by at least 30% and increases the yield by 22% (Choi et al., 1984).

14.1.6.3 CHEMICAL CONTROL

Seed-borne infection can be managed by treating the seed with thiram (0.3%) or Metalaxyl (0.1%). Secondary infection and further spread of the disease can be manage by three sprayings of Bordeaux mixture (0.8%), at an interval of 1 week after the appearance of the disease (Verma et al., 2005). Spraying of dithiocarbamate fungicides such as mancozeb (0.3%) or zineb (0.3%) and Fytolan (copper oxychloride) (0.3%) is reported to be effective in the control of the disease (Kalita et al., 2002).

14.1.6.4 BIOLOGICAL CONTROL

The several species of *Pseudomonas*, *Bacillus*, and *Streptomyces* are most active at 25°C–27°C at field capacity moisture level, can be suppressive to *Phytophthora* spp. in soil. The antagonistic fungi *viz. Trichoderma viride*, *T. harzianum,* and *Pseudomonas fluorescence* are the effective for the management of the disease (Chattopadhyay et al., 2016).

14.2 FUSARIAL WILT

14.2.1 INTRODUCTION

Fusarial wilt of sesame was reported for the first time from North America in 1950 (Armstrong and Armstrong, 1950). Since then, the disease is reported to occur in Egypt, Colombia, Greece, India, Iran, Israel, Japan, Korea, Malawi (formerly Nyasaland), the United States, former Venezuela and Soviet Union. Similar disease has been reported from Pakistan, Peru, Puerto Rico, and Turkey (Verma et al., 2005). The disease can be devastating on susceptible varieties of sesame, but many local varieties have been found to have some degree of resistance to local races of the fungus. Epiphytotic occurrence of the disease was reported in 1961 and 1964 in the United States and in 1959 in Venezuela.

14.2.2 SYMPTOMS

Plants get infected at any stage of the crop development including the damping-off phase in the seedling stage. During later stages of the plant, yellowing of the leaves is the first noticeable symptom of the wilt in the field. Leaves become yellowish, droop, and desiccate. Sometimes such leaves show inward rolling of the edges and eventually dry up. The terminal portion dries up and becomes shrunken and bent over. In a severe stage, the infected plant becomes defoliated and dry. In a less severe infection or when mature plants are infected, only one side of the plant may develop symptoms, resulting in partial wilting, and a half stem rot symptom has been reported (Cho and Choi, 1987). A blackish discoloration in the form of streaks appears on infected plants. Discoloration of the vascular system is conspicuous in the roots. Roots in the later stages show rotting, wholly or partially corresponding with that side of the plant showing disease symptoms. Numerous pink pinhead-sized sporodochia (containing macroconidia of the fungus) may be seen scattered over the entire dried stem. The capsules of wilted plants also show numerous sporodochia.

14.2.3 CAUSAL ORGANISM

Fusarium oxysporum

The pathogen is *Fusarium oxysporum* (Schell.) f. sp. *sesami* Jacz. isolation of the causal fungus could be obtained more easily from the infected dry sample (dry sample isolation) compared to conventional direct isolation technique from freshly infected sesame plants (Su et al., 2012). The fungus produces profuse mycelial growth on potato dextrose agar. The mycelium is arid, hyaline, septate, and richly branched, turning light pink when old. The micro-conidia are formed abundantly. They are hyaline, ovoid to ellipsoid, unicellular, and measure 8.5 x 3.25 μm in size. In the old culture, the macro-conidia are formed sparsely. They are 3–5, septate, and measure in the range of 35–49 x 4–5 μm in size. The macro-conidia are produced abundantly in sporodochia as they develop on affected plants. The chlamydospores are globose to subglobose, smooth, or wrinkled and measure 7–16 μm in diameter. Physiological studies on the pathogen have been made. The fungus grows best on Richard's medium. It grows at the temperature range of 10°C–30°C with an optimum temperature of 25°C. Nitrate nitrogen and pH 5.6–8 support the maximum growth of the fungus. Illumination inhibits spore germination (Liu et al., 2010).

14.2.4 DISEASE CYCLE

The fungus is restricted in its host range to sesame. Morphological differences and similarities have been reported in different isolates of *F. oxysporum* f. sp. *sesami*. Three strains have been reported in Venezuela on the basis of morphological differences, but these strains are reported to show a similar degree of pathogenicity. It is reported that there is a relationship between vegetative compatibility groups of the pathogen and geographic origin of the isolates collected from the different sesame-growing regions (Basirnia and Banihashemi, 2005). The pathogenic variation and molecular characterization of *Fusarium* species isolated from wilted sesame have been studied. The pathogen is reported to be seed and soil borne, and it may persist for many years in the soil. The amount of seed transmission of the pathogen varies in the range of 1.0–14.0% depending on the severity of systemically infected sesame plants (Basirnia and Banihashemi, 2006). It appears that the fungus penetrates the host through root hairs and causes trichomycosis. The most virulent isolates produce more cell-wall-degrading enzymes than the less virulent ones. The culture filtrates of *F. oxysporum* f. sp. *sesami* has been reported to have an inhibitory effect on sesame. Shoot and root growth is also inhibited by culture filtrate of the fungus indicating the production of toxic substances by the pathogen. Some elements like vanadium, zinc, boron, molybdenum, and manganese are highly inhibitory to *F. oxysporum* f. sp. *sesami*.

14.2.5 EPIDEMIOLOGY/FAVORABLE CONDITIONS

High soil temperature to a depth of 5–10 cm and 17.0–27.0% water-holding capacity during dry periods is favorable for the disease development. Drought stress in the sesame plants predisposes the plants to infection and development of wilt and influences the host genotype reaction to the disease (Kayak and Boydak, 2011). The Fusarium wilt of sesame is reported to be associated with nematode attack in South America and with *M. phaseolina* in Egypt, India, and Uganda (Kolte, 1985). The density of the fungus becomes higher in soil under continuous cropping.

14.2.6 MANAGEMENT

Several microbial antagonists, i.e., *Trichoderma viride*, *Gliocladim virens* (El-Bramawy and El-Sarag, 2012), *Bacillus polymyra*, *B. subtilis*, *Enterobacter cloacae* (Abdel-Salam et al., 2007), *Pseudomonas aeruginosa* (Abdel-Salam et al., 2007), *P. putida,* and *P. fluorescens*, *Streptomyces bikiniensis*, and *S. echinoruber* (Chung and Ser, 1992) are inhibitory to the growth of *F. oxysporum* f. sp. *sesami* and show high potential for their use in the management of Fusarium wilt of sesame. *Trichoderma* species grown on cow dung slurry and cow dung are the most effective in the control of the wilt disease of sesame (Sangle and Bambawale, 2004). Fusarium wilt of sesame can be controlled with application of plant growth promoting rhizobacteria, and this practice offers a potential nonchemical means for disease management. A combination of *Azospirillum brasilense* based Cerialin and *Bacillus megaterium*-based Phosphoren biofertilizers plus Topsin (100 ppm) has been found to be effective in reduction of Fusarium wilt incidence, with increased morphological characteristics and plant yield (Ziedan et al., 2012). Similarly, a mixture of *P. putida* plus *P. fluorescens* treatment together (Fusant) as biocide and biofertilizer gives better control of the wilt disease with higher sesamum crop yield. Fertilizer adaptive variant tetracycline resistant strain TRA2 of Azotobacter chroococcum, an isolate of wheat rhizosphere, has been found to show plant growth promoting attributes and strong antagonistic effect against sesame wilt and charcoal rot pathogens. Seed bacterization with the strain TRA2 results in significant decrease in Fusarium wilt disease incidence and increase in vegetative growth of sesame plants (Maheshwari et al., 2012). *Glomus* spp. (VAM) protects the sesame plants by colonizing the root system and consequently reduces colonization of fungal pathogens in sesame rhizosphere by stimulation of bacteria belonging to the Bacillus group. These bacteria show high antagonistic potential, and this significantly reduces Fusarium wilt incidence in sesame. Extracts of leaves of thyme, eucalyptus, and garlic reduce the incidence of Fusarium wilt disease of sesame. Extract of peppermint (*M. piperita*) leaves not only reduces the wilt incidence but also increases the yield of sesame plant (Sidawi et al., 2010).

14.3 ALTERNARIA LEAF SPOT

14.3.1 INTRODUCTION

Alternaria leaf spot of sesame was first reported by Kvashnina (1928) from the North Caucasus region in the former Soviet Union. Mohanty and Behera (1958) from India reported Alternaria blight of sesame and found the causal organism to closely resemble as *M. sesami*. However, it differed from *M. sesami* in that some of the spores were catenulate. On the basis of the catenulation, the fungus was keptin the genus *Alternaria* and renamed as *A. sesami* (Kawamura) Mohanty and Behera. In India and in the United States, it was earlier referred only by the name *Alternaria* sp. The Alternaria leaf spot is now reported to occur in most of the tropical and subtropical areas of the world. Epiphytotic occurrence of the disease has been reported from the Stoneville area in Mississippi in 1962, the Tallahassee area in Florida in 1958, and the coastal area of Orissa in 1957 and Maharashtra in India in 1975 (Kolte, 1985). The amount of damage to the sesame plant is dependent on the stage of the plant growth and environmental conditions. Disease severity is negatively correlated with the seed yield, 1000 seed weight, and seeds/capsule (Ojiambo et al., 2000b).

The disease causes 20.0–40.0% loss in sesame crop in the state of Uttar Pradesh in India (Kumar and Mishra, 1992). It is, however, reported that about 0.1–5.7 g seeds/100 fruits are lost due to the disease under Karnataka conditions in India (Kolte, 1985).

14.3.2 SYMPTOMS

Disease symptoms appear mainly on leaf lamina as small, brown, round-to-irregular spots, varying from 1.0 to 8.0 mm in diameter. These spots later become larger and darker with concentric zonations demarcated with brown lines inside the spots on the upper surface. On the lower surface, the spots are light brown in color. Further, these spots often coalesce and may involve large portions of the blade, which become dry and are shed. Dark brown, spreading, water-soaked lesions can be seen on the entire length of the stem. The lesions also occur on the midrib and even on veins of leaves. In very severe condition, plants may be dried within a very short period after symptoms are first noted, while milder attacks cause defoliation. Occasionally, seedlings, and young plants are killed exhibiting pre- and post-emergence damping-off.

14.3.3 CAUSAL ORGANISM

The pathogen is *Alternaria sesami* (Kawamura) Mohanty and Behera. The conidiophores of the fungus are pale brown, cylindrical, simple, erect, 0–3 septate, and not rigid, arise singly and mea-sure 30–54 x 4–7 and bears conidia at the apex. The conidia are produced singly or in chains of two. They are straight or slightly curved, obclavate, and yellowish brown to dark or olivaceous brown in color and measure 30–120 x 9–30 μm (excluding the beak). The conidia have 4–12 transverse septa.

14.3.4 MANAGEMENT

14.3.4.1 CULTURAL CONTROL

Salt density at 2–5% concentration can be used to sort out the infected seed from the seed lots to maintain healthy nucleus seed after further washing and drying the seed (Enikuomehin, 2010). Seeds floated at 2 and 5% salt conc.

are characteristically discolored, malformed, infected, and lightweight. Experimental evidence has been presented in Nigeria that intercropping sesame with maize in a single alternate row (1:1) arrangement can be useful in reducing the severity of Alternaria leaf blight of sesame (Enikuomehin et al., 2010).

14.3.4.2 CHEMICAL CONTROL

Two sprays of mancozeb at 0.25% (Mudingotto et al., 2002) or a combination of mancozeb at 0.25% plus methyldemeton at 1 ml/l or mancozeb at 0.25% plus streptocycline at 0.025% (Shekharappa and Patil, 2001a) have been found to be effective in the management of Alternaria leaf spot of sesame with increase in yield of sesame crop.

14.4 CHARCOAL ROT

14.4.1 INTRODUCTION

The charcoal rot of sesame has been reported from most of the sesame growing areas in the world (Kolte, 1985; Verma et al., 2005). The disease is particularly reported to be quite serious, limiting the cultivation of the seasmum crop in Ismailia Governorate Region in Upper Egypt (Abdou et al., 2001); Southeastern Anatolia Region in Turkey (Sager et al., 2009); in the Portuguesa state in Venezuela (Cardona and Rodriguez, 2006; Martinez-Hilders et al., 2013); In India, the disease has been reported from Chandrapur district of Vidarbha region of Maharashtra, the Gwalior Division of Madhya Pradesh, and in the states of Haryana and Chhattisgarh (Deepthi et al., 2014); and in Pakistan (Akhtar et al., 2011). Seedling mortality due to seed-borne infection aggravates the problem by reducing the plant stand per unit area resulted reduction of the yield. About 5–100% yield loss due to the disease is reported from different areas. At 40.0% disease incidence estimated yield loss of 57.0% is reported by Maiti et al. (1988). However, in favorable and severe condition the losses may be higher. In Venezuela, losses in sesame due to charcoal rot have been evaluated resulting up to 65% weight reduction of the seed in affected plants. The importance of the charcoal rot lies not only in affecting the yield and causing quantitative and qualitative losses (Sagar et al., 2009) but also in increasing soil infestation with the causal fungus. For example, sclerotia of *Macrophomina phaseolina* in Venezuelan soils of sesame production areas have been estimated to be up to 200 per gram of soil. If the disease appears simultaneously with Phytophthora blight or with Fusarium wilt, the losses in yield usually are very high.

14.4.2 SYMPTOMS

Sesame plants may be attacked immediately after the sowing of the crop. The germinating seeds may become brown and rot. In the seedling stage, the roots may become brown and rot, ultimately the affected plants may be dried. If the plants survive, the older plants are affected at the base of the stem indicating the formation of lesion that later spreads to the middle portion of the stem and becomes ashy, causing drooping of leaves from top of the plants. Such plants if survive make poor growth and remain stunted. The mycelium of the fungus progresses upward in the stem, and as the stem dries, pycnidia are formed as minute black dots. The stem may break off, and the blackening may extend upward on the stem. The capsules are also affected. Such capsules open prematurely, exposing shriveled and discolored seeds. Seeds may also show the presence of sclerotia on the surface.

14.4.3 CAUSAL ORGANISM

Macrophomina phaseolina

The pathogen *Macrophomina phaseolina* (Tassi) Goid belongs to Deutromycota and produces septate mycelium. The conidia are produced in black dots like fruiting bodies known as pycinidia. It is revealed that *M. phaseolina* populations in all the major sesame production regions in China (Wang et al., 2011), Iran (Bakhshi et al., 2010; Mandizadeh et al., 2012), Mexico (Munoz-Cabanas et al., 2005), and Venezuela (Martinez-Hilders and Laurentin, 2012) are highly genetically diverse based on genomic data. High level of genotypic variability is likely due in part to the exposure of the pathogen to diverse environment and a wide host range within these countries. However, no clear association between geographical origin and host of each isolate has been found, though isolates from the same location show a tendency to belong to their respective closer groups indicating closer genetic relatedness (Bakshi et al., 2010).

14.4.4 MANAGEMENT

14.4.4.1 HOST RESISTANCE

Different sesame germplasm lines and cultivars that have been found tolerant or little susceptible against charcoal rot are ORM 7, ORM 14, and ORM 17 (Subrahmaniyan et al., 2001; Thiyagu et al., 2007); TLC-246, TL6-279, and TLCCCCC-281 (John et al., 2005); ZZM0565, ZZM0570, Xiangcheng dazibai, Xincai Xuankang, Shangshui farm species, and KKU 3 (Zhao et al., 2012); mutants NS 13 Pl, NS 163-1, NS 270 P1, and NS 26004 (Akhtar et al., 2011); CMGS, OF 4A, and alpha-tubulin (Liu et al., 2012); UCLA-1, EXP-1, and DV-9 and Aceteru-M, Adnan (5/91), Taka 2, B 35, and mutation 48 (El-Fiki et al., 2004b). The sources of resistance appear to differ in the mechanism of resistance. Factors such as morphological traits like single stem medium branch numbers (El-Bramawy et al., 2009); creamy or white seed color (El-Bramawy et al., 2009); antifungal nutritional components such as phytin, trypsin inhibitor, and tannins (El-Bramawy and Embaby, 2011); certain biochemical factors as faster rate of activity of polyphenol oxidase enzymes in Chinese sesame cultivar Yuzhi 11 (Liu et al., 2012); and different isoenzyme band patterns (Zhang et al., 2001) have been linked with resistance to charcoal rot in sesame. The mature plant reaction, through hybridization studies, indicated that susceptibility in the mature plant is dominant over tolerance, and it is governed by 1, 2, or 3 pairs of genes (Kolte, 1985).

14.4.4.2 CULTURAL MANAGEMENT

The average charcoal rot incidence can be reduced by choice of sowing date and levels and time of irrigation depending on the local conditions in a particular geographical area. Early sowing by June 10 in Egypt and following hills-over-furrows method of sowing and giving only one irrigation during the whole growing season to a crop fertilized with N at 65 kg, P at 200 kg, and K at 50 kg/feddan (0.42 ha) result in significant reduction in charcoal rot incidence (Shalaby and Bakeer, 2000). Lowering concentration of Ca, Na, Mg, and Fe and increasing concentration of K, Cu, and Zn in the soil by applying chemical fertilizers and organic manure may reduce very much the charcoal root rot incidence (Narayanaswamy and Gokulakumar, 2010). Sesame crop sown as mixed or intercropped with green gram in 1:1 is useful in the management of the charcoal rot and results in higher sesame yield in the arid region. Six weeks of soil solarization of infested crop field sites in the summer months result in good sesame seed germination and better disease management under Indian conditions (Chattopadhyay and Sastry, 2001).

14.4.4.3 CHEMICAL CONTROL

Seed treatment with carbendazim (0.1%–0.3%) gives complete control of seed-borne infection of *M. phaseolina* in sesame when used as seed treatment fungicide (Shah et al., 2005; John et al., 2010). Other seed treatment fungicides are thiophanate methyl (John et al., 2010), Benlate or Rizolex T at 3 g/kg seed (El-Fiki et al., 2004a), mancozeb (Mudingotto et al., 2002), thiram, captan, and carboxin (Verma et al., 2005). Soil treatment with fungicides is effective but impracticable. The integration of fungicide carbendazim seed treatment (0.1%) with carbendazim-tolerant strain of T. viride (Tv-Mut) as induced by mutating the native strain of the fungus by UV irradiation and soil supplemented with 20 kg P and 15 kg K/ha show the highest reduction (91.7%) in sesame stem root rot incidence caused by *M. phaseolina* (Chattopadhyay and Sastry, 2002). Aminobutylic acid and potassium salicylate can effectively control charcoal rot in sesame by induction of host resistance against *M. phaseolina* and increasing plant height, indole acetic acid (IAA) content, and peroxidase (PO) activity (Shalaby et al., 2001).

14.4.4.4 BIOLOGICAL CONTROL

Effect of antagonistic fungi and bacteria isolated from the rhizosphere of sesame is reported to be efficiently more effective in managing the root rot and stem rot of sesame caused by *M. phaseolina*. Sesame seed treatment with (a) *T. viride* at 4 g/kg of seed (Zeidan et al., 2011), (b) *T. harzianum* (Moi and Bhattachrya, 2008), (c) *P. fluorescens* (Jayshree et al., 2000; Moi and Bhattachrya, 2008), and (d) *Bacillus subtilis* (Elewa et al., 2011) has been found effective in the management of charcoal rot disease. Green manure of *Crotolaria* mixed with *Trichoderma* constitutes a viable alternative for the control of charcoal rot of sesame (Cardona, 2008). A combination of seed treatment and soil application of the antagonists through the application of clay granules impregnated with *T. harzianum* or *P. fluorescent* at sowing time appears to be much more effective in the control of the charcoal rot (Cardona and Rodriguez, 2002). Application of vesicular-arbuscular mycorrhizae (VAM), namely, *Glomus* spp., together with biocontrol agents *T. viride* or *B. subtilis* significantly helps in efficient management of root rot (*M. phaseolina*) and Fusarium wilt diseases of sesame than individual application of either VAM or antagonists (Elewa et al., 2011; Zeidan et al., 2011). Soil solarization in combination with fungal antagonists *T. pseudokoningii* and *Emericella nidulans* singly or in mixed inocula reduces charcoal rot incidence in sesame significantly (Ibrahim and Abdel-Azeem, 2007). Seed treatment with *Azotobacter chrococum* and seed + soil treatment with *Azospirillum* also reduce the disease by about 30% (Maheshwari et al., 2012).

14.4.5 EFFECT OF PLANT EXTRACTS

Extracts of *Thevetia neriifolia* (Bayounis and Al-Sunaidi, 2008a), *Azadirachta indica*, *Datura stramonium*, *Nerium oleander*, *Eucalyptus camaldulensis* (Bayounis and Al-Sunaidi, 2008b), and *Helichrysum* flower (Shalaby et al., 2001) show inhibitory effect on the growth of *M. phaseolina*, indicating their potential use in the management of the disease. The extracts of Eucalyptus (*Eucalyptus rostrata*, *E. camaldulensis*), peppermint (*Mentha piperita*), and thyme (*Thymus serpyllum*), when used in sand culture or under *in-vitro* conditions in growth media and inoculated with *M. phaseolina*, have been found to show increase in sesame seed germination despite the presence of *M. phaseolina* in the culture, indicating potential usefulness of these extracts (Sidawi et al., 2010).

14.5 PHYLLODY

14.5.1 INTRODUCTION

Prevalence of the sesamum phyllody erroneously named leaf curl is traced since 1908 in Mirpur Khas area of India (now in Pakistan). Detailed historical evidence of the occurrence and causal pathogen of the disease has been reviewed earlier by Kolte (1985). It has been reported from India, Iran, Iraq, Israel, Myanmar, Sudan, Nigeria, Tanzania, Pakistan, Ethiopia, Thailand, Turkey, Uganda, Upper Volta, Venezuela, and Mexico (Kolte, 1985; Akhtar et al., 2009). The first evidence of association of mycoplasma-like organism (now known as phytoplasma) with the disease was observed in Upper Volta by Cousin et al. (1971). Affected plants remain completely or partially sterile, resulting in total loss in yield. The yield loss due to phyllody in India is estimated to about 39%–74%. The losses in plant yield, germination, and oil content of sesame seeds may be as high as 93.66%, 37.77%, and 25.92%, respectively. It is estimated that a 1% increase in phyllody incidence decreases the sesame yield by 8.4 kg under Coimbatore conditions in India.

14.5.2 SYMPTOMS

The initial symptoms of the phyllody is observed on the inflorescence and characterized by all floral parts are transformed into green leafy structures followed by abundant vein clearing in different flower parts. The plant is stunted with reduced internodes and abnormal branching (Figure 14.1). However, when infection takes place at later stages, normal capsules are formed on the lower portion of the plants, and phylloid flowers are present on the tops of the main branches and on the new shoots that are produced from the lower portions. The calyx becomes polysepalous and shows multi-costate venation compared to its gamosepalous nature in healthy flowers. The sepals become leaf like but remain smaller in size. The phylloid flowers become actinomorphic in symmetry, and the corolla becomes poly-petalous. The corolla may become deep green, depending upon the stage of infection. The veins of the flowers become thick and quite conspicuous. The stamens retain their normal shape, but they may become green in color. Sometimes, the filaments may, however, become flattened, showing its tendency to become leaf like. The anthers become green and contain abnormal pollen grains. In a normal flower, there are only four stamens, but a phylloid flower bears five stamens. The carpels are transformed into a leaf outgrowth, which forms a pseudo-syncarpous ovary by their fusion at the margins. This false ovary becomes much enlarged. In Sudan, red varieties of sesame have been found to be affected to the extent of 100%. Inside the ovary, instead of ovules, there are small petiole-like outgrowths, which later grow and burst through the wall of the false ovary producing small shoots. These shoots continue to grow and produce more leaves and phylloid flowers (Figure 14.2). The stalk of the phylloid flowers is generally elongated, whereas the normal flowers have very short pedicels. Increased IAA content appears to be responsible for proliferation of ovules and shoots. Sometimes, these symptoms are found to be accompanied with yellowing, cracking of seed capsule, germination of seeds in capsules, and formation of dark exudates on the foliage (Akhtar et al., 2009). Normal-shaped flowers may be produced on the symptomless areas of the plants, but such flowers are usually dropped before capsule formation, or the capsules are dropped later leaving the stalk completely bared.

FIGURE 14.1 Phyllody symptoms before flowering

FIGURE 14.2 Flowers are converted into leafy structure.

14.5.3 CAUSAL ORGANISM

Mycoplasma like organism (MLO)

The pathogen is now reported to be phytoplasma (formerly referred to as mycoplasma-like organism, wall-less bacteria belonging to the class Mollicutes). Light microscopy of hand-cut sections treated with Dienes stain shows blue areas in the phloem region of phyllody-infected sesame plants (Akhtar et al., 2009). The phytoplasma pleomorphic bodies are reported to be present in phloem sieve tubes of affected sesame plants. Electron microscopy has revealed that the big pleomorphic bodies. Generally, the phytoplasmas are round, but some may be 1500 nm long and 200 nm wide. Bodies with beaded structures can also be noticed. The phytoplasmas are bounded by a single unit membrane as is typical for the Mollicutes and show ribosome-like structure and DNA-like strands within. Phytoplasma cells contain one circular double-stranded DNA chromosome with a low G + C contents (up to only 23%), which is thought to be the threshold for a viable genome (Bertaccini and Duduk, 2009; Weintraub and Jones, 2010). They also contain extra-chromosomal DNA such as plasmids. Since phytoplasmas cannot be grown in axenic culture, advances in their study are mainly achieved by molecular techniques. Molecular data on sesame phytoplasmas have provided considerable insight into their molecular diversity and genetic interrelationships, which has in turn served as a basis for sesame phytoplasma phylogeny and taxonomy. Classification of phyllody phytoplasma associated with sesame has been attributed to at least three distinct strains worldwide including aster yellows, peanut witches' broom, and clover proliferation group (Al-Sakeiti et al., 2005; Khan et al., 2007). Based on restriction fragment length polymorphism (RFLP) analysis of polymerase chain reaction-amplified 16S rDNA, sesame phyllody phytoplasma infecting sesame in Myanmar (termed as SP-MYAN) belongs to the group 16SrI and subgroup 16SrI-B. Sequence analysis has confirmed that SP-MYAN is a member of Candidatus Phytoplasma asteris and it is closely related to that of sesame phyllody phytoplasma from India (DQ 431843) with 99.6% similarity (Khan et al., 2007; Win et al., 2010).

14.5.4 MANAGEMENT

14.5.4.1 HOST PLANT RESISTANCE

Selections of disease-resistant sesame lines, which would flower within 40–50 days after sowing (DAS), appear to be desirable and important from the yield viewpoint under Indian conditions (Kolte, 1985; Selvanarayanan and Selvamuthukumaran, 2000). From India, a considerable number of genotypes such as RJS 78, RJS 147, KMR 14, KMR 29, Pragati, IC 43063, and IC 43236 (Singh et al., 2007), SVPR-1 (Saravanan and Nadarajan, 2005), AVTS-2001-26 (Anandh and Sevanarayanan, 2005), Swetta-3, RT-127, No.171 (Dandnaik et al., 2002), TH-6 (Anwar et al., 2013), and three wild species, i.e., *S. alatum*, *S. malabaricum*, and *S. yanaimalaiensis*, are resistant to phyllody with mean incidence below 5%, which can be utilized as donor parents in resistance breeding to phyllody disease (Saravanan and Nadarajan, 2005; Singh et al., 2007). A single recessive gene governs resistance in cultivated varieties (KMR 14 and Pragati), whereas wild species possess a single dominant gene conferring resistance to phyllody (Singh et al., 2007).

Phyllody resistance in a land race of sesame is reported to be under the control of two dominant genes with complementary (9:7) gene action (Shindhe et al., 2011). Some genotypes in India have not been observed to show phyllody symptoms. Such genotypes are Ny-9, Sirur, Local, NKD-1037, K-50, TC-25, RT-15H, OCP-1827, No.5, No.16, No.17, No.18, No.21, No.23, and No.24 (Dandnaik et al., 2002). Interspecific hybrids between *S. alatum* and *S. indicum* are, however, moderately resistant to phyllody. Advanced phyllody disease-resistant sesame mutant lines with earliness, more capsules, and high harvest index have been developed in Pakistan under the series NS11-2, NS11 P2, NS100 P2, NS 103-1, and NS240 PI and phyllody disease-resistant sesame. These mutant lines can be of great potential use in breeding for disease resistance. Some other sesame lines as JT-7, IT-276, and N-32, though not resistant to the disease, have been found useful to escape the disease (Kolte, 1985).

14.5.4.2 CULTURAL CONTROL

An appropriate sowing date may be useful in avoiding severe occurrence of the disease. The incidence of the disease is reported to be reduced considerably by sowing the crop in early August under Indian conditions. The reduced population of the vector in the growth period of sesame plants is perhaps important in keeping the disease under check (Mathur and Verrna, 1972; Nagaraju and Muniyappa, 2005). Destruction of recent identified collateral host bhang *(Cannabis sativa* subsp. *Sativa)*. The phytoplasma survive on both the plants at same time (Singh et al., 1918).

14.5.4.3 CHEMICAL CONTROL

Insect vector management is the method of choice for limiting the outbreaks of phytoplasmas in sesame. At the time of sowing, soil may be treated with Thimet® 10 G at the rate of 10 kg/ha or with Phorate 10 G at the rate of 11 kg/ha or with Temik® 10 G at the rate of 25 kg/ha to get the management of the disease through vector control (Nagaraju and Muniyappa, 2005). An effective degree of management is obtained if the aforementioned treatment is combined with spraying of the crop with Metasystox (0.1%) or with any other effective chemical (Misra, 2003). Tetracycline sprays at 500 ppm concentration at the flower initiation stage have proved to be effective against phyllody, but recovery is temporary. A possibility of biochemical control by spraying manganese chloride has been indicated. It appears that manganese chloride oxidizes the phenol and protects or inhibits the enzymes, brining the auxin level to normal. Once hyperauxin is oxidized, the plant can gain its normal conditions (Chattopadhyay et al., 2016).

14.6 CERCOSPORA LEAF SPOT (CLS)/WHITE SPOT

14.6.1 SYMPTOM

The disease first initiate on the leaf as minute water soaked lesions in later stage these lesions enlarge to farm round to irregular spots up to 15 mm in size. The spotted area is light brown to white in the beginning later they converted dark gray in color. In severe conditions several spots may occur on a single blade, enlarging to coalesce and blight a large portion of a leaf. Often infection takes place on the leaf, petiole, stem, and pods producing leaniar dark color, deep seeded lesions. The damaged to the plant growth and yield depends on the severity of infection on the stem and pods on the stage which the infection takes place. Disease appears as small, angular brown leaf spots of 3 mm diameter with gray center and dark margin delimited by veins. In severity of the disease defoliation occurs. Under favorable conditions, the disease spreads to leaf petiole, stem, and capsules producing linear dark colored deep seated lesions.

14.6.2 CAUSAL ORGANISM

Cercospora sesami, C. sesamicola.

14.6.3 MANAGEMENT

Jeyalakshmi et al. (2013) reported that, most of the sesame diseases can be minimized by soil application of neem cake (250 kg/ha) along with seed treatment and soil application (2.5 kg/ha) of *Trichoderma viride* followed by foliar spray of azadirachtin @ 3 ml/l at 30 and 45 DAS.

Spray Mancozeb 2.5 gm/l in severe condition two spraying of bayleton 1.0 gm/l at 20 days interval is effective for managing the disease.

14.7 BACTERIAL BLIGHT

Xanthomonas axenopodis p. v. *sesami.*

14.7.1 SYMPTOM

Plants may be affected at all growth stages. The disease is characterized water soaked, small, and irregular spots are formed on the leaves which later increases and turn brown, under favorable conditions leaves become dry and brittle, severely infected leaves defoliate.

14.7.2 CAUSAL ORGANISM

Bacterium *Xanthomonas axenopodis* pv. *sesami.*

The bacterium is rod shape, motile by one polar flagellum. Colony color is yellowish white.

14.7.3 SURVIVAL AND SPREAD

Seed can carry the pathogen up to a period of 16 months. A weed plant, *Acanthospermum hispidum*, is observed to be susceptible against *X. axenopodis* pv. *sesami*. This host acts as a source of survival of the bacterium in its dried leaves from year to year. The bacterium does not survive in sesame plant debris in soil from year to year. The bacterium enters the host primarily through stomata and quickly becomes vascular. The secondary spread is by spattering rains. High temperature and humidity favor the disease. Seedling infection of sesame is most severe at soil temperature of 20°C. Infection does not take place when soil temperature is 40°C. The disease also becomes severe when the soil moisture is 30–40% and relative humidity (RH) is 75–87%.

14.7.4 MANAGEMENT

Crop rotation with non-host plants helps in reducing the build of primary inoculum. Three sesame genotypes SG-34, SG-22, T-58, and Sg-55 are resistant against bacterial blight. Chemical or antibiotic seed treatment or hot-water treatment of seed and antibiotic sprays to check secondary spread. Streptocycline seed treatment for 2 h followed by three sprays at 10-day interval of streptocycline plus copper oxychloride effectively controls the disease or Foliar spray of Streptocycline (500 ppm) as soon as symptoms are observed in the plants. Repeat the second spraying at 15 days interval (Chattopadhyay et al., 2016).

KEYWORDS

- **alternaria leaf spot**
- **bacterial blight**
- **cercospora leaf spot**
- **charcoal rot**
- ***Macrophomina phaseolina***
- **phyllody**

REFERENCES

Abdel-Salam, M. S., El-Halim, M. M. A., & El-Hamshary, Q. I. M., (2007). Enhancement of *Enterobacter cloacae* antagonistic effects against the plant pathogen *Fusarium oxysporium*. *J. Appl. Sci. Res., 3*, 848–852.

Abdou, E., Abd-Alta, H. M., & Galal, A. A., (2001). Survey of sesame root rot/wilt disease in Minia and their possible control by ascorbic and salicylic acids. *Assiut. J. Agric. Sci., 1*(23), 135–152.

Akhtar, K. P., Sarwar, G., & Arshad, H. M. I., (2011). Temperature response, pathogenicity, seed infection, and mutant evaluation against *Macrophomina phaseolina* causing charcoal rot disease of sesame. *Arch. Phytopathol. Plant Prot., 44*, 320–330.

Akhtar, K. P., Sarwar, G., Dickinson, M., Ahmad, M., Haq, M. A., Hameed, S., & Iqbal, M. J., (2009). Sesame phyllody disease: Symptomatology, etiology and transmission in Pakistan. *Turk J. Agric., 33*, 477–486.

Al-Sakeiti, M. A., Al-Subhi, A. M., Al-Saady, N. A., & Headman, M. L., (2005). First report of witches'-broom disease of sesame (*Sesamum indicant*) in Oman. *Plant Dis., 89*(5), 530.

Anandh, G. V., & Selvanarayanan, V., (2005). Reaction of sesame germplasm against shoot Webber *Antigastra catalaunalis* Duponchel and phyllody. *Indian J. Plant Prot*., *33*, 35–38.

Anonymous, (2004). Agricultural data. In: "*Agricultural Statistics Databases.*" Organization of the United Nations, Rome, Italy. http://faostat. fao. org (accessed on 13 January 2020).

Anwar, M., Hasan, E., Bibi, T., Mustafa, H. S. B., Mahmood, T., & Ali, M., (2013). TH-6: A high yielding cultism of sesame released for general cultivation in Punjab. *Life Sci*., *1*(1), 44–57.

Armstrong, J. K., & Armstrong, G. M., (1950). A *Fusarium* wilts of sesame in United States. *Phytopathology, 40*, 785.

Bakhshi, A., Elebarian, H. R., Aminian, B. A., & Ebrahimi, M., (2010). A study on genetic diversity of some isolates of *Macrophornina phaseolina* using molecular marker; PCR-RFLP and RAPD. *Iranian J. Plant Prot. Sci., 41*, 103–112.

Basirnia, T., & Banihashcmi, Z., (2006). Seed transmission of in *Sesamum indicum* in Fars province. *Iranian J. Plant Pathol*., *42*, 117–123.

Basirnia, T., & Banihashemi, Z., (2005). Vegetative compatibility grouping in *Fusarium oxysporum* f. sp. *sesami* the causal agent of sesame yellows and will in Pars province. *Iranian J. Plant Pathol., 41*, 243–255.

Bayounis, A. A., & Al-Sunaidi, M. A., (2008a). Efficacy of some plant powders in protecting Sesame seeds against *Macrophomina phaseolina* in greenhouse. *Univ. Aden. J. Nat. Appl. Sci*., *12*, 233–243.

Bayounis, A. A., & Al-Sunaidi, M. A., (2008b). Effect of some plant extracts on growth inhibition of *Macrophomina phaseolina. Univ. Aden J. Nat. Appl. Sci., 12*, 469–480.

Bertaccni, A., & Duduk, B., (2009). Phytoplasma and phytoplasma diseases: A review of recent research. *Phytopathol Mediterr*., *48*, 355–378.

Butler, E. J., (1918). *Fungi and Disease in Plants*. Thacker Spink and Co, Calcutta, India.

Cardona, R., & Rodriguez, H., (2002). Evaluation of *Trichoderma harzianum* for the control of *Macrophomina phaseolina* on sesame. *Fitopatol. Venez*., *15*, 21–23.

Cardona, R., & Rodriguez, H., (2006). Effects of *Trichoderma harzianum* fungus on the incidence of the charcoal rot disease on sesame. *Rev. Fac. Agron. Univ. Zulia, 23*, 44–50.

Cardona, R., (2008). Effect of green manure and *Trichoderma harzianum* on sclerotia population and incidence of *Macrophomina phaseolina* on sesame. *Rev. Fac. Agron. Univ. Zulia, 25*, 440–454.

Chattopadhyay, C., & Sastry, R. K., (2001). Potential of soil solarization in reducing stem-root rot incidence & increasing seed yield of sesame. *J. Mycol. Plant Pathol*., *31*, 227–231.

Chattopadhyay, C., & Sastry, R. K., (2002). Combining viable disease control tools for management of sesame stem-root rot caused by *Macrophomina phaseolina* (Tarsi) Goid. *Indian J. Plant Prot., 30*, 132–138.

Chattopadhyay, C., Kolte, S. J., & Waliyar, F., (2016). *Diseases of Edible Oilseed Crops* (pp 232–293). CRC Press, Boca Raton, Florida.

Cho, E. K., & Choi, S. H., (1987). Etiology of a half stem rot in sesame caused by *Fusarium oxysporum*. *Korean J. Plant Prot*., *26*, 25–30.

Choi, S. H., Cho, E. K., & Chae, Y. A., (1987). An evaluation method for sesame (*Sesamum indicum* L.) resistance to *Phytophthora nicotianae* var. *parasitica*. *Korean J. Crop Sci., 32*, 173–180.

Choi, S. H., Cho, E. K., & Cho, W. T., (1984). Epidemiology of sesame phytophthora blight in different cultivation types. *Res. Rep. Rural Dec. Korea Republic, 26*, 64–68.

Chung, B. K., & Hong, K. S., (1991). Biological control with *Streptomyces* spp. of *Fusarium oxysporum* f. sp. *vasinfectum* causing sesame wilt and blight. *Korean J. Weal., 19*, 231–237.

Chung, B. K., & Ser, S. O., (1992). Identification of antagonistic *Streptomyces* sp. on *Phytophthora nicotianae* var. *parasitica* and *Fusarium oxysporum* f. sp. *vasinfectum* causing sesame wilt and blight. *Korean J. Myco., 20*, 65–71.

Cousin, M. T., Kartha, K. K., & Delattre, (1971). Sur to presence d'organisms de type mycoplasme dans tubes cribles de *Sesamum orientale* L. atteient de phyllodie. *Rev. Plant Pathol., 50*, 430.

Dandnaik, B. P., Shinde, S. V., More, S. N., & Jangwad, N. P., (2002). Reaction of sesame lines against phyllody. *J. Maharashtra Agric. Univ., 27*, 233.

Deepthi, P., Shukla, C. S., Verma, K. P., & Sankar, R. E., (2014). Yield loss assessment and influence of temperature and relative humidity on charcoal rot development of sesame (*Sesamum indicum* L.). *Bio Scan, 9*, 193–195.

Dubey, A. K., Gupta, S., & Singh, T., (2011). Induced vivipary in *Sesamum indicum* L. by seed borne infection of *Phytophythora parasitica* var. *sesame*. *Indian J. Fundam. Appl. Life Sci., 1*, 185–188.

El-Bramawy, M. A. E. S. A., & Embaby, H. E., (2011). Anti-nutritional factors as screening criteria for some diseases resistance in sesame (*Sesamum indicum* L.) genotypes. *J. Plant Breed. Crop Sci., 3*, 352–366.

El-Bramawy, M. A. E. S., El-Hendawy, S. E., & Shaban, W. I., (2009). Assessing the suitability of morphological and phenological traits to screen sesame genotypes for *Fusarium* wilt and char coal disease resistance. *Plant Prot. Sci., 45*, 49–58.

Elewa, L. S., Mostafa, M. H., Sahab, A. F., & Ziedan, E. H., (2011). Direct effect of biocontrol agents on wilt and root-rot diseases of sesame. *Arch. Phytopathol. Plant Prot., 44*, 493–504.

El-Fiki, A. I. I., El-Deeb, A. A., Mohamed, F. G., & Khalifa, M. M. A., (2004b). Controlling sesame charcoal rot incited by *Macrophomina phaseolina* under field conditions by using the resistant cultivars and some seed and soil treatments. *Egypt. J. Phytopathol., 32*, 103–118.

El-Fiki, A. I. I., El-Deeb, A. A., Mohamed, F. G., & Khalifa, M. M. A., (2004a). Some applicable methods for controlling sesame charcoal rot disease (*Macrophomina phaseolina*) under greenhouse conditions. *Egypt. J. Phytopathol., 32*, 87–101.

Enikuomehin, O. A., (2010). Seed sorting of sesame (*Sesamum indicum* L.) by salt density and seed-borne fungi control with plant extracts. *Arch. Phytopathol. Plant Pmt., 43*, 573–580.

Enikuomehin, O. A., Aduwo, A. M., Olowe, V. I. O., Popoola, A. R., & Oduwaye, A., (2010). Incidence and severity of foliar diseases of sesame (*Sesamum indicum* L.) intercropped with maize (*Zea mays* L.). *Arch. Phytopathol. Plant Prot., 43*, 972–986.

Ibrahim, M. E., & Abdel-Azeem, A. M., (2007). Towards an integrated control of sesame (*Sesamum indicate* L.) charcoal rot caused by *Macrophomina phaseolina*. *Arab Univ. J. Agric. Sci., 15*, 473–481.

Jeyalakshmi, C., Rettinassababady, C., & Nema, S., (2013). Integrated management of sesame diseases. *Journal of Biopesticides, 6*(1), 68–70.

Jin, U. H., Lee, J. W., Chung, Y. S., Lee, J. H., Yi, Y. B., Kim, Y. K., et al., (2001). Characterization and temporal expression of a w-6 fatty acid desaturate cDNA from sesame (*Sesamum indicum* L.) seeds. *Plant Sci., 161*, 935–941.

John, P., Tripathi, N. N., & Kumar, N., (2010). Efficacy of fungicides against charcoal rot of sesame caused by *Rhizoctonia bataticola* (Taub.) Butler. *Res. Crops, 11*, 508–510.

Kalita, M. K., Pathak, K., & Barman, U., (2002). Yield loss in sesamum due to phytophthora blight in Barak Valley Zone (BVZ) of Assam. *Ann. Biol., 18*, 61, 62.

Kayak, H., & Boydak, E., (2011). Trends of sudden wilt syndrome in sesame plots irrigated with delayed intervals. *Af J. Microbial. Res., 5*, 1837–1841.

Khan, A. J., Bottner, K. Al-Saadi, N., Al-Subhi, A. M., & Lee, I. M., (2007). Identification of phytoplasma associated with witches' broom and virescence diseases of sesame in Oman. *Bull. Insectol., 60*, 133–134.

Kolte, S. J., (1985). *Diseases of Annual Edible Oilseed Crops* (Vol. II, pp. 83–109). Rapeseed-mustard and sesame diseases. CRC Press, Boca Raton, London, New York.

Kumar, P., & Mishra, U. S., (1992). Diseases of *Sesamum indicum* in Rohilkhand: Intensity and yield loss. *Indian Phytopathol., 45*, 121, 122.

Liu, L. M., Liu, H. Y., & Tian, B. M., (2012). Selection of reference genes from sesame infected by *Macrophomina phaseolina. Acta Agron. Sin.38*, 471–478.

Liu, Z. H., Huang, X. Y., Liu, Yang, M. H., Sun, J., & Zhao, Z. H., (2010). Biological characteristics of *Fusarium oxysporum* of the sesame stem blight causal agent. *J. Shenyang Agric. Univ., 41*, 417–421.

Maheshwari, D. K., Dubey, R. C., & Aeron, A., (2012). Integrated approach for disease management and growth enhancement of *Sesamum indicum* L. utilizing *Azotobacter chroococcum* TRA2 and chemical fertilizer. *World J. Microbiol. Biotechnol., 28*, 3015–3024.

Maiti, S., Hegde, S. R., & Chattopadhyay, S. B., (1988). *Handbook of Annual Oilseed Crops* (pp. 109–137). Oxford and IBH Publishing Co. Pvt. Ltd., New Delhi, India.

Mandizadeh, V., Safaie, N., & Goltapeh, E. M., (2012). Genetic diversity of sesame isolates of *Macrophomina phaseolina* using RAPD and ISSR markers. *Trakia J. Sci., 10*, 65–74.

Martinez-Hilders, A., & Laurentin, H., (2012). Phenotypic and molecular characterization of *Macrophomina phaseolina* (Tassi) Goid. coming from the sesame production zone in Venezuela. *Bioagrn, 24*, 187–196.

Martinez-Hilders, Mendoza, A., Y. Peraza, D., & Laurentin, H., (2013). Genetic variability of *Macrophomina phaseolina* affecting sesame: Phenotypic traits, RAPD markers, and interaction with crop. *Res. J. Recent Sci., 2*, 110–115.

Mathur, Y. K., & Verma, J. P., (1972). Relation between date of sowing and incidence of sesamum phyllody and abundance of its cicadellid vector. Indian). *Entomol., 34*, 74, 75.

Melean, A. S., (2003). Resistance of white seeded sesame (*Sesamum indicam* L.) cultivars against charcoal rot (*Macrophomina phaseolina*) in Venezuela. *Sesame Safflower News, 18*, 72–76.

Misra, H. P., (2003). Efficacy of combination insecticides against tit leaf Webber and pod borer, *Antigastra catalaunalis* (Dupon.), and phyllody. *Ann. Plant Prot. Sci., 11*, 277–280.

Mohanty, N. N., & Behera, B. C., (1958). Blight of sesame (*Sesame orientate* L.) caused by *Alternaria sesami* (Kawamura). *Comb. Cure Sci., (India), 27*, 492.

Moi, S., & Bhattacharyya, P., (2008). Influence of biocontrot agents on sesame mot rot. *J. Mycopathol. Res., 46*, 97–100.

Mudingotto, P. J., Adipala, E., & Mathur, S. B., (2002). Seed-borne mycoflora of sesame seeds and their control using salt solution and seed dressing with Dithane M-45. *Muarik Bull.*, *5*, 35–43.

Munoz-Cabanas, R. M., Hernandez-Delgado, S., & Mayek-Perez, N., (2005). Pathogenic and genetic analysis of *Macrophomina phaseolina* (Tassi) Goid. on different hosts. *Re, Mexi. Fitopatol.*, *23*, 11–18.

Nagaraju, & Muniyappa, V., (2005). Viral and phytoplasma diseases of sesame. In: Saharan, G. S., Mehta, N., & Sangwan, M. S., (eds.), *Diseases of Oilseed Crops* (pp. 304–317). Indus Publishing Co, New Delhi, India.

Narayanaswamy, R., & Gokulakumar, B., (2010). ICP-AES analysis in sesame on the root rot disease incidence. *Arch. Phytopathol. Plant Prot.*, *43*, 940–948.

Ojiambo, P. S., Ayiecho, P. O., Narla, R. D., & Mibey, R. K., (2000). Tolerance level of *Altemaria sesami* and the effect of seed infection on yield of sesame in Kenya. *Exp. Agric.*, *36*, 335–342.

Sagar., P., Sagir, A., & Sopt, T., (2009). The effect of charcoal rot disease (*Macrophomina phaseolina*), irrigation, and sowing date on oil and protein content of some sesame lines. *J. Turk. Phytopathol.*, *38*, 33–42.

Sangle, U. R., & Bambawale, O. M., (2004). New strains of *Trichoderma* spp. strongly antagonistic against *Fusarium oxysporum* f. sp. *sesame*. *J. Myco. Plant Pathol.*, *34*, 107–109.

Saravanan, S., & Nadarajan, N., (2005). Pathogenicity of mycoplasma like organism of sesame (*Sesamum indicum* L.) and its wild relatives. *Agric. Sri. Dig.*, *25*, 77, 78.

Sarwar, G., & Akhtar, K. P., (2009). Performance of some sesame mutants to phyllody a leaf curl virus disease under natural field conditions. *Pak. J. Phytopathol.*, *21*, 18–25.

Sehgal, S. P., & Prasad, N., (1966). Variation in the pathogenicity of single-zoospore isolates of sesamun phytophthora: Studies on the penetration and survival. *Indian Phytopathol.*, *19*, 154–158.

Sehgal, S. P., & Prasad, N., (1971). Instability of pathogenic characters in the isolates of sesamum phytophthora and effect of host passage on the virulence of isolates. *Indian Phytopathol.*, *24*, 295–298.

Selvanarayanan, V., & Selvamuthukumaran, T., (2000). Field resistance of sesame cultivars against phyllody disease transmitted by rosins albicinctus distant. *Sesame Safflower Newsl.*, *15*, 71–74.

Shah, R. A., Pathan, M. A., Jiskani, M. M., & Qureshi, M. A., (2005). Evaluation of different fungicides against *Macrophomina phaseolina* causing root rot disease of sesame. *Pak. J. Agric. Eng. Vet. Sci.*, *21*, 35–38.

Shalaby, I. M. S., El-Ganainy, A. M. A., Botros, S. A., & El-Gebally, M. M., (2001). Efficacy of same natural and synthetic compounds against charcoal rot caused by *Macrophomina phaseolina* of sesame and sunflower plants. *Assiut J. Agric. Sci.*, *32*, 47–56.

Shalaby, O. Y. M., & Bakeer, A. T., (2000). Effect of agricultural practices on root rot and will of sesame in Fayoum. *Ann. Agric. Sci. Moshtohor.*, *38*, 1399–1407.

Shekharappa, G., & Patil, P. V., (2001). Chemical control of leaf blight of sesame caused by *Alternaria sesami*. *Karnataka J. Agric. Sci.*, *14*, 1100–1102.

Shim, K. B., Kim, D. H., Park, J. H., Lee, S. W., Kim, K. S., & Rho, J. H., (2012). A new black sesame variety Kanghcuk' with lodging and phytophthora blight disease resistance, and high yielding. *Korean J. Breed. Sci.*, *44*, 38.

Shindhe, G. G., Lokesha, R., Naik, M. K., & Ranganath, A. G. R., (2011). Inheritance study on phyllody resistance in sesame (*Sesamum indicum* L.). *Plant Arch.*, *11*, 775, 776.

Sidawi, A., Abou Ammar, G., Alkhider, Z., Arifi, T., Alsaleh, E., & Alalees, S., (2010). Control of sesame will using medicinal and aromatic plant extracts. *Julius-Kuhn-Archie, 428*, 117.
Singh, A. K., Gopala, R. A., Goel, S., & Rao, G. P., (2018). Identification of '*Candidatus* phytoplasma asteris' causing sesame phyllody disease and its natural weed host in Jammu, India. *Indian Phytopathology.* https://doi. org/10.1007/s42360–018–0013–3 (accessed on 28 January 2020).
Singh, P. K., Akram, M., Vajpeyi, M., Srivastava, R. L., Kumar, K., & Naresh, R., (2007). Screening and development of resistant sesame varieties against phytoplasm. *Bull. Insectol., 60*, 303, 304.
Su, Y. L., Miao, H. M. Wei, L. B., & Zhang, H. Y., (2012). Study on separation and purification techniques of *Fusarium oxysporum* in sesame (*Sesamum indicum* L.). *J. Henan Agric. Sri., 41*, 92–95.
Subrahmaniyan, K., Dinakaran, D., Kalaiselven, P., & Arulmozhi, N., (2001). Response of root rot resistant cultures of sesame (*Sesamum indicum* L.) to plant density and NPK fertilizer. *Agric. Sci. Dig., 21*, 176–178.
Thiyagu, K., Kandasamy, G., Manivannan, N., Muralidharan, V., & Manoranjitham, S. K., (2007). Identification of resistant genotypes to root rot disease (*Macrophomina phaseolina*) of sesame (*Sesamum indicum* L.). *Agric. Sri. Dig.*, *27*, 34–37.
Verma, M. L., & Bajpai, R. P., (2001). Effect of bioinoculants on phytophthora blight and macrophomina root/stem rot of sesame (*Sesamum indicum* L.). Abstr. In: *National Symposium on Plant Protection Strategies for Sustainable Horticulture* (p. 113). Society of plant protection science, SKAUAT, Jammu, India.
Verma, M. L., (2002). Effect of soil types on diseases and yield of sesame (*Sesamum indicum* L.). Abstr. In: *Asian Congress of Mycology and Plant Pathology and Symposium on Plant Health for Food Security* (p. 53). University of Mysore, Mysore, India.
Verma, M. L., Mehta, N., & Sangwan, M. S., (2005). Fungal and bacterial diseases of sesame. In: Saharan, G. S., Mehta, N., & Snwan, M. S., (eds.), *Diseases of Oilseed Crops* (pp. 209–303). Indus Publishing Co., New Delhi, India.
Wang, L. H., Zhang, Y. X., & Li, D. H., (2011). Variations in the isolates of *Macrophomina phaseolina* from sesame in China based on amplified fragment length polymorphism (AFLP) and pathogenicity. *Afr J. Microbial. Res., 5*, 5584–5590.
Weintraub, P. G., & Jones, P., (2010). *Phytoplasmas: Genomes, Plant Hosts and Vectors* (p. 147). CAB International, Oxford shire, U. K.
Win, N. K. K., Back, C. G., & Jung, H. Y., (2010). Phyllody phytoplasma infecting sesame (*Sesamum indicum*) in Myanmar. *Trop. Plant Pathol.*, *35*, 310–313.
Zhao, H., Miao, H. M., Gao, H. T., Ni, Y. X., Wei, L. B., & Liu, H. Y., (2012). Evaluation and identification of sesame germplasm resistance to *Macrophomina phaseolina*. *J. Henan Agric. Sci., 41*, 82–87.
Ziedan, E. H., Elewa, L. S., Mostafa, M. H., & Sahab, A. F., (2011). Application of mycorrhizae for controlling root diseases of sesame. *J. Plant Prot. Res.*, *51*, 355–361.
Ziedan, E. H., Mostafa, M. H., & Elewa, I. S., (2012). Effect of bacterial inoculation *Fusarium oxysporum* f. sp. *sesami* and their pathological potential on sesame. *J. Agric. Technol.*, *8*, 699–709.

CHAPTER 15

Sunflower *(Helianthus annuus* L.) Diseases and Their Management by Integrated Approach

KOTHAKOTA VENKATARAMANAMMA,[1] LINGAN RAJENDRAN,[2] and C. GOPALAKRISHNAN[2]

[1]*Scientist (Plant Pathology), Regional Agricultural Research Station, ANGRAU, Nandyal–518502, Andhra Pradesh, India*

[2]*Department of Plant Pathology, TNAU, Coimbatore–3, Tamil Nadu, India, E-mail: rucklingraja@gmail. com (L. Rajendran)*

Sunflower (*Helianthus annuus* L.) is the most important edible oil seed crop in world occupying third place, grown in an area of 2.4 million hectares in India with a production and productivity of 1.44 million tonnes and 608 kg/ha. Large-scale cultivation has started with the introduction of high yielding open pollinated varieties. Among the states, the four namely Andhra Pradesh, Karnataka, Maharashtra, and Tamil Nadu contribute 90% of total area (Chander rao et al., 2015). The crop is infected by many fungal, bacterial, and viral diseases (Gulya et al., 1994) with an average yield loss of 25–40% (Shankergoud et al., 2006). Among them, the following are the most important diseases cause drastic yield losses under field conditions.

15.1 FUNGAL DISEASES

15.1.1 ALTERNARIA LEAF SPOT OR BLIGHT

15.1.1.1 INTRODUCTION/ECONOMIC IMPORTANCE

It is widely distributed wherever sunflower grown and is considered as a most destructive one. The leaf spot prevalent in North America, Africa, Argentina,

Brazil, Japan, Australia, India, Serbia, and Montenegro (former Yugoslavia), Romania, and France (Anahosur, 1978; Zimmer and Hoes, 1978; Davet et al., 1991; Pereyra and Escande, 1994). Due to this disease, seed yield (27–80%) and oil yield (17–35%) reduction has been reported. Further, a negative correlation between disease intensity and seed yield, oil content has been observed. The parameters, *viz.*, number of seeds per head and seed yield per plant are mostly affected and also affects the quality of seeds by affecting germination and initial vigor of the seedlings.

15.1.1.2 SYMPTOMS

The symptom appears in all stages in varieties and hybrids and starts to appear twenty days after sowing (DAS) on leaves further spreads to petiole, stem, bracts, and flower heads. Initial symptoms of the disease appears in the form of small spots, circular to oval, ranging from 0.2 to 0.5 mm in diameter having yellow halos in the older leaves. This enlarges in size having concentric rings in the center and becomes irregular in shape. Stem lesions appear as long, narrow lesions which coalesce to form large blackened areas resulting in stem breakage. It causes severe premature defoliation (Anahosur, 1978; Almeida et al., 1981; Davet et al., 1991). The disease reduces the seed yield (27–80%) and oil yield (17–33%) (Balasubramanaym and Kolte, 1980). High humidity and moderate to warm temperatures favors the disease.

Being a seed borne disease, it reduces the seed germination, initial vigor of plants and seedling survival.

15.1.1.3 CASUAL ORGANISM

Alternaria helianthi/Alternaria alternata/Alternariaster helianthi

The fungus *Alternaria helianthi* produces pale grey-yellow colored cylindrical conidiophores which are straight or curved, geniculate, simple or branched, septate, and bear single conidium (straight or slightly curved, cylindrical to long ellipsoid, multi septate, pale grey-yellow to pale brown).

15.1.1.4 DISEASE CYCLE

The fungus survives on seed, infected plant debris as primary inoculum and wind born conidia act as a secondary spread.

FIGURE 15.1a Leaf spot symptoms on 20 days crops.

FIGURE 15.1b Leaf spot symptom on mature leaf.

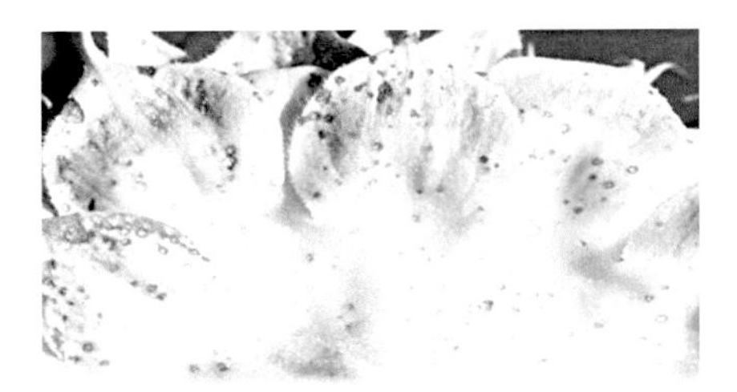

FIGURE 15.1c Leaf spot symptoms on bracts.

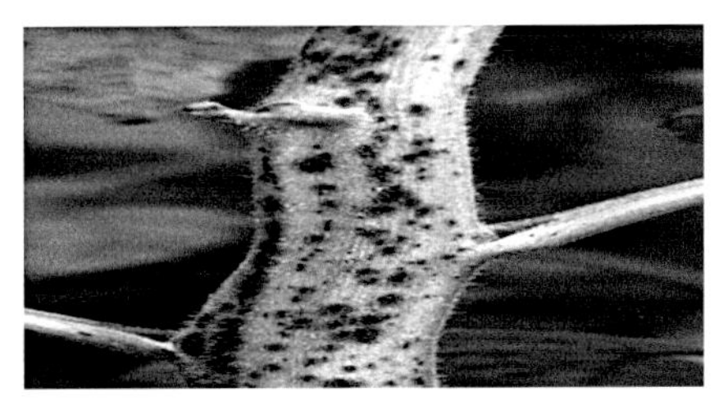

FIGURE 15.1d Leaf spot symptoms on stem.

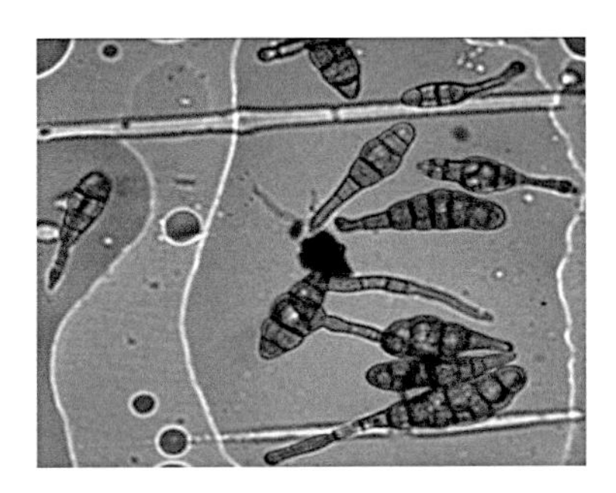

FIGURE 15.1e Microscopic view of conidia *of Alternaria alternata.*

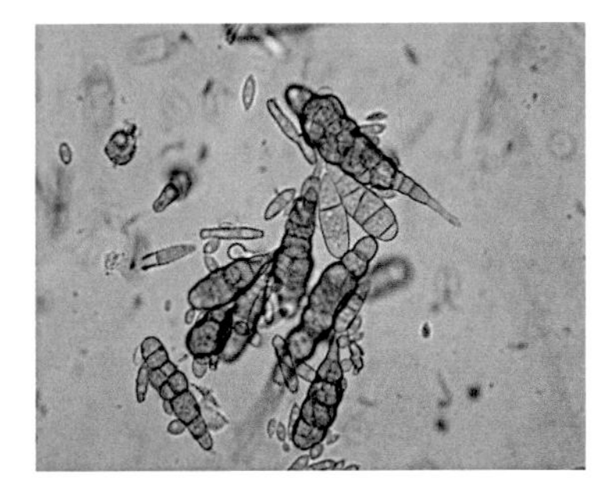

FIGURE 15.1f Microscopic view of conidia of *Alternaria helianthi.*

15.1.1.5 EPIDEMIOLOGY

Hot weather and frequent rain during the flowering and seed filling stages favor the infection.

15.1.1.6 MANAGEMENT

- Selection of healthy seeds and seed treatment with the fungicide namely iprodione 25% + carbendazim 25% or Mancozeb 63% + carbendazim 12% @ 2 g per kg of seed during the sowing
- Providing adequate spacing (60 x 30 cm or 45 x 30 cm) and the crop rotation may be followed in the continuous cropping
- Foliar application of iprodione 25% + carbendazim 25% @ 0.2% or Propiconazole 25% EC @ 0.1% or mancozeb @ 2.5 g/l for two times at 30 and 45 DAS.

15.1.2 POWDERY MILDEW

15.1.2.1 INTRODUCTION/ECONOMIC IMPORTANCE

It is a more common disease and occurs under dry conditions towards the end of winter months during flowering or seed filling stage. It is distributed worldwide, but occurs in greater intensity in tropical areas (Zimmer and Hoes, 1978; Gulya et al., 1997).

15.1.2.2 SYMPTOMS

The disease appears as white to grey powdery patches (white mold) on the upper surfaces of older leaves progresses from lower to upper leaves. These patches enlarge, coalesce, and cover most of the plant parts and turn grey in color during disease progress. When the crop reaches maturity, fungus forms sexual reproductive bodies (chasmothecium) and appears as dark dots, reduces seed yield. This disease is severe in both rainfed and irrigated crops. Yield loss was found to be proportional to the disease severity and stage of the crop (Diaz Franco, 1983; Zimmer and Hoes, 1978; Almeida et al., 1981).

15.1.2.3 CASUAL ORGANISM

Erysiphe cichoracearum

The fungus produces hyaline, ectophytic mycelium and conidiophores are short, bears conidia in chain. These are oval or barrel shaped, single celled and hyaline. The chasmothecium are dark, globose with the hyaline or pale brown simple, hypha like appendages. The asci are ovate and produce ascospores, which are thin walled, elliptical, and pale brown in color (Kapoor, 1967).

(a) (b)

FIGURE 15.2 (a) Plant affected with powdery mildew. (b) Powdery mildew symptoms as patches on leaf.

15.1.2.4 DISEASE CYCLE

The fungus survives on plant debris and produce resting structure known as chasmothecium. The next spring, sexual spores (ascospores) are released from the chasmothecium and carried by air current to leaves of plant where new infection will start. The primary infection is by the ascospores and the secondary spread is through wind-borne conidia.

15.1.2.5 EPIDEMIOLOGY

The temperature (25°C) and relative humidity (RH) (95%) are optimum conditions for the infection. Cool temperature with low RH is reported to cause severe epidemics (Kapoor, 1967; Zimmer and Hoes, 1978; Kolte, 1985).

15.1.2.6 MANAGEMENT

- Avoiding high humidity area and having full sunlight most of the day.
- Providing morning irrigation and maintaining optimum population.

- Following crop rotation and over dose application of nitrogen fertilizer may be avoided.
- Application of wettable sulfur 80 WP @ 3 g or Dinocap 48% EC@ 1 ml or Difenconazole @ 0.05% for two times at 45 and 60 DAS.

15.1.3 DOWNY MILDEW

15.1.3.1 INTRODUCTION/ECONOMIC IMPORTANCE

It is a major disease potentially very destructive originating in North America, with the movement of sunflower around the world, the pathogen is currently endemic in all sunflower growing areas (Gulya et al., 1997). Most countries have specific regulations to prevent the introduction or spread of the pathogen (Pereyra and Escande, 1994). In India disease is reported from Maharashtra, Karnataka, and Andhra Pradesh and reported to reduce the seed yield ranging from 5 to 60%.

15.1.3.2 SYMPTOMS

This pathogen can produce different types of symptoms depending on the plant age, genotype, and humidity, temperature conditions and amount of inoculum prevailing (Zimmer and Hoes, 1978; Almeida et al., 1981; Davet et al., 1991; Pereyra and Escande, 1994; Gulya et al., 1997). The affected plants are stunted during the early stage of infection and produce white fungal growth on the lower surface of the leaves corresponding upper surface become chlorotic, yellow discoloration and covers large area leads to systemic infection. Further, the flower heads becomes sterile, stiff, and face upwards without the seed formation.

15.1.3.3 CASUAL ORGANISM

Plasmopara halstedii

It is obligate parasite, which is systemic produces mycelium (intercellular) and sporangiophore and sporangia arise on the lower side of the leaves. The sporangia are thin tending to release biflagellate zoospores (Zimmer and Hoes, 1978; Davet et al., 1991).

FIGURE 15.3 Downy mildew symptom.

15.1.3.4 DISEASE CYCLE

The primary source of inoculum is by sporangia, oospore, and the secondary spread is through wind borne sporangia and zoospores. The sexual spore (oospore) acts as a resting spore.

15.1.3.5 EPIDEMIOLOGY

The temperature (15–18°C), RH (more than 95%) with high rainfall conditions favors the disease (Davet et al., 1991).

15.1.3.6 MANAGEMENT

- Pre sowing irrigation followed by one irrigation at 10 or more DAS.
- Rouging the infected seedlings, removal, and destruction.
- Seed treatment with metalaxyl-M 31.8% ES @ 2 g/kg and metalaxyl 35% WS 6 g/kg of seed.
- Foliar pray of Ridomil MZ@ 2.5 ml/l twice at 30 and 45 DAS.

15.1.4 CHARCOAL ROT

15.1.4.1 INTRODUCTION/ECONOMIC IMPORTANCE

It is a soil-borne disease of many crops in arid and semiarid areas in warmer region.

15.1.4.2 SYMPTOMS

It usually begins to appear after flowering. The first symptoms are wilting of the plant during the hot sunny hours followed by a recovery in the evenings as temperatures decline. A gray discoloration at the base of the stalk of infected plants. The pith decays internally leaving only the water conducting vascular bundles. This gives the internal stem a shredded appearance. Further, the vascular bundles may cover with small, black sclerotia. The disease restricts the flow of water and nutrients, reduced seed size and light test weight usually happen. The disease can be economically important in the High Plains during hot, dry seasons.

FIGURE 15.4 Diseased stalk due to charcoal rot.

15.1.4.3 CASUAL ORGANISM

Macophomina phaseolina (Tassi) Goid.

15.1.4.4 DISEASE CYCLE

The sunflower root comes into contact with small sclerotia like bodies which infect the plant. The fungus overwinters in the soil but can also be seed borne.

15.1.4.5 EPIDEMIOLOGY

High soil temperature and moisture stress favors the disease.

15.1.4.6 MANAGEMENT

- Crop rotation with cotton for 2 to 3 years.
- Early season and short duration cultivars may be chosen for sowing.
- Irrigate during the hot and dry season to avoid water stress.
- Collect and destroy residues of past crops before planting for next season.
- Spot drenching with Carbendazim @ 1 g/l.
- Seed treatment with *Trichoderma viride* @ 4 g/kg or *Pseudomonas fluorescens* 10 g/kg of seed.
- Soil application of native *Pseudomonas fluorescens* (Pf1) or *Trichoderma viride* (Tv1) –2.5 kg/ha + 500 kg of well decomposed FYM or vermicompost or sand at 30 DAS.

15.1.5 HEAD ROT

15.1.5.1 INTRODUCTION/ECONOMIC IMPORTANCE

Head rot generally affects the crop when there is intermittent rain or drizzling during the crop heading stage. Almost total loss may result from this disease because of poor filling and loss of seeds.

15.1.5.2 SYMPTOMS

The initial symptom appears as brown, irregular water soaked spots on the back of ripening head, usually adjacent to the flower stalk. They become big in size and become soft and pulpy and covered by a loose greyish fungal spore mass. It leads to rotting of sunflower heads and drop off (Mishra et al., 1972) and seeds are transformed into a black, powdery mass. Injury to the flower head is necessary for infection.

Heliothis larvae are reported to pre dispose for infection due to creation of small wounds. It is important in wet weather and causes significant yield losses and in warm humid weather, the disease spread is rapid. The most typical disease symptoms include rotting of sunflower head with a loose cover of greyish fungal spore mass.

FIGURE 15.5 Head rot symptom.

15.1.5.3 CASUAL ORGANISM

Rhizopus spp. (mostly *Rhizopus arrhizus*)

It produces many aerial stolons and rhizoids with dark brown or black colored, non-septate hyphae. Sporangia are globose and black in color with a central columella having aplanate, dark colored, and ovoid sporangiospores.

15.1.5.4 DISEASE CYCLE

Primary inoculum is by infected plant debris and secondary spread is through wind born conidia.

15.1.5.5 EPIDEMIOLOGY

The damage caused by insects and caterpillars along with the prolonged rainy weather at flowering stage.

15.1.5.6 MANAGEMENT

- Avoid caterpillar infestation on heads.
- Foliar application of Fenthion 1 ml/l and wettable sulfur @ 3 g/l of water or mancozeb @ 2.5 g/l at 10 days interval.

15.1.6 SCLEROTIUM WILT OR COLLAR ROT

15.1.6.1 INTRODUCTION/ECONOMIC IMPORTANCE

This is most important pathogen in the world and is distributed under varied climate, *viz*., temperate, tropical or subtropical (Gulya et al., 1997). The losses depend on the part of the plant affected and quickly kill the infected plants at seedling stage. Seeds of infected head can fall, resulting in total loss of production (Zimmer and Hoes, 1978; Davet et al., 1991).

15.1.6.2 SYMPTOMS

Rotting starts when the mycelium of the fungus, originating from sclerotia comes in contact with the lateral roots. The first symptom observed is a sudden wilting of the plant. The infected plant can recover turgidity after a rainfall, but within a few days, this symptom becomes irreversible. A light brown and soft injury appears at ground level, the lesion may be covered by white mycelium. The fungus develops internally and destroys the internal tissues of the stem. Many sclerotia are found within the colonized portion of the stem, but few are found in the root and in the outdoor area. Diseased plants can lodge easily. A white mycelium can cover the lesion, and sclerotia are observed within the stem. Plants can break at the lesion site (Punja, 1985).

15.1.6.3 CASUAL ORGANISM

Sclerotium rolfsii.

15.1.6.4 DISEASE CYCLE

The fungus survives as sclerotia in soil and plant debris. The secondary spread occurs through sclerotia by implements and irrigation water.

15.1.6.5 EPIDEMIOLOGY

Alternate periods of high soil moisture and water stress conditions predispose the disease.

15.1.6.6 MANAGEMENT

- Deep summer plowing.
- Seed treatment with carboxin @ 3 g/kg of seed or *Trichoderma viride* @ 4 g/kg or *Pseudomonas fluorescens* 10 g/kg of seed.
- Drenching the affected plants with Copper oxy chloride @ 3 g/l at the base of the plants.
- Soil application of *Pseudomonas fluorescens* or *Trichoderma viride* –2.5 kg/ha + 500 kg of well decomposed FYM or vermicompost or sand at 30 DAS.

15.1.7 RUST

15.1.7.1 INTRODUCTION/ECONOMIC IMPORTANCE

Rust is the wide spread, most severe and more common in temperate and sub-tropical region. It causes a considerable yield loss (10–30%) when it infects in early stage.

15.1.7.2 SYMPTOMS

Small reddish brown pustules (uredospores) appear on the lower surface of the leaves and slowly they spread to other parts. This uredosori are slowly replaced by teliosori/teliospores (black) when the crop reaches maturity.

FIGURE 15.6 Rusty spots on leaves.

15.1.7.3 CASUAL ORGANISM

Puccinia helianthi

This is an autoecious rust. The uredospores are round or elliptical, dark cinnamon-brown color with minute echinulate whereas teliospores are elliptical or oblong, bi-celled, smooth walled and chestnut brown with long, colorless pedicel. The pycnial and aecial stages occur on volunteer crops grown during off-season.

15.1.7.4 DISEASE CYCLE

The primary inoculum is by sporidia of teliospores, or volunteer plants at high altitudes and comes through wind currents.

15.1.7.5 EPIDEMIOLOGY

Temperatures (25.5–30.5°C), RH (86–92%) favors the disease. The incidence of rust increases with age (75 days old plants).

15.1.7.6 MANAGEMENT

- Clean cultivation, removal of volunteer plants and growing resistant varieties.
- Crop rotation may be followed.
- Foliar application of mancozeb or zineb @ 2.5 g/l for 2–3 times at 10 days interval. The first spray should be given as soon as the disease is noticed or 35 DAS.

15.2 VIRAL DISEASES

15.2.1 NECROSIS DISEASE

15.2.1.1 INTRODUCTION/ECONOMIC IMPORTANCE

The appearance of sunflower necrosis disease (SND) was observed for the first time during 1997 at Bangalore which later spread to other parts of Karnataka, Tamil Nadu, Andhra Pradesh, and Maharashtra. The incidence ranged from 5–70%, which was observed on all stages of crop growth both in Kharif and Rabi.

15.2.1.2 SYMPTOMS

This disease is observed in all stages of crop growth from seedling to harvesting stage (Chander Rao et al., 2002). Initial symptoms appear on leaves as small irregular, necrotic patches on leaf lamina more near to the midrib. It results in twisting of the leaf as the disease progresses. It spreads to petiole and stem through one side of leaf lamina and finally it reaches to the tip of the shoot and reaches to flowers and leads to paralytic symptom. Sometimes stem bending and twisting (S shaped curve) beneath the flower head is seen. The infected plants will die at seedling stage itself when the infection at early stage. Further, the early infection leads to stunting, weak, and death of the plant. If the infection before flowering, flower formation is affected and flowers fails to open. Whereas after flowering, seed filling is affected and in severe cases seed setting will not occur. Disease incidence is higher in kharif and summer, where as low in Rabi season. Highest disease incidence is observed during prolonged dry spells immediately after heavy rains.

Necrosis of leaves

FIGURE 15.7a Necrosis symptoms of leaves by Tobacco streak virus

Stem bending

FIGURE 15.7b Stem bending symptoms by Tobacco streak virus

Necrosis of leaves

FIGURE 15.7c Necrosis symptoms of leaves by Tobacco streak virus.

Necrosis of terminal bud

FIGURE 15.7d Necrosis symptoms on terminal bud by Tobacco streak virus.

FIGURE 15.7e Stunted plant symptoms by Tobacco streak virus.

Stunted plant

FIGURE 15.7f Stunted plant symptoms by Tobacco streak virus.

FIGURE 15.7g Stunted plant symptoms by Tobacco streak virus.

15.2.1.3 CASUAL ORGANISM

Tobacco streak virus (TSV).

15.2.1.4 DISEASE CYCLE

Pollen of Parthenium plays a major role in spreading the disease as it acts as a symptom less carrier and provides continuous source of inoculum for infection. Other weed hosts such as *Commelina, Achyranthus, Euphorbia, Tridox* sps were also found infected with virus. The main source of inoculum identified as the pollen grains of infected crop plants or weeds.

15.2.1.5 TRANSMISSION

Thrips spp. helps in spreading the disease by carry the infected pollen and not by acquiring the virus in the body.

15.2.1.6 EPIDEMIOLOGY

Dry weather (July–August) with moderate temperature (30–32°C) and RH (55–75%) is conducive to thrips incidence (Singh, 2005).

15.2.1.7 MANAGEMENT

- Rouging of the virus infected plants as early as they have been identified.
- Removal of Parthenium from bunds, adjoining areas as well as from fields.
- Growing of 5–7 rows of jowar or maize as a border to sunflower crop as they attracts thrips population and acts as a barrier in spreading of wind borne thrips and also to the infected Parthenium pollen.
- Seed treatment with imidachloprid (Gaucho 70WS) @ 5 g/kg of seed helps the crop from early stage infection.
- Foliar application of imidachloprid @ 0.4 ml or Thiomethaxam @ 0.25 g/l of water at 30 and 45 DAS will manage the thrips population.

15.2.2 LEAF CURL

15.2.2.1 INTRODUCTION/ECONOMIC IMPORTANCE

Leaf curl virus belongs to Begomo virus group was first time observed in Raichur, Karnataka. This disease causes yield losses up to 40%.

15.2.2.2 SYMPTOMS

The initial symptoms were first noticed on 30 days old plants (Deepa et al., 2015) and the common symptoms are like shortened growth, curling, thickening, and brittling of leaves, vein thickening and reduction of leaf size and enations (Govindappa et al., 2011). Leaf cupping from the margins and upward curling of the leaves are also seen. Leaves show yellow discoloration and significantly affect diameter of the head, 100 seed weight, oil content and all yield components depending on the growth stages at which first symptom appears.

15.2.2.3 CASUAL ORGANISM

Sunflower leaf curl virus.

Leaf thickening

FIGURE 15.8a Leaf thickening symptoms.

Brittle leaves

FIGURE 15.8b Brittle leaves symptoms.

Leaf curling

FIGURE 15.8c Leaf curling symptoms.

Leaf curl affected field

FIGURE 15.8d Leaf curl symptoms: A view in affected fields.

15.2.2.4 TRANSMISSION

This disease is transmitted through white flies, i.e., *Bemisia tabaci.*

15.2.2.5 MANAGEMENT

- Avoid growing of sunflower nearby cotton areas.
- Roguing of infected plants and foliar application of insecticide metasystax or rogor @ 2 ml/l of water at 10 days interval.

KEYWORDS

- **charcoal rot**
- **head rot**
- **leaf curl**
- **necrosis disease**
- **Sclerotium wilt**
- **tobacco streak virus**

REFERENCES

Almeida, A. M. R., Machado, C. C., & Carrão-Panizzi, M. C., (1981). *Doenças do girassol: Descrição de Sintomas e Metodologia para Levantamento* (p. 24). Londrina: Embrapa-CNPSo, (Embrapa-CNPSo. Circular técnica, 6).

Anahosur, K. H., (1978). *Alternaria helianthi. CMI Descriptions of Pathogenic Fungi and Bacteria, 582*, 1–2.

Balasubramanyam, N., & Kolte, S. J., (1980). Effect of alternaria blight on yield components, oil content, and seed quality of sunflower. *Indian Journal of Agricultural Sciences*, *50*(9), 701–706.

Chander Rao, S., Sujatha, M. Karuna, K., & Varaprasad, K S., (2015). Powdery mildew disease in sunflower: A review. *Journal of Oilseeds Research, 32*(1), 111–122.

Chander, R. S., Prasada, R. R. D. V. J., Singh, H., & Hegde, D. M., (2002). *Sunflower Necrosis Disease and Its Management.* Information Bulletin, DOR, Hyderabad.

Davet, P., Pérès, A., Regnault, Y., Tourvieille, D., & Penaud, A., (1991). *Les Maladies du Tournesol* (p. 72). Paris: CETIOM.

Deepa, G. S., Govindappa, M. R. Naik, M. K., & Suresh, S. R., (2015). Estimation of yield loss in sunflower due to new sunflower leaf curls virus disease at different stages of crop growth. *International Journal of Plant Protection*, *8*(1), 138–141.

Govindappa, M. R., Shanker, G. I., Shankarappa, K. S., Wickramarachchi, W. A. R. T., Anjaneya, R. B., & Rangaswamy, K. T., (2011). Molecular characterization and partial characterization of begomovirus associated with leaf curl disease of Sunflower (*Helianthus annus*) in southern India. *Plant Pathology Journal*, 1–7.

Gulya, T. J., Rashid, K. Y., & Masirevic, S. M., (1997). Sunflower diseases. In: Schneiter, A. A., (ed.), *Sunflower Technology and Production* (pp. 263–379). Madison: American Society of Agronomy.

Kapoor, J. N., (1967). *Erysiphe Cichoracearum. CMI Descriptions of Pathogenic Fungi and Bacteria, 152*, 1, 2.

Kolte, S. J., (1985). *Diseases of Annual Edible Oilseed Crops* (pp. 9–96). III CRC Press, Florida.

Mishra, R. P., Kushwaha, U. S., Khare, M. N., & Chand, J. N., (1972). Rhizopus rots of sunflower in India. *Indian Phytopathology*, *25*, 236–239.

Pereyra, V., & Escande, A. R., (1994). *Enfermedades del girasol en la Argentina: Manual de Reconocimiento* (p. 113). Balcarce: INTA.

Punja, Z. K., (1985). The biology, ecology, and control of *Sclerotium rolfsii. Annual Review of Phytopathology*, *23*, 97–127.

Shankergoud, I., Parameshwarappa, K. G., Chandranath, H. T., Pramod, K., Mesta, R. K., Go lasangi, B. S., et al., (2006). *Sunflower and Castor Research in Karnataka: An Overview* (p. 21). University of Agricultural Sciences, Bangalore/Dharwad.

Singh, H., (2005). Thrips incidence and necrosis disease in sunflower. *Journal of oil Seed Research, 22*(1), 90–92.

Zimmer, D. E., & Hoes, J. A., (1978). Diseases. In: Carter, J. F., (ed.), *Sunflower Science and Technology* (pp. 225–262). Madison: American Society of Agronomy.

CHAPTER 16

Diseases of Taramira (*Eruca sativa* Mill.) and Their Management: Indian Perspective

K. K. SHARMA[1] and J. N. SRIVASTAVA[2]

[1]*Regional Research Station (PAU), Ballowal Saunkhri, P. O. Takarla, Tehsil-Balachaur, SBS Nagar, Punjab, India*

[2]*Department of Plant Pathology, Bihar Agricultural University, Sabour–813210, Bhagalpur, Bihar, India*

Taramira or Arugula or garden rocket or Safed Sarson (*Eruca sativa* syn: *E. vasicaria* sub sp. Sativa (Miller) Thell., *Brassica eruca* L. syn: *Brassica erucoides* Roxb.) is an important oilseed crop belongs to family Brassicaceae which is cultivated in arid and semi-arid regions of the world. It is believed to be a native of southern Europe and North Africa (Bailey, 1949) and Central Asia (Ugur et al., 2010). This crop is cultivated in many countries like Mexico, Portugal, Spain, Turkey, China, Japan, Netherlands, France, and Morocco including India. In India, Taramira crops grown in Rajasthan, Haryana, Punjab, M. P. and U. P. states. Rajasthan has the highest area amounting 236, 200 hectares of production under Taramira (India Oilseed and product update, 2017). In Uttar Pradesh, the leading areas of cultivation of Taramira are Meerut, Agra, Jhansi, Rohilkhand, Allahabad, and Banaras while in Rajasthan Alwar, Dholpur, Bharatpur, Bhilwara, Jaipur, Jodhpur, Kota, Udaipur, and Shri Ganganagar. In Punjab, Taramira is grown on 4.9 thousand hectares of land and sub-montanous zone or kandi area of the state is well known for its cultivation (POP for Rabi crops of Punjab, 2016–17).

It is an annual herb with diploid ($2n = 22$) number of chromosomes. Plant is usually 2 to 4 feet tall with erect, simple or branched with hairy stem. The basal leaves are stalked while the upper ones are sessile and the *flowers* are pedicellate, pale yellow or whitish, 1.5–2.0 cm across. The fruit are shorter and stout and seeds are round to oval, yellowish brown or reddish of 1.5–2 mm size, arranged in 2 rows in each compartment of the fruit (Miyazawa et al., 2002). This crop has smaller requirements of fertilizers and other inputs than those of wheat and can be grown well on sandy and loamy-sand soils under rainfed areas. It is well known for its drought tolerance. Its strong and fast penetrating root system allowing absorption deep-seated water in soil layers and hence requires very less preparatory tillage. In rainfed areas, one or two plowings with desi plow or a cultivator each followed by planking are sufficient. Sowing was done during whole October at the seed rate of 1.5 kg seed/acre with rows 30 cm apart and 4–5 cm deep. Plant to plant distance of 15 cm is maintained. It is raised as an alternative crop in dry land or rainfed areas and light textured soils even with conserved soil moisture (Gupta et al., 1998). The crop matures in about 150 days and is ready for harvest when pods turn yellow (Figure 16.1).

FIGURE 16.1 Taramira (*Eruca sativa*).

Taramira is also known for its medicinal properties viz., astringent, diuretic, digestive, tonic, laxative, aphrodisiac, anti-inflammatory for colitis and rubefacient properties (Yaniv, 1998; Bajilan and Al-Naqeeb, 2011). The oil content ranges from 31.6–41.31% which is used skin diseases and to get rid of lice and dandruff. Oil is also mixed with mustard oil to enhance its pungency and the oil cake is a nutritious feed for the cattle.

The losses in oilseed crops due to biotic stresses is about 19.9% world-wide, out of which diseases cause severe yield reduction at different growth stages. In India, the Alternaria blight, White rust (WR)/Downy mildew cause significant yield losses (up to 47%) in oilseed crops followed by Sclerotinia rot (SR) (up to 40%) and Powdery mildew (up to 18%) diseases (Kumar et al., 2008).

Taramira or Arugula or garden rocket or Safed Sarson (*Eruca sativa)* is affected by many diseases. These diseases as well as pathogen either alone or in combination causes substantial damage to crop resulting in heavy economic losses every year. This communication deals with important diseases of Taramira and their management strategies. Especially, Alternaria blight and SR or Sclerotinia wilt are the major constrains in its production and productivity. Downy mildew and WR may also affect the production. But Alternaria blight at the time of pod formation stage supposed to be a major constrain in both the quality and quantity. Besides this, Fusarium wilt and Root rot, Bacterial blight and Club-root disease are also reported from other countries but either not found or minor disease in India.

16.1 ALTERNARIA BLIGHT

This disease of *E. sativa* is very common and become severe under favorable conditions at the time of pods formation if not managed. It is prominent in all rapeseed-mustard growing areas of India including Rajasthan, Punjab, Haryana, and U. P. (Kadian and Saharan, 1983; Sangwan et al., 2002) and also reported from other countries (Gilardi et al., 2013; Abdalla M. El-Alwany, 2015). Losses caused by *A. brassicae* in pod and seed weight were reported by Jain (1992) amounting 4.11–51.06% and 6.67–55.85%, respectively.

16.1.1 SYMPTOMS

The disease appears first on the lower leaves and characterized by small circular dark brown to brownish-black necrotic spots with concentric rings which slowly increase in size; center of these spots dries up and falls down later. Many spots coalesce to form large patches results in blightening of leaves and defoliation may occur in severe cases. Circular to linear, dark brown lesions also develop on stems which become elongated at later stage. The pods are also affected during the flowering and pod formation stage and these pods produce undersized, discolored, and shriveled seeds. The diseased pods turn black in color and may also rot under severe disease outbreak. In case of severe attack, the upper parts of the stem and pods wither (Figure 16.2).

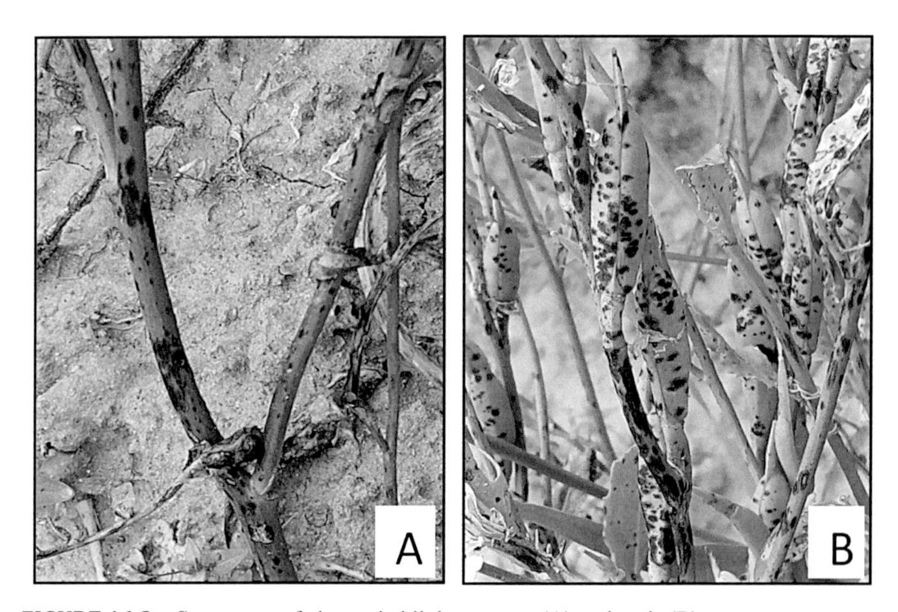

FIGURE 16.2 Symptoms of alternaria blight on stem (A) and pods (B).

16.1.2 CASUAL ORGANISM

Alternaria brassicae or *A. brassicicola.*

16.1.3 SURVIVAL AND SPREAD

The pathogen survives through spores or mycelium in diseased plant debris or weed. The disease is seed born in nature (externally and internally). The weather conditions like temperature in range of 21–25°C and relative humidity (RH) of more than 70% (moist and warm weather) with intermittent rains favor disease development. The *Alternaria* blight on pods of *E. sativa* was favored and developed at temperature ranged from 21–24°C and 63–70% RH under field conditions in Bawal of Haryana, India.

16.1.4 MANAGEMENT

- First of all farmers should preferably use the disease resistant or tolerant or disease escaping cultivars if available.
- Grower should use of healthy disease free seeds for sowing to avoid disease risk.
- Collect and destroy diseased debris after the harvest of the crop.
- Choosing the sowing (early sowing) time is very important as it affects the disease incidence significantly. Practices like deep plowing, timely weeding and maintenance of optimum plant population can avoid the disease risk at some extent.
- Avoidance of irrigation at flowering and pod formation stages may help to manage the disease.
- Seed treatment with biocontrol agents viz., *T. viride*, *G. virens* or botanicals like *Allium sativum* bulb extract (1% w/v) and spray of *Trichoderma viride* at 45 and 75 days after sowing (DAS) was found effective to reduce infection. Use of biocontrol agents is advantageous as they are often effective against a wide range of soil-borne pathogens. Moreover, they are ecofriendly, cost effective and their use avoids the risk of development of resistance in the pathogen towards the control agent.
- Seed treatment with carbendazim @ 0.1% a. i. or mixture of carbendazim with Apron 35 SD at the rate of 6 g/kg (Kumar et al., 2008). As soon as the symptoms start appearing, the crop should be sprayed with Mancozeb 75 WP at the rate of 2 kg in 1000 liters of water per hectare at 10–15 days interval (3–4 Spray).
- PAU has recommended three spray to the crop with Blitox or Indofil M-45@ 250 g in 100 liters of water starting at 75 days old crop followed by second spray of Score 25EC@ 100 ml and third spray with Blitox or Indofil M-45@ 250 g in 100 liters of water at 15 days interval (POP for Rabi crops of Punjab, 2017).

16.2 SCLEROTINIA STEM ROT

Shaw and Ajrekar (1915) reported *S. sclerotiorum* first time in India. The pathogen has a wide host range able to infect about 408 plant species

(Boland and Hall, 1994) including rapeseed-mustard. None of the fungicide and source of resistance was found to prove its effectiveness for managing this disease and hence, it has become a serious problem in rapeseed and mustard growing areas of India (Kang and Chahal, 2000; Shivpuri et al., 2000; Ghosolia et al., 2004). Yield losses up to 39.9% were reported in rapeseed-mustard in India according to Chattopadhyay et al. (2003). *Eruca sativa* was found highly susceptible (45% disease incidence and >50% disease severity) to SR and symptoms were also appeared early (45–52 DAS) as compare to other Bressica species (Rathi and Singh, 2009a).

16.2.1 SYMPTOMS

Disease symptoms can be observed as elongated water soaked lesions, appear particularly on base near to the crown region or at internodes of the stem, covered with white cottony mycelial growth. The affected plants become whitish and may be seen easily from distance. Later on, dark brown to black sclerotia are produced on and/into stem under unfavorable conditions (nutrition and moisture depletion), which may also see on other infected plant parts. Under severe infection plant are showed, defoliation, shredding of stem, wilting, and drying of the plants. Affected pants either fail to produce or reduced number of shriveled seed in siliquae. Plant looks whitish or creamish white from distance (Figure 16.3).

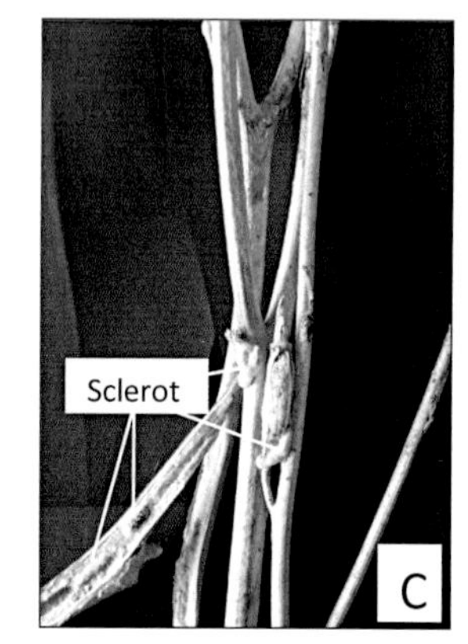

FIGURE 16.3 Symptoms of sclerotinia rot and wilt in standing crop on stem and pods (A & B), sclerotia on and inside the stem (C).

16.2.2 CASUAL ORGANISM

Sclerotinia sclerotiorum.

16.2.3 SURVIVAL AND SPREAD

The pathogen may survive in infected plant parts or crop residue or soil surface or as a contaminant of seed in the form of mycelium and sclerotia. Coley-Smith and Cooke (1971) reported myceliogenic and carpogenic survival of fungal sclerotia. The average temperature in the range of 18–25°C and high humidity (90–95%) with windy weather favors the disease development. Due to short height of the plant covering the soil increases the humidity within the crop canopy provides more favorable microclimate of crop canopy for the development of disease (Ghasolia and Asha Shivpuri, 2007). Prolonged cool and moist weather during flowering period of crop, which is susceptible stage for disease, leads to epidemic development.

16.2.4 MANAGEMENT

- First of all farmers should preferably use the disease resistant or tolerant or disease escaping cultivars if available.
- Grower should use of healthy disease free seeds for sowing to avoid disease risk.
- Collect and destroy diseased debris after the harvest of the crop.
- The affected plants (early ripened) could be rogue out prior to sclerotia formation to reduce soil inoculum.
- Neither fungicide was found to control this disease after appearance in field nor recommended till date. Use of agrochemical is not feasible and economical to spray as success rate is varied drastically and disease appears generally late at pod formation stage (Mehta et al., 2005; Rathi and Singh, 2009b). However, disease was significantly managed through systemic fungicides as prophylactic sprays reported under field conditions (Rathi et al., 2012).
- Either soil application of 2.5 kg *Trichoderma harzianum* + *T. viride* mixture colonized on 50 kg of farm yard manure (FYM) in one hactare at the time of preparation of field and/or seed treatment with above mixture @10 g/kg seeds and/or foliar spray of the same mixture @0.2% just after appearance of Sclerotinia stem rot at 20

days intervals was reported to prevent the spread of Sclerotinia stem rot disease.

16.3 DOWNY MILDEW

This disease may cause considerable loss when mixed infection of downy mildew and WR is developed in affected field. The disease has been reported from India (Sharma et al., 1991; Sastry, 2003) and other countries like California (Koike, 1998), Northern Italy (Minuto et al., 1997), Japan (Satou et al., 2004), Argentina (Romero and Zapata, 2005), Argentina (Larran et al., 2006), and Korea (Choi et al., 2010).

16.3.1 SYMPTOMS

The disease affects all aerial parts of the plants although; its severity is more on foliage and inflorescence. Small, irregular light green or grayish white necrotic lesions develop on the lower surface of the leaves; later on lesions increase in size and become purplish brown. A brownish white cottony fungal growth may also be seen on the spots under favorable conditions. Subsequently, affected leaf portion shrivels, dries up and tears easily. Infection of inflorescence results in thickening of peduncle due to hypertrophy and hyperplasia results in deformed floral organs which is most distinct symptom of the disease known as stag head stage. Affected inflorescence either fails to produce siliquae or abnormal siliquae without seeds.

16.3.2 CASUAL ORGANISM

Peronospora brassicae/Hyaloperonospora brassicae.

16.3.3 SURVIVAL AND SPREAD

The pathogen survives as oospores on the affected plant tissues in soil and on weed hosts. The environmental conditions like atmospheric temperature in the range of 10–20°C and RH more than 90% favors disease development.

16.3.4 MANAGEMENT

- First of all farmers should preferably use the disease resistant or tolerant or disease escaping cultivars if available.
- Grower should use of healthy disease free seeds for sowing to reduce the risk of disease.
- Destroy diseased debris from the previous crop.
- Application of recommended doses of N, P, and K fertilizers with split application of N may be helpful for the management of disease.
- Apply 3–4 spray of 250 g of Blitox or Indofil M-45 in 100 liters of water at 15 days interval starting at about 75 days old crop.

16.4 WHITE RUST (WR)

It is less likely to occur on available varieties of Taramira but indidence may be high on susceptible/tolerant variety under favorable environmental conditions. There are some records from India (Vardhana, 2017a and b) and other countries like Califonia (Scheck, 1997), Northern Italy (Minuto et al., 1997), Argentina (Zapata et al., 2005), Australia (Kaur et al., 2011), South Africa (Mangwende, 2015) where the occurrence of this disease was observed.

16.4.1 SYMPTOMS

Plant may be infected locally and systemically both. Local infection is easily observed as white to creamy white raised, scattered pustules appear on the under surface of the leaves which later coalesce to form patches. In case of systemic infection, swelling, and distortion of floral heads (Secondary infection) due to hypertrophy and hyperplasia and 'stag head' structures can be seen in affected field. The flowers also get malformed and become sterile; petals become green and stamens are transformed into leaf-like structures, which become thick and club shaped. During humid weather, mixed infection of WR and downy mildew is often.

16.4.2 CASUAL ORGANISM

Albugo candida/Cystopus candida.

16.4.3 SURVIVAL AND SPREAD

The pathogen survives through oospores in affected host tissues and soil. Secondary infection is carried out by sporangia and zoospores which produce new infection. Favorable environmental conditions like high moisture (>70% RH) and cool weather (12–20°C) aggravate the development of disease.

16.4.4 MANAGEMENT

- First of all farmers should preferably use the disease resistant (Bansal et al., 1997) or tolerant or disease escaping cultivars if available.
- Grower should use of healthy disease free seeds for sowing to reduce the risk of disease.
- Destroy diseased debris from the previous crop.
- Follow crop rotation with non-cruciferous crop varieties to prevent disease build up.
- Application of recommended doses of N, P, and K fertilizers with split application of N may be helpful for the management of disease.
- Apply 3–4 spray of 250 g of Blitox or Indofil M-45 in 100 liters of water at 15 days interval starting at about 75 days old crop.

Besides the above described diseases, other diseases like Clubroot caused by *Plasmodiophora brassicae* (Lima et al., 2004), *Fusarium leaf spot* (Garibaldi et al., 2010), *Bacterial blight,* and *rot* have also been reported but not causing significant yield loss so far in Taramira production and productivity although, they may become serious problem in future.

KEYWORDS

- **alternaria blight**
- **downy mildew**
- ***Eruca sativa***
- ***Fusarium leaf spot***
- **sclerotinia stem rot**
- **white rust**

REFERENCES

Abdalla, M., & El-Alwany, (2015). Plant pathogenic *Alternaria* species in Libya. *Open Access Library Journal*, *2*, 1–8.

Bailey, L. H., (1949). *Botanical Review*, *22*, 81–86.

Bajilan, S. I., & Al-Naqeeb, A. A., (2011). Effect of the aqueous extract of rocket (*Eruca sativa*) leaves on the histological structure of some organs in male mice. *Journal of the College of Basic Education*, *17*(70), 13–21.

Bansal, V. K., Tewari, J. P., Tewari, I., Gomez-Campo, C., & Stringam, G. R., (1997). Genus Eruca: A potential source of white rust resistance in cultivated Brassicas. *Plant Genet Res Newslett.*, *109*, 25–26.

Boland, G. J., & Hall, R., (1994). Index of plant hosts of *Sclerotinias clerotiorum*. *Can. J. Plant Pathol.*, *16*, 94–108.

Chattopadhyay, C., Meena, P. D., Kalpana, S. R., & Meena, R. L., (2003). Relationship among pathological and agronomic attributes for soil borne diseases of three oilseed crops. *Indian J. Plant Protect*, *31*, 127, 128.

Choi, Y. J., Park, M. J., Kim, J. Y., & Shin, H. D., (2010). An unnamed *Hyaloperonospora* sp. causing downy mildew on arugula (rocket) in Korea. *Plant Pathology*, *59*(6), 1165.

Coley-Smith, J. R., & Cooke, R. C., (1971). Survival and germination of fungal sclerotia. *Ann Rev Phytopathol.*, *9*, 65–92.

Garibaldi, A., Gilardi, G., Bertoldo, C., & Gullino, M. L., (2011). First report of leave spot of rocket (*Eruca sativa*) caused by *Fusarium equiseti* in Italy. *Plant Dis.*, *95*(10), 1315.

Ghasolia, R. P., & Shivpuri, A., (2007). Morphological and pathogenic variability in rapeseed and mustard isolates of *Sclerotinia sclerotiorum*. *Indian Phytopathl.*, *60*(1), 76–81.

Ghasolia, R. P., Shivpuri, A., & Bhargava, A. K., (2004). Sclerotinia rot of Indian mustard (*Brassica juncea*) in Rajasthan. *Indian Phytopathol.*, *57*, 76–79.

Gilardi, G., Gullino, M. L., & Garibaldi, A., (2013). New Diseases of Wild and Cultivated Rocket in Italy. Retrieved from: http://agris. fao. org/agris-search/search. do?recordID= US201400148929 (accessed on 14 January 2020).

Gupta, A. K., Agarwal, H. R., & Dahama, A. K., (1998). Taramira: A potential oilseed crop for the marginal lands of Rajasthan, India. In: Bassam, N. E. I., et al., (eds.), *Sustainable Agriculture for Food, Energy, and Industry Strategy Towards Achievement* (pp. 687–691). James and James (Science Publishers) Ltd., London (U. K.).

India Oilseed and Product Update, (2017). Retrieved from: http://Oilseed_and_Product_Update_New Delhi_India_11-28-2017. pdf (accessed on 14 January 2020).

Jain, S. C., (1992). Assessment of losses in taramira (Eruca sativa Lam) due to alternaria blight disease. *Madras Agricultural Journal*, *79*(1), 57–58.

Kadian, A. K., & Saharan, G. S., (1983). Symptomatology, host range, and assessment of yield losses due to *Alternaria brassicae* infection in rapeseed and mustard. *Indian J. Mycol. & Pl. Pathol.*, *13*, 319–323.

Kang, I. S., & Chahal, S. S., (2000). Prevalence and incidence of white rot of rapeseed and mustard incited by *Sclerotiniasclerotiorum* in Punjab. *Pl. Dis. Res.*, *15*, 232, 233.

Kaur, P., Sivasithamparam, K., & Barbetti, M. J., (2011). Host range and phylogenetic relationships of *Albugo candida* from cruciferous hosts in Western Australia, with special reference to *Brassica juncea*. *Plant Disease*, *95*(6), 712–718.

Koike, S. T., (1998). Downy mildew of arugula, caused by *Peronospora parasitica* in California. *Plant Disease*, *82*, 1063.

Kumar, A., Premi, O. P., & Thomas, L., (2008). *Rapeseed - Mustard Cultivation in India-An Overview*. New letter published by national research center on rapeseed—Mustard (NRCRM) Sewar, Bharatpur.

Larran, S., Ronco, L., Mónaco, C., & Andreau, R. H., (2006). First report of *Peronospora parasitica* on rocket (*Eruca sativa*) in Argentina. *Australasian Plant Pathology, 35*, 377.

Lima, M. L. P., Cafe-Filho, A. C., Nogueira, N. L., Rossi, M. L., & Schuta, L. R., (2004). First report of clubroot of *Eruca sativa* caused by *Plasmodiophora brassicae* in Brazil. *Plant Disease*, *88*, 573.

Mangwende, E., Kalonji-Kabengele, J. B., Truter, M., & Aveling, T. A. S., (2015). First report of white rust of rocket (*Eruca sativa*) caused by *Albugo candida* in South Africa. *Plant Disease*, *99*(2), 290.

Mehta, N., Sangwan, M. S., & Saharan, G. S., (2005). Fungal diseases of rapessed-mustard. In: Saharan, G. S., Naresh, M., & Sangwan, M. S., (eds.), *Diseases of Oilseed Crops* (p. 15–86.). Indus Publishing Co. New Delhi, India.

Minuto, G., Minuto, A., & Garibaldi, A., (1997). Eruca Sativa: Chemical Control of Downy Mildew and White Rust. Retrieved from: https://www. cabdirect. org/cabdirect/abstract/19971006865 (accessed on 14 January 2020).

Miyazawa, M., Maehara, T., & Kurose, K., (2002). Composition of the essential oil from the leaves of *Eruca sativa*. *Flavor and Fragrance Journal*, *17*(3), 187–190.

POP for Rabi Crops of Punjab, (2016–17). *Alternaria Blight. Package of Practices for Crops of Punjab, Rabi (2016–2017)* (p. 59). Punjab Agricultural University, Ludhiana.

Rashtra, V., (2017a). Plant's diseases of district Meerut and adjacent areas. *Plant Archives*, *17*(1), 385–390.

Rashtra, V., (2017b). Plant's diseases of district Ghaziabad and adjacent areas. *Plant Archives*, *17*(1), 727–732.

Rathi, A. S., & Singh, D., (2009a). Comparison of Sclerotinia rot incidence and sclerotial formation in different rapeseed-mustard species. *Paper Presented in 16th Australian Research Assembly on Brassica Held at Ballarat Mercure Hotel* (pp. 40–44). Ballarat, Victoria, Australia, Conference proceedings.

Rathi, A. S., & Singh, D., (2009b). Integrated management of alternaria blight and white rust in Indian mustard. *Paper Presented in 16th Australian Research Assembly on Brassica Held at Ballarat Mercure Hotel* (pp. 51–54). Ballarat, Victoria, Australia, Conference proceedings.

Rathi, A. S., Sharma, S., & Singh, D., (2012). Efficacy of carbendazim as prophylactic control of sclerotinia rot in Indian mustard. *Paper Presented in 1st National Brassica Conference on "Production Barriers and Technological Options in Oilseed Brassica" Held at CCS HAU* (Abstracts p. 130). Hisar.

Romero, A. M., & Zapata, R., (2005). First report of downy mildew of arugula caused by *Peronospora parasitica* in Argentina. *Plant Disease*, *89*(6), 688.

Sangwan, M. S., Mehta, N., & Gandhi, S. K., (2002). Some pathological studies on *Alternaria raphani* causing leaf and pod blight of radish. *J. Mycol. Pl. Pathol., 32*, 125, 126.

Sastry, E. V. D., (2003). Taramira (*Eruca sativa*) and its improvement: A review. *Agricultural Review*, *24*(4), 235–249.

Satou, M., Uematsu, S., Nishi, K., & Kubota, M., (2004). Downy mildew of rocket (Erucavesicaria ssp. sativa) and leaf mustard (Brassica juncea) caused by Peronospora parasitica. *Annual Report of the Kanto-Tosan Plant Protection Society* (No.51, pp. 21–24). Retrieved from: https://www. cabdirect. org/cabdirect/abstract/20053050227 (accessed on 14 January 2020).

Scheck, H. J., (1999). First occurrence of white rust of arugula caused by *Albugo candida*. *Plant Disease*, *83*(9), 877.
Sharma, R. K., Agrawal, H. R., & Sastry, E. V. D., (1991). *Taramira: Importance, Research, and Constraints*. S K N College of Agriculture, Jobner (Mimeo).
Shaw, F. J. W., & Ajrekar, S. L., (1915). The genus rhizoctonia in Indian. *Dept. Agri. Nom. Bot. Ser., 7*, 177–194.
Shivpuri, A., Sharma, K. B., & Chhipa, H. P., (2000). Some studies on the stem rot (*Sclerotinia sclerotiorum*) disease of rapeseed/mustard in Rajasthan. *J. Myco. Pl. Pathol.*, *30*, 268.
Uğur, A., Süntar, I., Aslan, S., Orhan, I.e., Kartal, M., Şekeroğlu, N., Esiyok, D., & Sener, B., (2010). Variations in fatty acid compositions of the seed oil of *Eruca sativa* Mill. Caused by different sowing periods & nitrogen forms. *Pharmacognosy Magazine*, *6*(24), 305–308.
Yaniv, Z., Schafferman, D., & Amar, Z., (1998). Tradition, uses, and biodiversity of rocket (*Eruca sativa*, Brassicaceae) in Israel. *Econ. Bot.*, *52*, 394–400.
Zapata, R., Romero, A. M., & Maseda, P. H., (2005). First report of white rust of arugula caused by *Albugo candida* in Argentina. *Plant Disease*, *89*(2), 207.

CHAPTER 17

Current Status of Smut Disease of Sugarcane and Its Management in India

GEETA SHARMA[1] and JAI SINGH[2]

[1]*Department of Plant Pathology, College of Agriculture, G. B. P. U. A. T., Pantnagar-263145, Uttarakhand, India*

[2]*JNKVV Krishi Vigyan Kendra, Sidhi-486661, Madhya Pradesh, India*

The current status of Smut disease of Sugarcane and the integrated disease management (IDM) programme in India to control them are discussed in this review. The Smut of sugarcane caused by *Sporisorium scitamineum* (*Ustilago scitaminea* Syd.), is a disease worldwide and present in all the sugarcane growing areas of the world. Losses can be range from very minor to complete losses. Smut infected plants results not only in fewer canes but also in reduced sugar content. Losses due to smut are greater with primary infection vs. secondary infection, early vs. late season, and with ratoon vs. plant crops. Infection of planted cuttings occurs by chlamydospores present in the soil or in irrigation water and by diseased planting stock. Standing cane becomes infected mainly in the buds by wind-borne spores. Temperature and relative humidity (RH) required for maximum spore dispersal are 22–24°C and 50–60%, respectively. Disease severity depends on the environmental conditions and the resistance of the sugarcane varieties grown. Integration of cultural practices, chemical control and biological control are effective tools for management smut in Sugarcane.

17.1 HISTORY AND DISTRIBUTION

Sugarcane smut is one of the important disease caused by fungus *Sporisorium scitamineum* (*Ustilago scitaminea*) (Sydow, 1924). The disease incidence was first reported from Natal, South Africa in 1877 as reported by Luthra et al. (1940). In India the disease was reported by Butler in the year 1906.

By 1930s, it was a serious problem in India and other countries of Asia, viz., Sri Lanka, Bangladesh, Taiwan, Pakistan, Philippines, and Japan. In Okinawa (Japan), it spread from 0.4% affected area to 22% in merely 6 years, i.e., from 1972 to 1978 (Yamauch, 1978). Then it spread to neighboring countries, like Brazil, Paraguay, and Bolivia and it was well established up to 1957 in these countries. In 1971, the disease spreaded to Hawaii (USA) and later in 1974 in Guyana. By 1981, smut threatened cane production in most of the Caribbean, Northern region of South America, Central America and the Continental United States. It was first discovered in the USA in Florida during 1978 and in Louisiana, Puerto Rico, and in Texas in 1981. Sugarcane smut was identified for the first time in Australia in the Ord River Irrigation Area (ORIA), in the semi-arid tropics of Western Australia, in July 1998 (Riley et al., 1999).

Due to these outbreaks a great deal of experimental work on sugarcane smut started up (Heinz, 1987) and subsequently, the occurrence of sugarcane smut in Morocco and Iran was established. The incidence of the disease was widespread covering several countries in East Africa, the Pacific and the Caribbean islands, wherein a severe outbreak of the disease resulted in devastating loss to the sugarcane plantations. Lovick (1978) comprehensively reviewed on various aspects of sugarcane smut viz. symptoms, yield reduction, causal organism, physiological races of the smut fungus, epidemiology, host resistance and management.

In India, smut had been a serious problem in the sustenance of superior varieties like Co 419, Co 740, Co 975, Co 1158, Co 1148 and CoS 767, CoS 510. An unprecedented epidemic of smut had broken out during 1942–43 in Bihar affecting 66% of cane area (Chona, 1956) and once again during 1950–52 affecting Co 213, Co 453, Co 513 and BO 11 varieties. The epidemic caused huge losses in Darbhanga and Saharan districts of Bihar. In tropical India, Subramanian, and Rao, 1951 reported that during 1947–48 smut severity was so high, particularly in the variety Co 419 in the district Bellary (Karnataka) and the disease had to be controlled by banning ratoons under the Madras pest and Disease Act. After that the disease has established in almost all the sugarcane growing states of the country, especially Maharashtra, Karnataka, Andhra Pradesh, Tamil Nadu, Uttar Pradesh, Bihar, and Orissa in traces to severe form.

17.2 RATOON CROP OF SUGARCANE

Ratooning is an important feature of sugarcane crop as it grows perennially and the root system or ratoon that remains in the ground will re-sprout from each stalk. Consequently ratoon crops grow faster than the plant crop.

Ratoon ability is an important criterion in all sugarcane screening programs. A variety which possesses multi-ratooning potential is beneficial as it can save the operational costs and of planting material for many years. With the less production cost, the profitability of sugarcane cultivation can be improved. Although several ratoon crops are possible, but due to impact of pests and diseases, ratoon crop eventually lead to declining in yield. Subba Raja et al. (1972) reported the effect of smut incidence on plant and ratoon crops considering the parameters of yield and quality of the cane.

17.3 ECONOMIC IMPORTANCE

The severity of smut as reported from different sugar producing countries has varied widely. Economic losses have ranged from negligible proportion to the serious levels which is enough to threaten the agricultural economy of the area. Disease severity is associated with hot dry climate where crop is suffering to water stress. Crop age and cycle at the time of infection is also important. The disease becomes more severe as the number of ratoons increases (Ferreira and Comstock, 1989). According to Antoine (1961), damage due to smut usually depends on three factors;

- Type of infection (primary or secondary).
- Type of crop (plant or ratoon).
- Time of infection (early or late).

The losses on account of smut can broadly be grouped into two categories:

- Quantitative losses; and
- Qualitative losses.

17.3.1 QUANTITATIVE LOSSES

In smut affected seed cane, the nodal buds lost vigor and in severe cases, the bud dies. Luthra and Sattar, (1942) obtained only 16% germination in sets derived from diseased canes as compare to 44% in healthy canes as the millable canes usually become slender and short in height (Kirtikar and Verma, 1962). Antoine (1961) reported more than 50% yield reduction, while Lovick (1978) reported loss estimates ranging from negligible to 73%. In India, nearly 60% loss in cane weight has been reported in 22 varieties

(Khanna, 1947). Chona (1956) recorded 29 and 23% reduction in the yield of sugarcane from varieties Co312 and Co313, respectively. In these varieties, Vasudeva (1954) also observed 30.6 and 40.5% reduction in cane yield and Kirtikar and Verma (1962) reported 69.2% losses. Subba Raja et al. (1972) recorded more than 67% yield losses in third ratoon of Co 419 in which disease incidence was 40.4%. They found similar trend of losses in other smut susceptible varieties.

17.3.2 QUALITATIVE LOSSES

Smut reduces the cane quality parameters like Brix, Sucrose, and Purity index. Decreases in sugar extractability and recovery was also estimated because of reduction in juice quality. The ratio of sugar to fiber is reduced and Kirtikar and Verma (1962) also reported drop in sucrose from 17.66 to 11.71%. Apparently, smut affected cane contains nearly 10% less juice and 3 to 7% less sucrose (Agnihotri, 1990). Sankpal and Nimbalkar (1980) reported that infected plants contain more reducing sugars but less non-reducing and total sugars than healthy ones. Kumar (2002) assessed the loss in the yield parameters and juice quality due to sugarcane smut, caused by *Ustilago scitaminea* in sugarcane cultivars BO 128 and Co 1158. He conducted the experiment with 3 types of planting materials (I) whip bearing canes, (II) healthy canes from smutted stool, and (III) healthy canes from healthy stool. Reduction in sett germinability, weight of cane and its juice, girth, and height of cane, polarity, brix, purity, commercial cane sugar, non-reducing sugars and ascorbic acid was comparatively of higher magnitude in whip bearing cane than that of healthy cane of diseased stool. Similarly, accumulation of phenol, reducing sugars and amino acid was higher in whip bearing cane in comparison to healthy canes.

17.4 SYMPTOMS

The most recognizable diagnostic feature of a smut infected plant is the emergence of a "smut whip." A smut whip is a curved, pencil-thick growth, gray to black in color that emerges from the top of the affected cane plant. These whips arise from the terminal bud or from lateral shoots on infected stalks (Ferreira and Comstock, 1989). They can vary in length from a few inches to several feet long. The whip is composed partly of host plant tissue and partly of fungus tissue. Whips begin emerging from infected cane by

2–4 months (primary infection) of age with peak whip growth occurring at the 6th or 7th month (secondary infection). Other smut symptoms may be evident before the characteristic whip is seen. Spindle leaves are erect before the whip emerges. Affected sugarcane plants may tiller profusely with the shoots being more spindly and erect with small narrow leaves (i.e., the cane appears as grass-like). Less common symptoms are leaf and stem galls, and bud proliferation (Rott et al., 2000; Agnihotri, 1983).

The smut whip has generally been considered a modified inflorescence. However, the true morphological nature of the abnormal growth has not been established with certainty. Butler (1918) was the first to suggest that the whip was modified inflorescence. Chona (1943) found that smutted clumps produced mummified arrows in which the lower portion consisted of a normal inflorescence with typical flowers while the upper portion of the rachis of the inflorescence was converted into a typical smut whip. Fawcett (1944) mentioned that in some cases the axis of the inflorescence had been found enclosed in the appendage with the chlamydospores. McMartin (1945) considered the 'curious organ' to be an abnormal arrow. Sharma (1956) considered the whip-like structure as a modified stem instead of a converted inflorescence. Besides patent symptoms, occasionally some abnormal symptoms are also observed. These include gall formation, bud proliferation, knife-cut (Byther and Steiner, 1974), shortening of anther stalks, presence of two ovaries in flower, presence of six feathery stigma instead of one bilobed feathery stigma (Sharma, 1956); and production of a typical ovaricolous smut (Bhombe and Somani, 1978). In the north Indian plains, two clear cut flushes of the diseased whips can be observed in a year (Agnihotri, 1983). The first flush appears in May-June due to primary infection and the second one in October-December as a result of secondary infection. However, in tropical India, emergence of smutted whips can be observed round the year.

17.5 CAUSAL ORGANISM

Sugarcane smut is caused by *Sporisorium scitamineum or Ustilago scitaminea* H&P. Sydow, a basidiomycetous fungus (Rott, 2000) that exists in several physiologic races (Grishan, 2001; Agnihotri, 1983; Anon., 1984). Using modern analytical methods, as well as pathogenicity studies with different sugarcane cultivars, researchers have suggested at least six races of the fungus exist worldwide.

17.6 TAXONOMY AND NOMENCLATURE OF PATHOGEN

Sporisorium scitamineum which was earlier known as *Ustilago scitaminea* belongs to the family Ustilaginaceae, order Ustilaginales and the class Basidiomycotina. The pathogen was first described and identified in 1870 by Mundkur in 1939 as *Ustilagosacchari* Rabenh. Sydow (1924) firstly concluded that the fungus causing the true culmicolous smut of sugarcane was quite distinct from *U. sacchari.* And he named it *Ustilago scitaminea* Sydow.

Piepenbring et al. (2002) regrouped the generic position of the sugarcane smut pathogen and renamed it as *Sporisorium scitamineum*. Three species of smut fungi (Ustilaginales, Basidiomycota) of economic importance, *Ustilagomaydis* on corn, *U. scitaminea* on sugarcane, and *U. esculenta* on *Zizanialatifolia*, were investigated in order to define their systematic position using morphological characteristics of the sori, ultrastructure of teliospore walls, and molecular data of the LSU rDNA analysis, which suggested that *U. scitaminea* belong to the genus *Sporisorium*. The sugarcane smut fungus develops sori with whip-shaped axes corresponding to columellae and henceforth, *U. scitaminea* is called *S. scitamineum.*

17.6.1 SYSTEMATIC POSITION OF PATHOGEN

Kingdom: Fungi
Phylum: Basidiomycotina
Sub-Phylum: Ustilaginomycotina
Class: Ustilaginomycetales
Sub-Class: Ustilaginomycetidae
Order: Ustilaginales
Family: Ustilaginaceae
Genus: *Ustilago (Sporisorium)*
Species: *scitaminea (scitamineum)*

17.6.2 CULTURAL AND MORPHOLOGICAL CHARACTERS OF PATHOGEN

The smut fungus can grow luxuriantly on potato dextrose and Richard's media (Galloway, 1936). It can also grow on Czapek's media (Fawcett, 1942) and can utilize in organic nitrogen sources as well as peptone. Singh and Budhraja (1962) found maximum growth of the fungus on asparagines-yeast agar medium at 28°C. Triaone (1980) successfully produced mature teliospores of smut fungus in auxenic culture. The pathogen is characterized by its dark brown; minutely punctate teliospores (5.5 to 7.5 μm diameter) with a thin epispore. Alexander et al. (1985) reported that diameter of spores, collected from different locations in India, varied from 16.5 to 17.04 μm The fungus is pathogenic to *Saccharum officinarum* and other Saccharum species, e.g., *S. spontaneum*, *S. barberi*, *S. sinense,* and *S. robustum* and also to related genera of *Miscanthus, Sclerostachya,* and *Imperata*.

17.7 TELIOSPORES

Teliospores germinate readily under moist conditions, each giving rise to a promycelium of variable dimensions averaging 16 μm long by 3–4 μm wide and usually divided transversely into three or four cells. Each of these cells is capable of producing sporidia (basidiospores), sometimes five or six at a time. The sporidia are hyaline and oval shaped, tapering towards their ends, and measure approximately 6 x 2 μm. The sporidia usually germinate by means of long, septate hyphae or, it may again produce sporidiaunder favorable nutrient conditions. The promycelium is also capable of producing several, long, septate, branched hyphae, each originating from one of the original cells of the promycelium. After hyphal and/or sporidial fusions occur, the dikaryon is formed and becomes the infectious hypha of the organism (Hirschorn, 1943).

Several environmental conditions like temperature and nutrition affect teliospore germination (Saxena and Khan, 1963, 1971; Block, 1964). Temeraturebelow 30°C and decreased sucrose concentration leads teliosporegermination more mycelial and as temperature rises above 30°C and sucrose concentration increased, sporidial germination takes place, therefore germination of teliospores is never completely sporidialnor mycelial.

Although teliospores of *U. scitaminea* have no dormancy so they can remain viable for considerable duration depending upon the storage conditions. Luthra et al. (1938) reported longevity of spores which ranged from 56 to 1306 days at 5°C. Alexander and Ramakrishnan (1978) found viability up to 11 years. Khan and Saxsena (1964) noticed that the temperature and RH were important for determining longevity of teliospores. They found viability of teliospores up to 110–148 days at 0–10°C; 105–128 days at 25°C; 68–90 days at 30°C; 35–60 days at 35°C and 1018 days at 40°C.

At 100 and 98% RH the viability was adversely affected on second and tenth day of storage, respectively. The teliospores readily germinateunder moist conditions (RH 100%). The optimum temperature for the germination of teliospores is 25 to 35°C (Luthra et al., 1938; Appalanarasayya, 1964). Rafay and Agrawal (1959) reported that smut spores could germinate even at 6°C. Waller (1969) also reported that the presence of water was necessary for spore germination. He also noted increase in spore germination by fresh sugarcane debris and washing from sugarcane leaves.

17.8 DISEASE DEVELOPMENT

Young buds are most vulnerable to infection (Chona, 1956). Like all species of *Ustilago, U. scitaminea* is a parasite of young meristematic tissues. Infection is initiated when teliospores are deposited on lateral buds of standing cane. The optimum temperature for production of infection hyphae, promycelia, and sporidia is 31°C (Bock, 1964). On cane surface, germ tube elongates considerably (Waller, 1969), the infectious hyphae penetrate the basal bud scales and invade the meristematic region of the bud (Dastur, 1920; Bock, 1964; Waller, 1970). Till the bud germination the fungus lies dormant and as soon as the meristem becomes active during bud germination, the fungus starts its association with the meristem. During shoot development each bud primordium gets infected with the fungus, whereas, internodal tissues remain free from hyphae (Fawcett, 1944). Then the mycelium become intercellular in meristematic tissues, sends haustoria into the cells to draw the nutrition and establish there (Antoine, 1961).

The apical meristem assumes an intercalary function and acts as a basal meristem for the development of smut whip. The vegetative phase of the fungus changes into reproductive one. The spores initially are not connected directly to hyphae. Later immature teliospores develop and mature in the sorus. The period from infection to whip production is about six months under field conditions (Waller, 1969).

The inoculum that deposits into the soil is carried by irrigation or rain water and on coming in contact with the buds in the basal portion of standing cane, infects them (Chona, 1956). The soil temperature of 25°C is the optimum for infection in the buds (Lambat et al., 1967). Irrigation in the infected field increases the severity of disease. Fresh sugarcane debris and washings from sugarcane leaves stimulate the spore germination at the surface of soils (Waller, 1968).

Alexander (1982) studied the disease progress in smut susceptible and resistant varieties. He observed that the disease incidence in ratoon crop of susceptible variety increases 5–6 times than the resistant one. Due to death of infection foci in the first ratoon of resistant variety, the disease levels in first and second ratoon crops remained at par.

17.9 SURVIVAL OF PATHOGEN IN SOIL

Survival of teliospores is relatively short in field soil under normal soil moisture regimes. James (1969) found that spores can remain viable in dry soil for 16–32 days. In soil of plant crop, the spores were viable for at least 64 days. In wet soils (without crop), spores retained viability for 4–8 days only, whereas under similar moisture conditions in soil with plant crop, the viability extended up to 16 days.

17.10 PERPETUATION OF THE DISEASE

The smut disease perpetuates by plantingsetts of infected shoots and ratooning of smut affected plants (Luthra and Sattar, 1942; Chona, 1956). In plant cane, smut infection usually remains latent in the buds of underground stubble. When such buds are harvested, the shoots that come up bear smutted whips (Agnihotri, 1983). Primary transmission of the smut fungus occurs through planting diseased seed cane while secondary spread is through windblown spores. Spores in or on soil are carried to different fields by rain or irrigation water where they can cause new infections to cane (Agnihotri, 1983).

17.11 HOST RANGE

Sugarcane smut has been reported to infect several other members of poaceae family, e.g., *Imperata arundinacea* (Dabh grass) and *Erianthus saccharoides* (McMartin, 1945), *Saccharums pontaneum* (Chonaand Gattani, 1950), *Sclerostachya fusca* (Khuree or Khilut) and *Saccharum barberi* (Indian cane) (Mundkur and Thirumalachar, 1952), *Zea mays* (Hirschhorn, 1963), *S. robustum* (Robust cane), *S. sinense* (Chinese cane) and *Narenga* spp. (Srinivasan and Alexander, 1965), *Sorghum bicolor* (Shatter cane) (Hutchinson, 1972), *Rottboeilia cochinchinensis* (Itch grass) (Latiza, 1980), *Cyperus dilatatus* (Olufolaji, 1987) and *S. munja* (Munj grass) (Rao et al., 1990). In India, *S. spontaneum* (Kans grass) occurs commonly and is considered a potential source of infection for sugarcane.

17.12 DISPERSAL OF PATHOGEN

Waller (1969) estimated that from infected sori, on an average 8 x 10^6 spores per day are released. Teliospores of the pathogen are blown by wind, some come into contact with standing canes whereas others fall on the soil. According to Sreeramulu (1973), the daytime dispersal of spores are maximum. The maximum dispersal of spores takes place at 24 to 27°C at 50 to 60% RH. The inoculum trapped below the height of the canopy serves as a source for secondary infection. Infection on standing cane may take place by the wind borne spores because young buds are more vulnerable to infection (Chona, 1956) and the spores already present in the soil or carried in by irrigation water are the main source of infection on planted cuttings (McMartin, 1948).

Fawcett (1942) has shown experimentally that spores in the soil die within a few months. According to Chona (1956), the spores of *U. scitaminea* do not have a long resting period, but lost their viability after three to four months. McMartin (1948) referring to experiment carried out in India, mentioned that if the spores are kept under dry condition then more than 70% of spores remain viable even after 210 days, whereas, under moist soil conditions, almost all the spores starts germination within 48 hrs.

Therefore, under moist soil conditions, the chlamydospores would germinate soon after shedding and the fungus may soon die in absence of suitable host plant in the immediate vicinity for infection. Under dry soil conditions the spores may remain viable for few months and germinate when sufficient moisture became available.

Some insects like Endomychid beetles (Francis, 1938), *Brachytarsus zeae*, *Anthicus bifasciatus*, and *Phalacrus* spp. are constantly associated with the smut whips (Hayward, 1943). Bowler et al. (1975) reported that the members of the Acrididae, Anthribidae, Cerambycidae, Chelisochidae, Tettigoniiadae, and Cucujidae feed on whips although these insects play a minor role in teliospore dissemination.

17.13 VARIATION IN THE PATHOGEN AND RACES

Physiologic races of the pathogen exist in nature. Mundkur (1939) studied 73 collections of the fungus and divided them into two species and two varieties.

In Argentina, Hirschhorn (1943) grouped smut into six and Hirschhorn and Astizgasso (1988) suggested at least three races. There are two races in Hawaii denoted as A and B (Comstock and Heinz, 1977), at least two in Brazil (Toffano, 1976; Da Silva and Sanguino, 1978), two races in Taiwan (Leu and Teng, 1974) and five races in Pakistan (Muhammed and Kausar, 1962). Six races using five varieties have been reported by Gillaspie et al. (1983). Peros (1985) compared isolates from different geographic regions for the electrophoretic character of three enzymes viz., esterases, acid phosphatases and leucin amino peptidases. He noted that the bands were similar in number and position. Alexander (1979) concluded that in India there is only one race with eleven biotypes. Variable smut reactions have been reported for the same varieties in the different countries, suggesting that different races may exists in different countries.

Mundkur and Thirumalachar (1952) observed the following distinctive characteristics of the sugarcane smut:

1. ***U. Consimilis* Syd:** Spores chestnut, epispore thick with smooth surface, 3.5 to 6 m in diameter, attacking *Sclerostachya fusca* and *Saccharum spontaneum* L.
2. ***U. scitaminea* Syd:** Spores Rood's—brown, epispore thin, minutely punctuate, 5.5 to 9.5 m in diameter, attacking *Saccharum spontaneum* L.
3. ***U. scitaminea* var. *sacchari*: barberi Mundkur:** Spores mummy brown, epispore thick, minutely verruculose, 5 to 8 m in diameter, attacking *S. barberi, S. officinarum* and *S. spontaneum.*
4. ***U. scitaminea* var. *sacchari*: *officinarum* Mundkur:** Spores Vandyke-brown, epispore medium to thick, 6.5 to 11 m in diameter, attacking *S. officinarum.*
5. ***U. scitaminea*,** like all *Ustilago* species, is a parasite of young meristematic tissues, entering the host exclusively through the lower part of the bud, below the scales (Fawcett, 1944).

17.14 ENVIRONMENTAL FACTOR

17.14.1 EFFECT OF TEMPERATURE

Every organism requires an optimum range of temperature for its activities, *U. scitaminea* can grow at temperatures ranging from 15 to 35°C, with an optimum range from 25 to 35°C (Jones, 1922). Maire (1898) found that the spores of the smut fungus germinate more quickly if the temperature is little raised (20°C to 25°C) and that the optimum temperature for sporidial and filamentous development is 25°C and 30°C. Piemeisel (1917) reported optimum conditions for the propagation of the fungus and budding of sporidia lie between 20°C and 26°C, the maximum at about 40°C and the thermal death point approximately at 46°C. He also found that incubation temperature of 24°C to 38°C did not influence the rate or amount of germination of the smut spores while Appalanarasayya (1964) claimed optimum temperature range 25°C to 35°C for the germination of teliospores.

17.14.2 EFFECT OF PH

Herrera et al. (1995) reported that neutral pH (7.0) is optimum for growth of fungus showing budding yeast like cells and at acidic pH (6.0) it develops into mycelial form while slightly acidic media favored the growth of the pathogen.

17.15 INTEGRATED MANAGEMENT

17.15.1 CULTURAL METHOD

17.15.1.1 ROUGING DISEASED SHOOTS

Rouging of diseased shoots has been widely recommended as a useful and effective means of controlling smut in the field (Antoine, 1961; Lovick, 1978). James (1972) reported that rouging of infected cane resulted in higher yields of cane and sugar than from non-rogued plots and whip removal as the more efficient and economic technique. Francis (1938) reported that good control was obtained through removal of diseased plants from the field and destroying it by immersing them in boiling water for 15 minutes. Chona (1943) obtained effective control of the disease in India through systematic rouging of entire clumps, not by removing individual smutted canes, and careful selection of seed material for three consecutive seasons.

17.15.1.2 PLANTING OF HEALTHY SETTS

This method was used for controlling smut and it became successful in countries where the disease does not become severe or where moderately resistant varieties were planted (Antoine, 1961; Lovick, 1978). To obtain disease free plants, it is essential to take them from certified seed nurseries which is free from disease. Planting material should be taken from a seed crop raised from heat-treated plants (preferably moist hot air at 54°C for 4 hr). The seed crop should be inspected regularly and smutted clumps, if any, should be rogued out. Avoiding ratooning of a disease crop means any plant crop having more than 10% infection should not be maintained for ratoon. In highly susceptible varieties like Co 740 and Co1158, the practice of multiple ratooning should be discouraged to avoid dispersal of the disease. Luthra et al. (1940) stated that the control of smut disease was to avoid the ratooning of diseased crops.

17.15.1.3 ADOPTING SUITABLE CROP ROTATION

For control of smut, crop rotation was first recommended by Robinson (1959) and then by Antoine (1961). Starving smut spores in the soil through crop rotation has been recommended as one control measure. Fawcett (1941) advised a rotation with lucerne, maize or some other non-susceptible crop. Robinson (1959) considered that the rotation which includes a fallow or green manure crop can give good control of smut.

17.15.1.4 USE OF RESISTANT GENOTYPES

Use of resistant varieties is most effective, reliable, and cheap method of disease control. Considerable genetic variability for resistance has been noted in different *Saccharum* species. Resistance against smut in *S. sinense* and *S. robustum* is more than *S. barberi*. Using these proven materials, resistant varieties have been developed time to time and these have been employed to control smut epidemics in various sugarcane growing countries. In India, Co 6806 (Alexander, 1975) and CoLk8102 (Sinha, 1995) had been found highly resistance to smut.

17.15.2 THERMO-THERAPY

Presently, two methods, viz. hot water and moist hot air treatment are being used to control smut. These methods are more effective in eliminating external infections, i.e., killing the spores adhering to the host surface or spores entrapped within bud scales. However, the internal infection of bud is partially or fully control by hot water treatment (50°C for 30 min to 2 hr). Joshi (1954) obtained complete control at 52°C for 18 min. In other words, there is an auto lamination of infected buds during this treatment. The work of various investigators suggests that no single method is effective in controlling the smut.

17.15.3 CHEMICAL METHODS

Treatment of setts with fungitoxicantsis an important tool to manage the disease. In the beginning Organomercurials were widely used (Kirtikar and Srivastava, 1962; McMartin, 1945; Muthusamy and Raja, 1972). Besides Organomercurials, treatment of setts with 0.1% Formalin (Luthra et al., 1940) and Dithane Z-78 (Joshi, 1954) has also been found effective in reducing smut incidence. Benomyl and Oxycarboxin completely suppress the systemic smut disease in sugarcane (Mameghmay, 1984). Conversely, Agnihotri et al. (1973) did not achieve control of smut with vitavax because it was rapidly degraded in plant system.

Muthusamy and Raja (1973) found that Agallol at 0.5% and Dithane-Z-78 at 0.3% both gave satisfactory control of *U. scitaminea* when used as dip treatments for setts before planting. Bailey (1978) found that setts treated for 2 hrs at 50°C with 500 ppm of triadimefon had a significantly lower rate of smut infection. Natarajan and Muthusamy (1981) reported that systemic infection of smut could be reduced by soaking two budded setts in triadimefon (0.1%) for 5 min. Goyal et al. (1983) found that 2 hr dip of 2-budded setts in 0.2% Bayleton [triadimefon] gave the best control of the pathogen (93.2% over the control) even in a highly susceptible cv. Co.1158. Padmanabhan et al. (1987) reported that hot water treatment of setts at 52°C for 30 min. mixed with carbendazim or bayleton offered elimination of systemic infection. Gul et al. (1989) reported that Agallol gave the best control and the highest cane and sugar yield in both plant and ratoon crops.

Sharififar and Kazemi (1999) reported effective control of smut infection through the use of propiconazole as sett dip. Vijaya (2000) assessed 5 fungicides viz., carbendazim @ 0.1%, triadimefon @ 0.1%, chlorothalonil @ 0.2%, thiophanate-methyl @ 0.1% and tridemorph @ 0.1% against settborne smut infection and the results indicated that all the fungicide treatments increased sett germination in comparison with the control. Fungicidal dip in tridemorph for 4 h gave complete control of smut. Satyanarayana et al. (2001) also found propiconazole sett dip treatment as promising method for disease reduction.

Wada (2003) selected suitable sugarcane varieties for resistance to the smut pathogen grown in Nigeria. Plant setts were dipped in smut spore suspension and then in solutions or suspension of six different fungicides for ten minutes and planted in a glass house and field trials and found thatmancozeb, carbendazim + maneb, metalaxyl + carboxin + furathiocarb, pyroquilon, benomyl, and chlorothalonil all wereeffective against *U. scitaminea.* Nageswara Rao and Patro (2006) managed the smut incidence by sett treatment with different fungicides. Pre-inoculation and post inoculation sett dip treatment with propiconazole (0.1%), hexaconazole (0.2%), carbendazim (0.1%) or tridemorph (0.1%), had shown radial reduction in smut incidence. McFarlane et al. (2007) bayleton 25% WP has been used effectively to limit smut development in susceptible cultivars for many years and has been useful in protecting germinating sugarcane from smut infection after hot water treatment. Agar plate spore germination assays were used to evaluate the efficacy of a range of antifungal formulations against *U. scitaminea.* Those that prevented the germination of smut spores were then tested for phytotoxicity to sugarcane in greenhouse experiment. Three fungicides inhibited smut spore germination on agar plates at all evaluated concentrations and had no marked effect on the germination of seed cane in greenhouse experiments. These three formulations were included in an inoculated field trial at the Pongola Research Station. One formulation from Syngenta (fludioxonil: mefenoxam: azoxystrobin) provided acceptable smut control and gave consistently good yields using all methods of application.

17.15.4 PLANT EXTRACTS

Judicious use of potentially hazardous pesticides has been matter of growing concern of both environmentalists and public health authorities. Plant products have traditionally been used by farmers to overcome the problems of pests long before chemical pesticides were invented. So instead of fungicides, plant extracts are the best alternatives available today. The presence of antifungal compounds in higher plants has long been recognized as an important factor to disease control (Mahadevan, 1982). More than two lakhs plant species of the known higher plants are there, approximately, out of which 1500–2000 are said to possess medicinal properties in the Indian system of medicine (Tiwari et al., 1998).

Most commonly used plant species for pest control are *Azadirachata indica*, *Nicotiana tabacum, Annona squamosa, Chrysanthemum cinerasiaefolium*, Derris root, *Sabadila officinals*, *Ryania speciosa, Quasia amara,* etc.

(Buckinghum, 1993). Lal et al. (2002) reported that *in vitro* leaf extracts of several plant species were found effective in inhibiting the mycelial growth of *U. scitaminea.* They also reported the control of smut by using botanicals. Leaf extracts of *Calendula officinalis*, *Solanum nigrum* inhibited mycelial growth and teliospore germination of *U. scitaminea*, they also reported that there was delaying in germination of teliospore by 48 hrs in case of neem extract. Plant extracts from different species are being tried against plant pathogens causing foliar diseases in cereal, oilseed, and vegetable crops (Ganesan, 1994; Valarini et al., 1994; Sardrud et al., 1994). Sindham et al. (1999) suggested that all the plant extracts viz., Onion, Ginger, Neem, Garlic, Mint, Eucalyptus, Tulsi, Datura, Bougainvillea were inhibitory to the growth of *R. solani* and *M. phaseolina* even at 5% concentration.

17.15.5 ESSENTIAL OILS

Essential oils are concentrated, hydrophobic liquids containing volatile aromatic compounds extracted from plants (Isman, 2000). The antifungal activities of some essential oil viz; Palmarosa (Srivastava et al., 2003), Citronella, Lemongrass oil (Srivastava et al., 2001), Ocimum, Cardamom, Curcuma, and Mentha spp. were recorded by Handique and Singh (1990) and Singatwadia and Katewass (2001). Sokovic et al. (2009) from their experiment concluded that essential oils of Thymus and Mentha species possess great antifungal potential and could be used as natural preservatives and fungicides. Antifungal activity of essential oil of Eucalyptus (*Eucalyptus camaldulensis*) and found that it suppressed the mycelial growth of postharvest pathogenic fungi, *Penicillium digitatum*, *Aspergillus flavus*, *Colletotrichum gloeosporioides*, and soil borne pathogenic fungi, *Pythium ultimum*, *Rhizoctonia solani* and *Bipolaris sorokiniana* (Katooli et al., 2011).

17.15.6 BIOLOGICAL CONTROL

Biological control is the reduction of inoculums density or disease producing activities of pathogen or parasite in its active or dormant state by one or more organism, accomplished naturally or through manipulation of environment, host or antagonist or introduction of one or more antagonists (Baker and Cook, 1974). The bio control of plant pathogens can be achieved by either promoting natural bio control agents to reach density enough to support a pathogen or by introduced biocontrol agents. The microorganisms which are used in control of the plant diseases are called as biocontrol agents. Bio control agents include all classes of organism—fungi, bacteria, nematode, protozoa, viruses, and seed plants (Cook and Baker, 1983). Presently more emphasis is being given and use of antagonistic bacteria and fungi for bio control of plant diseases.

17.15.6.1 FUNGAL BIO CONTROL AGENTS

Fungi are the most extensively researched group of biocontrol agents and they have been used against aerial, root, and soil microbes (Whipps, 1993). Antibiosis and the mycoparasitism of thirty two strains *Trichoderma* sp. were tested to see their effect on the germination of teliospores of *U. scitaminea*. Most of these strains showed good antagonism and a strong inhibitory effect on the growth of fungi pathogenic to sugarcane (Martinez et al.1998). Srivastava et al. (2006) observed enhancement in germination and yield parameters of sugarcane in both plant and ratoon crops by the use of *Trichoderma* sp. Lal et al. (2009) reported the control of smut by using botanicals and bioagents (*Trichoderma viride*). Reduction in disease incidence and enhancement in growth parameters in various crops by the use of *Trichoderma* spp. may be due to suppression of deleterious root mycoflora on account of antifungal activity of the bioagent (Harman et al., 1993; Datnoff et al., 1995).

17.15.6.2 BACTERIAL BIOCONTROL AGENTS

There is an increasing attempt to introduce bacterial bio-control agents for managing plant pathogens. In this context, Fluorescent Pseudomonads has received much attention and has been used in lab and field experiments. This is mainly due to the production of variety of metabolites by Fluorescent Pseudomonads. That suppresses the pathogens directly or indirectly (Mathre et al., 1999).

17.15.6.3 ORGANIC AMENDMENT

Effect of amended soil extracts with different types of amendments against *U. scitaminea, in-vitro.* So far, none of the report is available on the effect of soil extracts amended with different types of amendments on the growth of

U. scitaminea, causing smut of sugarcane. Neelam (2010) tested 10 different concentrations of vermiwash against *U. scitaminea* and found that at lower concentration, vermiwash was less effective but as the concentration increases there was inhibition of mycelial growth of *U. scitaminea.*

17.16 VARIOUS CLIMATIC FACTORS AFFECTING SMUT INCIDENCE

India is the native place of sugarcane, standing second after Brazil in the world with respect of its production. Sugarcane is a crop of tropical wet regions requiring a temperature of 27°C and average rainfall above 100 cm, deep fertile, alluvial, and black soil is ideal for sugarcane cultivation. Weather conditions are critically important in the development and spread of the pathogen causing smut of sugarcane. Some of these can be utilized to form the basis of disease prediction. They may vary in their combinations in different agro-climatic zones and influence not only the pathogen but also the host.

Waller (1969) estimated that 8 x 10^6 spores per day are released by a smutted whip. According to Sreeramulu (1973), the daytime dispersal of spores is maximum. The maximum dispersal of spores takes place at 24 to 27°C and 50 to 60% RH. The inoculum trapped below the height of the canopy serves as a source for secondary infection. Infection of the buds may take place, on standing cane, by the wind borne spores. Young buds being more vulnerable to infection (Chona, 1956) and on planted cuttings, by spores already present in the soil or carried in by irrigation water (McMartin, 1948).

17.17 CONCLUSION

Smut is one of the important diseases of sugarcane caused by *Ustilago scitaminea* Syd. (Syn: *Sporisorium scitamineum*) which is causing significant yield reduction. The severity of smut, as reported from different sugar producing countries has varied widely. Economic losses have ranged from negligible proportion to levels serious. It reduces not only cane quality parameters like Brix, Sucrose, Purity, but also decreases sugar extractability and recovery. The continuous and indiscriminate use of chemicals to manage the disease, results in accumulation of harmful chemical residues in soil, water, and in the planting material. Therefore, management of smut disease of sugarcane is a difficult task by any single approach. Weather has a pivotal role which varied differently time to time and therefore it can be conclude that temperature and

RH play an important role in spread and development of the smut disease in field. An analysis of temperature and RH ranges prevailing at the time of maximum dispersal revealed that the optimum conditions ranges and lie between 22–24°C and 50–60%, respectively.

KEYWORDS

- **bacterial bio control agents**
- **essential oils**
- **fungal biocontrol agents**
- **organic amendment**
- **plant extracts**
- **thermo-therapy**

REFERENCES

Agnihotri, V. P., (1983). *Diseases of Sugarcane* (p. 363). Oxford & IBH Publishing Co., New Delhi.

Agnihotri, V. P., (1990). *Diseases of Sugarcane and Sugarbeet* (2nd edn., p. 483). New Delhi, Oxford, and IBH Publishing Co.

Agnihotri, V. P., Singh, K., & Budhraja, T. R., (1973). *Proc. Indian Nat. Sci. Acad., 39*, 561–568.

Alexander, K. C., & Ramkrishna, K., (1978). Studies on the smut disease (*Ustilago scitaminea)* of Sugarcane. Longevity and Viability of teliospores. *Indian J. Sugarcane Technol., 1*, 47–49.

Alexander, K. C., & Srinivasan, K. V., (1966). Sexuality in *U. scitaminea Syd. Curr. Sci., 35*, 603, 604.

Alexander, K. C., (1975). Studies on smut disease *U. Scitaminea Syd.* of sugarcane. *PhD Thesis* (p. 91). Calicut Univ., India.

Alexander, K. C., (1979). *Progress in Resistance Breeding for Red Rot and Smut* (pp. 89–91). Maharashtra Sugar August.

Alexander, K. C., (1982). Growth of foci of infection, secondary spread, and loss in yield in smut *(U. scitaminea Syd.*). *Disease of Sugarcane: Sugarcane Pathol. Newsl., 28*, 3–6.

Alexander, K. C., Mohanraj, D., & Joshi, R., (1985). *Annual Report, Div. of Pathology*. S. B. I., Coimbatore, India.

Allsopp, P., Samson, P., & Chandler, K., (2000). Pest management. In: Hogarth, M., & Allsopp, P., (eds.), "*Manual of cane Growing, " Bureau of Sugar Experimental Stations* (pp. 291–337). Indooroopilly, Australia.

Anonymous, (2009). *Sugarcane Agriculture: Chapter III*. Ministry of Food and Public Distribution, Government of India. Available at: http://fcamin. nic. in/ (accessed on 28 January 2020).

Anonymous, (2009). *United States Department of Agriculture, Food Safety* (p. 3).
Anonymous, (2010). *Deputy Director, Integrated Watershed Development Project*. Kotdwar Garhwal, 'Vermiwash': Production, use and general basic information (Pamphlet).
Anonymous, (2010). FAOSTAT. *FAO Statistics Division*. http://faostat. fao. org (accessed on 14 January 2020).
Anonymous, (2010). *Indian Sugar* (Vol. LIX).
Anonymous, (2011). *Handbook of Agriculture*. ICAR, New Delhi.
Anonymous, (2011). *Uttarakhand Sugarcane Department*. Kashipur.
Anonymous, (2012). *Scientific Report, 2011-12 Sugarcane Breeders and Pathologists Meet.* S. B. I., Coimbatore.
Ansari, M. M., (1995). Control of sheath blight of rice by plant extracts. *Indian Phytopath., 48*, 268–270.
Antoine, R., (1961). Smut. In: Martin, J. P., Abbott, E. V., & Hughes, C. G., (eds.), *Sugarcane Diseases of the World. I* (pp. 326–354). Amsterdam, Elsevier.
Antoine, R., (1964). Smut. In: Martin, J. P., Abbott, E. V., & Hughes, C. G., (eds.), *Sugarcane Disease of the World. I* (pp. 56–60). Amsterdam, Elsevier.
Appalanarasayya, P., (1964). Some physiological studies on sugarcane smut *(Ustilago scitaminea* Syd,). *Indian Phytopathol, 17*, 284–287.
Bailey, R. A., (1978). The effect of hot water treatment on ratoon stunting disease and moisture stress on the incidence of smut in sugarcane. *Proc. Int. Soc Sugarcane Technol., 16*, 327–335.
Baker, R. F., & Cook, R. J., (1974). *Biological Control of Plant Pathogens* (p. 443). W. H. Freeman and Co., San Francisco.
Bhombe, B. B., & Somani, R. B., (1978). *Indian Phytopathology*, *31*, 239, 240.
Bock, K. R., (1964). Studies on Sugarcane Smut (*U. scitaminea*) in Kenya. *Trans. Br. Mycol. Soc., 4*, 403–417.
Bowler, P. A., (1975). Insect feeding on Sugarcane in Hawaii. *Proc. Hawaii Entomol. Soc., 22*, 451–456.
Buckingham, J., (1993). *Dictionary of Natural Products* (p. 250). Chapman and Hall, London.
Bull, T., (2000). The sugarcane plant. In: Hogarth, M., & Allsopp, P., (eds.), "*Manual of Cane Growing"* (pp. 71–83). Bureau of sugar experimental stations, Indooroopilly, Australia.
Butler, E. J., (1906). Sugarcane disease in Bengal. *Mem. Deptt. Agric. (India). Bot. Ser.*, *1*(B), 1–53.
Butler, E. J., (1918). *Fungi and Diseases in Plants*. Calcutta Thacker, Spink & Co.
Byther, R. S., & Steiner, G. W., (1974). *Plant Dis. Reptr.*, *58*, 401–405.
Chona, B. L., & Gattani, M. L., (1950). Kans grass (*Saccharum spontaneum* L.): A collateral host for sugarcane smut in India. *Indian J. Agric. Sci., 20*, 359–362.
Chona, B. L., (1943). Sugarcane smut and its control. *Indian Fmg., 4*, 401–404.
Chona, B. L., (1956). Presidential address, pathology section. *Proc. Intern. Soc. Sugarcane Techno.9th Congr.* (pp. 975–986).
Comstock, J. C., & Heinz, D. J., (1977). A new race of culmicolous smut of sugarcane in Hawaii. *Sugarcane Pathol. Newsl., 19*, 24–25.
Comstock, J., Ferreira, S. A., & Tew, T. L., (1983). Hawaii's approach to control of sugarcane smut. *Plant Dis., 67*, 452–457.
Cook, R. J., & Baker, R. F., (1983). The nature and paradise of biological control of plant pathogens. *The Amr. Phytopathol. Soc*. (p. 539). Stpaul, Minn.
Cox, M., Hogarth, M., & Smith, G., (2000). Cane breeding and improvement. In: Hogarth, M., & Allsopp, P., (eds.), "*Manual of Cane Growing"* (pp. 91–108). Bureau of sugar experimental stations, Indoroopilly, Australia.

Croft, B., Magarey, R., & Whittle, P., (2000). Disease management. In: Hogarth, M., & Allsopp, P., (eds.), "*Manual of Cane Growing*" (pp. 263–289). Bureau of Sugar Experimental Stations, Indooroopilly, Australia.

Daniels, J., & Roach, T., (1987). In: Heinz, D. J., (ed.), *Sugarcane Improvement Through Weeding* (pp. 7–84). Amsterdam, Elsevier.

Dasilva, W. M., & Sanguino, A., (1978). Evaluating reaction of American cane varieties to *U. scitaminea* in Brazil. *Sugarcane Pathol. Newsl.*, *21*, 10–11.

Dastur, J. F., (1920). The mode of infection by smut in sugarcane. *Ann. Bot.*, *34*, 391–397.

Datnoff, L. E., & Pernezny, K. L., (1995). Biological control of *Fusarium* crown and root rot of tomato in Florida using *Trichoderma harzanium* and *Glomus intrardices*. *Biol. Contrl.*, *5*, 427–431.

Diner, A. N., Mott, R. L., & Amerson, H. V., (1984). Cultured cells of white pine showing genetic resistance to axenic Blister rust hyphae. *Science*, *24*, 407–408.

Durairaj, V., Natarajan, S., Ahmed, N. J., & Padmanabhan, D., (1972). Results of some experiments on smut of sugarcane in Tamil Nadu State, India. *Sugarcane Pathol. Newsl.*, *8*, 34, 35.

Fawcett, G. L., (1942). Notassobre el 'carbon' de la Cana de Azucar. *Circ. Estac. Exp. Agric.* (p. 114). Tucuman.

Fawcett, G. L., (1944). El 'carbon' de la cana de Azucar. *Bol. Estac. Exp. Agric.* (p. 47). Tucumar.

Fawcett, G. L., (1946). Departmento de Botanica y Fitopato-Logica. Ex Memoria annual Delano 1945. *Rev. Induster-Agric. Tucuman.*, *36*, 165, 166.

Ferreira, S. A., & Comstock, J. C., (1989). In: Ricaud, C., Egan, B. T., Gillaspie; A. G. J., & Hughes, C. G., (eds.), *Disease of Sugarcane* (pp. 211–229). Amsterdam. Elsevier.

Francis, C. B., (1938). Sugarcane smut. *Madras Agric. J.*, *26*, 468–474.

Galloway, L. D., (1936). Report of the imperial mycologists. *Sci. Rept. Agric. Res. Inst., Pusa*, *35*, 120–130.

Ganesan, T., (1994). Antifungal properties of wild plants. *Advances in Plant Sciences, 7*, 185–187.

Gangrade, S. K., Shrivastava, R. D., Sharma, O. P., Jain, N. K., & Trivedi, K. C., (1991). *In vitro* antifungal effect of essential oils. *Indian Perfumer*, *35*, 46–48.

Gillaspie, A. G., Mock, R. G., & Dean, J. L., (1983). Differentiation of *U. scitaminea* isolates in green house tests. *Plant Dis*., *67*, 373–375.

Goyal, S. P., Vir, S., Beniwal, M. S., & Bishnoi, S. S., (1983). Efficacy of fungicides in controlling Sugarcane smut (*Ustilago scitaminea Sydow*). *Indian-Sugar*, *33*, 463, 464.

Grover, R. K., & Moore, J. D., (1962). Toximetric studies of fungicides against brown rot organisms, sclerotiafructicola and *S. laxa*. *Phytopathol.*, *52*, 876–880.

Gul, F., & Hassan, S., (1989). Efficacy of different fungicides to control whip smut of sugarcane. *Sarhad. J. Agric.*, *5*, 87–89.

Handique, A. K., & Singh, H. B., (1990). Antifungal action of lemongrass oil on some soil borne plant pathogens. *Indian Perfumer*, 232–234.

Harman, G. E., Chet, I., & Baker, R., (1933). Factors affecting *Trichoderma harzianum* applied to seed as a biocontrol agent. *Phytopathology*, *71*, 569–572.

Hayward, K. J., (1943). El 'carbon' de li Cana y losansectos. *Circ. Estac. Agric. Tucuman*, 123.

Heinz, D. J., Krishnamurthy, M., Nickell, L. G., & Maretzki, A., (1987). Tissue and organ culture in sugarcane improvement. In: Reinert, J., & Bajaj, Y. P. S., (eds.), *Plant Cell, Tissue, and Organ Culture* (pp. 3–17). Berlin, Springer Verlag.

Herrera, J. R., (1995). Yeast-mycelial dimorphism of haploid and diploid strains of *Ustilago maydis*. *Microbiology*, *141*, 695–703.

Hirschhorn, E., & Astiz, G. M. M., (1988). Physiological specialization of sugarcane smut (*U. scitaminea Syd.*) in the Republic of Argentina. *Fitopatologia, 23*, 10–14.
Hirschhorn, E., (1943). Algunoscarateres del 'carbon' de la Cana de azucaren la Argentina (*Ustilago scitaminea* Syd,). *Notas Mus. La Platl., 8*, 23–39.
Holder, D. G., & Dean, J. L., (1979). Screening of sugarcane smut resistance in Florida. *Sugar J., 41*, 18–19.
Holliday, M. J., & Klarman, W. L., (1979). Expression of disease reaction types in soybean callus from resistance and susceptible plants. *Phytopathology, 69*, 576–578.
Hutchinson, P. B., (1970). A standardized rating system for recording varietal resistance to sugarcane disease. *Sugarcane Pathol. Newsl., 5*, 7.
Hutchinson, T. B., (1972). Alternate hosts for disease of Sugarcane. *Sugarcane Pahtol. Newsl., 8*, 36–38.
Isman, B. M., (2000). Plant essential oils for pest and disease management. *Crop Prot., 19*, 603–608.
James, G. L., (1969). Smut susceptibility testing of Sugarcane verities in Rhodesia. *Proc. S. Afr. Sugar Technol. Assoc., 43*, 85–91.
James, G. L., (1969). Viability of Sugarcane smut spores in the soil. *Sugarcane Pathol. Newsl., 3*, 10–12.
James, G. L., (1972). A summary of varietal resistance ratings to smut in Rhodesia 196371. *Sugarcane Pathol. Newsl., 8*, 14.
Jones, E. S., (1922). Influence of temperature on the spore germination of *Ustilagozeae. Journal of Agricultural Research, 24*, 303–305.
Joshi, N. C., (1954). Chemotherapy against sugarcane diseases. *Indian Sugar, 4*, 343.
Katooli, N., Raheleh, M., & Seyed, E. R., (2011). Evaluation of eucalyptus essential oil against some plant pathogenic fungi. *Journal of Plant Breeding and Crop Science, 3*, 41–43.
Khan, A. M., & Saxena, S. K., (1964). Effect of temperature and relative humidity on the viability of chlamydospores of *Ustilago scitaminea. Indian Sugarcane J., 9*, 55.
Khanna, K. L., (1947). Annual report. *Cent. Sugarcane Res. Stn. Pusa* (p. 117). Bihar.
Kirtikar, & Varma, H. S., (1962). *A Review on Effect of Sugarcane Diseases on Yield and Juice Quality in U. P. Indian Sugar, 12*, 103–108.
Kumar, B., Kumar, S., & Rai, B., (2002). *Annals of Agri-Bio Research, 7*, 69–71.
Lal, R. J., Sinha, O. K., Bhatnagar, S., Lal, S., & Awasthi, S. K., (2009). Biological control of Sugarcane smut through botanicals and *Trichoderma viride. Sugar Tech., 11*, 381–386.
Lal, R. J., Srivastava, S. N., & Sinha, O. K., (2002). *Fungitoxicity of Some Botanicals Against Fungal Pathogens of Sugarcane* (pp. 106, 107). International symposium on "food, nutrition, and security through diversification in sugarcane production and processing system" held at I. I. S. R., Lucknow.
Lambat, A. K., Chenulu, V. V., & Chona, B. L., (1967). Influence of soil temperature of Sugarcane by the smut fungus *Ustilago scitaminea. Indian Phytopathol., 19*, 237, 238.
Latiza, A. S., (1980). Host range of *U. scitaminea* Syd. in the Philippines. *Sugarcane Pathhol. Newsl., 24*, 11–13.
Leu, L. S., & Teng, W. S., (1974). Culmicolous smut of Sugarcane in Taiwan. V. Two pathogenic strains of *U. scitaminea. Proc. Intern. Soc. Sugar Cane Technol., 15th Cong.*, 275–279.
Leu, L. S., Teng, W. S., & Wang, Z. N., (1976). Culmicolous smut of sugarcane in Taiwan. IV. Resistance trial. *Taiwan Sugar Exp. Stn. Res. Rep., 74*, 37–45.
Lovick, G. L., (1978). Smut of sugarcane—*Ustilago scitaminea. Rev. Plant Pathol., 57*, 181–188.

Luthra, J. C., & Sattar, A., (1942). Control of Sugarcane smut. *Indian Fng., 3*, 594.
Luthra, J. C., Sattar, A., & Sandhu, S. S., (1938). Life history and modes of perpetuations of smut of Sugarcane (*U. scitaminea Syd.*). *Indian J. Agric. Sci., 8*, 849–861.
Luthra, J. C., Sattar, A., & Sandhu, S. S., (1940). Experiments on the control of smut of Sugarcane (*Ustilago scitaminea* Sydow). *Proc. Indian Acad. Sci. B., 12*, 118–128. (RAM 20:135).
Mahadevan, A., (1982). Biochemical aspects of plant disease resistance. *Part I, Performed Inhibitory Substances* (pp. 425–431). New Delhi: Today and tomorrow's Printers and Pub.
Maire, R., (1898). Note sur us développementsaprophytiqueet sur la structure cytologique des sporidies-levûres chez l' Ustilagomaydis. *In Bul. Soc. Mycol., 14*, 161–173.
Mameghmay, R. S., (1984). Chemotherapeutic effects of fungicides on sugarcane systemically infected by smut. *Sugarcane, 1*, 3.
Martinez, B., Gonzales, R., & Balance, C., (1998). Antagonism of *Trichoderma* spp. strains on some Sugarcane pathogens. *Phytopatol., 33*, 207–211.
Mathre, D. E., Cook, R. J., & Callum, N. W., (1999). From discovery to use: Transversing the world of commercializing biocontrol agents for plant disease control. *Plant Dis., 83*, 972–983.
McFarlane, K., Moodley, D., Chinnasamy, G., & McFarlane, S. A., (2007). *XXVI Congress, International Society of Sugar Cane Technologists* (pp. 1031–1035). ICC, Durban, South Africa.
McMartin, A., (1945). Sugarcane smut: Reappearance in natal. *S. Afr. Sugar. J., 29*, 55–57.
McMartin, A., (1948). Sugarcane smut: A report on visits to the Sugarcane estates of Southern Rhodesia and Portuguese East African, with general observation on the disease. *S. Afr. Sugar J., 32*, 737–749.
Morton, D. J., & Stroube, W. H., (1955). Antagonistic and stimulatory effect of soil microorganisms upon Sclerotium. *Phytopathology., 45*, 417–420.
Muhammed, S., & Kausar, A. G., (1962). Preliminary studies on the genetic of sugarcane smut, *Ustilago scitaminea* Sydow. *Biologia., 8*, 65–74.
Mundkur, B. B., & Thirumalachar, M. J., (1952). *Ustilaginales of India* (p. 84). Common wealth Mycological Institute, Kew England.
Mundkur, B. B., (1939). Taxonomy of Sugarcane smuts. *Kew Bull., 10*, 523–533.
Muthusamy, S., & Raja, K. T. S., (1973). Fungicides in the control of Sugar cane smut. *Sugarcane Pathol. Newsl., 10*, 6, 11–13.
Nageswara, R., & Patro, (2006). Management of Sugarcane smut through sett treatment with fungicide. *J. Mycol. Pl. Pathol., 36*, 456–459.
Piepenbring, M., Stoll, M., & Oberwinkler, F., (2002). The generic position of *Ustilago maydis, Ustilago scitaminea*, and *Ustilago esculenta* (Ustilaginales). *Mycol. Progress, 1*, 71–80.
Riley, I. T., Jubb, T. F., Egan, B. T., & Croft, B. J., (1999). First outbreak of sugarcane smut in Australia. *Proc. XXIII-ISSCT-Congress* (Vol.2, pp. 333–337). New Delhi India.
Rott, P., Bailey, R. A., Comstock, J. C., Croft, B. J., & Saumtally, A. S., (2000). *A Guide to Sugarcane Diseases.* Montpellier, France: CIRAD Publication Service.
Sankpal, S. D., & Nimbalkar, J. B., (1980). Physiological studies on the smut infected sugarcane variety Co.740. *Indian J. Expt. Biol., 18*, 95, 96.
Sokovic, M. D., Vukojevic, J., Marin, P. D., Brkic, D. D., & Vajs, V., (2009). Chemical composition of essential oils of *Thymus* and *Mentha* species and their antifungal activities. *Molecules, 14*, 238–249.
Sreeramulu, T., (1973). Aeromycological observations and their implications in the epidemiology of some diseases of Sugarcane. *Indian Natn. Sci. Acad. Bull., 46*, 506–510.

Subba Raja, K. T., Natrajan, S., Ahmed, J., & Padmanabhan, D., (1972). Effect on smut incidence in plant and ratoon crops on yield and quality. *Sugarcane Pathol. Newsl., 9*, 9–11.
Subramanian, T. V., & Lakshmipati, R. V., (1951). Infection and development of *Ustilago scitaminea* Sydow in sugarcane. *Proc. Sugarcane Res. Workers, India, 1st Conf.* (pp. 55–63).
Sydow, H., (1924). Notes on the *Ustilaginaceae*. *Ann. Mycol., 22*, 277–291.
Waller, J. M., (1969). Sugarcane smut in Kenya. I. Epidemiology. *Trans. Br. Mycol. Soc., 52*, 139–151.

CHAPTER 18

Diseases of Tobacco (*Nicotiana tabacum* Linn.) and Their Management

K. JAYALAKSHMI, [1] H. RAVINDRA, [2] and J. RAJU[3]

[1]*ICAR-NAARM, Hyderabad, Telangana–500030, India*

[2]*AINRP (Tobacco), ZAHRS, University of Agricultural and Horticultural Sciences, Shivmaogga, Karnataka–577201, India*

[3]*Plant Quarantine station, Ministry of Agriculture and Farmers Welfare, Government of India, Mangalore, Karnataka–575011, India*

Tobacco (*Nicotiana tobaccum* L.) is one of the major commercial crops earning sizeable foreign exchange and internal revenue is susceptible to several fungal, viral, bacterial, and nematode diseases both in nursery and field. It affects both the quality and yield of exportable leaf. The yield losses caused by several diseases may vary from 10–15%.

18.1 NURSERY DISEASES

18.1.1 DAMPING OFF

Damping-off is the major widespread nursery disease of tobacco causes pre-emergence, post-emergence damping off.

18.1.1.1 SYMPTOMS

- Pre-emergence damping off: Germinating seedlings are infected and wither before emergence from the soil.
- Post-emergence damping off: On stem region minute water soaked lesions appear and soon girdling appears leading to toppling over of the seedlings.

- Young seedlings in the nursery are destroyed in patches and infection spreads quickly and under favorable conditions, the entire seedlings in the nursery are killed within 3 to 4 days.

FIGURE 18.1 Damping off symptoms.

- On the surface of the soil, a thin mat of mycelium may be seen.
- The pathogen spreads quickly and affect the entire seed bed causing enormous loss of seedlings.

18.1.1.2 CAUSAL ORGANISM

Pythium aphanidermatum (Edson) Fitzp.

18.1.1.3 DISEASE CYCLE

1. **Primary Source of Inoculum:** By soil-borne oospores.
2. **Secondary Spread:** By sporangia and zoospores disseminated by wind and irrigation water.

18.1.1.4 EPIDEMIOLOGY

Excess number of seedlings, Ill drained nursery beds, Heavy shade in the nursery, 90–100% relative humidity (RH), High soil moisture, Low temperature (below 24°C) are condition of disease development.

18.1.1.5 MANAGEMENT

- Deep summer plowing of nursery area.
- Soil solaraization during April and may for four weeks
- Raised seed beds of 15–45 cm height preferable to avoid water stagnation.
- Using optimum seed rate to avoid overcrowding of seedlings
- Rabbing the seed bed with slow burning of agricultural wastes like paddy husk, bajra cobs, maize cobs.
- Regulated watering to avoid excessive dampness.
- Application of *Trichoderma viride* or *T. harzianum* 10 g/m^2.
- Soil application of 0.4% Bordeaux mixture or 0.2% Copper Oxychlorideor 0.2% Metalaxyl +Mancozeb, two days before sowing and 20 and 30 days after germination.

18.1.2 BLACK SHANK

Black shank, *Phytophthora parasitica* var. *nicotianae* is a major disease of tobacco results in considerable loss of the crop.

18.1.2.1 SYMPTOMS

- Black shank disease majorly affects the roots and base of the stem.
- Initially produces minute black spot on the stem region and spreads along the stem to produce irregular black patches and often girdling occurs.
- The coalesce of black spots leads to development of necrotic patches on the stems.
- The affected stem is split open, the pith region is found to be dried up in disc-like plates showing black discoloration.
- The pathogen also spreads to the leaves and causes blighting and drying of the bottom leaves.

FIGURE 18.2 Black Shank symptoms.

18.1.2.2 CAUSAL ORGANISM

Phytophthora parasitica var. *nicotianae.*

18.1.2.3 DISEASE CYCLE

The pathogen survives as a saprophyte on infected crop residues in the soil. Primary infection is through Oospores present in the soil and secondary spread takes place by sporangia or zoospores disseminated by wind or irrigation water and agriculture implements.

18.1.2.4 EPIDEMIOLOGY

Intermittent rainfall, cloudy weather, prolonged dampness and low temperature (below 22°C) favors for sudden out break of the disease and also high population of root knot nematode, *Meloidogyne incognita,* favors the disease leads disease complex.

18.1.2.5 MANAGEMENT

- Removal and destruction of the infected plants in the field.
- Using disease free seedlings for transplanting.
- Soil application of *Trichoderma viride or T. harzianum* @2.5 kg with Farmyard manure (FYM) (100 kg).
- Soil drenching or spot application of 0.2% Bordeaux mixture or Copper oxychloride or Metalaxyl + Mancozeb during planting and periodical sprays in 15 days interval gives good control.

18.2 LEAF SPOT DISEASES OF TOBACCO

18.2.1 FROG EYE LEAF SPOT

18.2.1.1 SYMPTOMS

- Initially minute small circular, brown or tan with grey center spots appears on lower leaves and gradually spread to upper leaves.
- The spot has a white center, by a dark brown margin, resembling the eyes of a frog hence the name as frog eye leaf spot.
- Under severe condition the spots coalesce to form large necrotic areas, causing the leaf to dry up from the margin and wither prematurely and also exhibit shot hole symptoms.
- Theses spots may affect both yield and quality.

18.2.1.2 CAUSAL ORGANISM

Cercospora nicotianae.

18.2.1.3 DISEASE CYCLE

1. **Primary Source of Inoculums:** By infected plant debris in the soil.
2. **Secondary Source of Inoculums:** By air-borne conidia.

FIGURE 18.3 Frog eye leaf spot symptoms.

18.2.1.4 EPIDEMIOLOGY

High RH (80–90%), temperature around 27°C, Close spacing of plants, using of high nitrogenous fertilizers favors the spread of the disease.

18.2.1.5 MANAGEMENT

- Destruction of affected plant debris.
- Avoid the application of excess nitrogenous fertilization.
- Spraying of 0.2% Bordeaux mixture or Thiophanate Methyl or Carbendazimor Zineb or 0.1% Benomyl 2–3 times at monthly interval.

18.2.2 BROWN LEAF SPOT

18.2.2.1 SYMPTOMS

- Small brown concentric circular lesions appear on older lower leaves.

- The spots spread to upper leaves, petioles, and stalks, and in severe condition the spots may coalesce resulting in leaf blight.

18.2.2.2 CAUSAL ORGANISM

Alternaria alternate.

18.2.2.3 DISEASE CYCLE

The fungus survives in the affected plant debris of tobacco stems, weeds, and other hosts. Under favorable weather condition, conidial production starts which infect the lowermost leaves as the condition progresses, repeated infection cycles of the fungus attack healthy tissues of all aerial parts of tobacco of any age under high RH and moisture. Fungus persists as a mycelium in dead tissue for several month. Secondary spread takes place through airborne conidia.

18.2.2.4 MANAGEMENT

- Removal and destruction of affected plant debris.
- Spraying of fungicides, 0.2% Bordeaux mixture or zinebor, Copper oxychloride or Mancozeb, two or three times at 15 days intervals.

18.2.3 ANTHRACNOSE

This disease can be seen in any stage of the crop. The disease can cause heavy loss when favorable conditions to the pathogen prevail.

18.2.3.1 SYMPTOMS

- Small, light green to white water soaked lesions develop on young leaves. These lesions enlarge in wet weather to form oily, circular spots. Further these spots dry up, become papery, thin, grey white surrounded by brown border.
- Affected leaves become wrinkled and distorted.
- In severe cases elongated, black or brown lesions on the midrib and petiole are seen.

- The disease spreads on the stem and produces several elongated or oblong sunken brown lesions.

18.2.3.2 CAUSAL ORGANISM

Colletotrichum tabacum.

18.2.3.3 DISEASE CYCLE

1. **Primary Source of Inoculums:** Affected plant debris of aerial parts left in the soil and dried plant debris.
2. **Secondary Source of Inoculums:** by air borne conidia.

18.2.3.4 EPIDEMIOLOGY

Optimum temperature 18°C and high RH are favorable for the outbreak of this disease.

18.2.3.5 MANAGEMENT

- Grow resistant varieties.
- Use healthy pathogen free seed for Sowing/planting.
- Seed treatment with Carbendazim @2 g/kg seed.
- Foliar spray, twice, with carbendazin @ 0.1% at 15 days interval or chlorothalonil @ 0.2% or Thiophanate-methyl @ 0.1% or mancozeb @ 0.2% at 10–15 days interval.

18.2.4 POWDERY MILDEW

18.2.4.1 SYMPTOMS

- Ash colored powdery mycelial mat appears on the upper surface of the leaf.
- In severe case the affected leaves turn brown and show a scorched appearance.

18.2.4.2 CAUSAL ORGANISM

Erysiphe cichoracearum.

18.2.4.3 DISEASE CYCLE

1. **Primary Infection:** By Cleistothecia which survive in the plant debris or in soil.
2. **Secondary Infection:** Air-borne conidia.

18.2.4.4 EPIDEMIOLOGY

Application of excessive nitrogen causing vigorous vegetative growth and close planting which increase the humidity around the plants and shade are predisposing factors for infection and development of the disease. The fungus normally attacks maturing plants in the field.60–75% RH and 16 to 23°C optimum temperature favors the infection.

18.2.4.5 MANAGEMENT

- Destruction of infected plant debris.
- Dusting sulfur powder at 40 kg/ha to soil between plant rows 6–8 weeks after planting. Ash or sand is to be mixed with sulfur for easy application. Care to be taken that sulfur does not fall on tobacco leaves. Recommended for black soils only.
- Spraying 0.2% Karathane or 0.05% Carbendizim just before the disease sets in and repeating the same at 10–12 days interval if necessary.

18.2.5 WILD FIRE DISEASE OF TOBACCO

The name of the wildfire disease, results from the burnt appearance of heavily infected tobacco plants. In India, this disease having major importance which can cause devastating losses, especially in wet seasons. The bacteria that cause wildfire and angular leaf spot are identical in all respects except that the wildfire bacteria produce a toxin.

18.2.5.1 SYMPTOMS

- Small brown or black water soaked lesion, surrounded by a broad chlorotic halo and has angular margins because the lesion is confined by the lateral veins.
- The lesions increase in diameter and may coalesce until the diseased tissue eventually falls out leaving ragged holes.
- Wildfire can be systemic in seedlings, causing distortion of the apical bud and leaves.

18.2.5.2 CAUSAL ORGANISM

Pseudomonas syringae pv. *Tabaci.*

18.2.5.3 DISEASE CYCLE

The bacteria are spread in wind and rain splash within the plants and within the field, and from infected weed hosts. These diseases spread by the seed also. Affected plant debris should always be destroyed at the end of the season, as they are sources of inoculum to infect overwintering weed hosts.

Secondary spread of the disease is usually observed after rain storms, with the direction of spread determined by the wind.

18.2.5.4 EPIDEMIOLOGY

Cloudiness, inclement wind with rain splashing are auspicious factors for infection and rapid spread of the disease. Excess soil moisture accompanied with high sultriness, excess nitrogen and low potassium alimentation withal enhance the disease incidence.

18.2.5.5 MANAGEMENT

- Removal and destruction of diseased debris and susceptible weeds from the seedbeds.
- Destruction of affected seedlings and using disease free seedlings for transplanting.

- Application of 1% Bordeaux mixture 1% in the nurseries. Under field condition the disease was effectively managed by two to three sprays with 0.5 g Streptocycline + 3 g copper oxychloride in one liter water at 15 days interval.

18.2.6 TOBACCO MOSAIC

Mosaic is the most important and very common virus disease on tobacco in India appearing in every tobacco growing tract of the country.

18.2.6.1 SYMPTOMS

- Disease observed in any stage of the crop.
- Primarily the disease appears as light discoloration along the veins of the youngest leaves.
- Soon the leaves develop a characteristic light and dark green pattern, the dark green areas are usually associated with the veins.
- The plants that become infected early in the season leads tunting with small, chlorotic, mottled, and curled leaves.

FIGURE 18.4 Tobacco mosaic virus symptoms.

- Under hot weather dark brown necrotic spots develop on leaves and this symptom is called "Mosaic burn" or "Mosaic scorching."
- The plants are infected in the early stages show the stunted growth reducing the yield and quality considerably. If the disease occurs when the crop is full grown, the tobacco yields are not affected.

18.2.6.2 CAUSAL ORGANISM

Tobacco mosaic virus *(Mormor tabaci)* or *Nicotiana* virus.

18.2.6.3 DISEASE CYCLE

The virus has a wide host range, affecting nearly 50 plant species belonging to nine different families. Virus produces different types of symptoms on several species of *Nicotiana*, tomato, brinjal, chilli, *Datura stramonium*, *Solanum nigrum,* and *Petunia.*

18.2.6.4 TRANSMISSION

TMV is contagious disease. This disease spreads through contact by labor and implements used for intercultivation, indiscriminately touching diseased and healthy plants, use of tobacco products by workers while working in field, presence of susceptible weeds and crop plants near fields are some of the factors responsible for disease development. The virus is sap-transmissible and enters the host through wounds.

18.2.6.5 MANAGEMENT

- Grow resistant varieties viz., Jayasree (MR) or VT 1158.
- TMV is a contagious disease, the following sanitary measures are suggested to prevent the disease spread.
- Washing hands in soap water before and after field operations, rouging the diseased plants early in the season and avoiding smoking during field operations.
- Prophylactic sprays with virus inhibitors of 1% neem leaf extracts on 30th, 40th, and 50th day of planting tobacco.

- Removal of mosaic affected plants within 3 weeks after planting and replaced with healthy seedlings.

18.2.7 TOBACCO LEAF CURL

18.2.7.1 SYMPTOMS

- Affected leaves show slight wrinkles and drooping. Under severe conditions the leaves and stem curls and leaves become thick and shrinkled and unfit for curing. Sometimes enations are also seen on under side of the upper leaves.
- Diseased leaves show vein clearing, puckering of leaves, downward curling of leaf margins; leaves become brittle; thickening of veins.

FIGURE 18.5 Tobacco leaf curl symptoms.

18.2.7.2 CAUSAL ORGANISM

Leaf curl virus (*Ruga tabaci*)

18.2.7.3 EPIDEMIOLOGY

Viruses are transmitted by whiteflies *(Bemisia tabaci)*. Since whiteflies spread disease, conditions such as too much bushy vegetation around tobacco fields serve as breeding places for whiteflies and thus spread the disease. Growing of brinjal, sunflower nearby tobacco fields also encourage build-up of whiteflies.

18.2.7.4 MANAGEMENT

- Removal and destruction of alternate weed hosts around nursery area.
- Installing Yellow-sticky traps (20 cm x 15 cm size galvanized iron sheet painted with yellow color coated with castor oil) at 5 per acre.
- If the whiteflies population reached ETL insecticides *viz*., Imidacloprid 200 S. L. @ 2.5 ml in 10 l of water or Thiamethoxam 25 WG @ 2 gm in 10 l of water may be sprayed at 10 days interval commencing from 4 weeks after germination.

18.2.8 ROOT-KNOT NEMATODE (RKN)

Root knot nematode, *Meloidogyne incognita* is prevalent in most of the light soil nurseries. Sandy soils and red sandy loams with adequate soil moisture favor the disease development. Several weed hosts growing in and around nurseries serve as alternate hosts of root knot nematodes. They are responsible for build-up of the nematodes population in the soil.

18.2.8.1 SYMPTOMS

- Yellowing of leaves, stunted growth of seedlings, wilting of plants, and premature death of seedlings leading to patches in seed beds.
- Several galls on roots were seen on the affected seedlings when pulled out from soil.

FIGURE 18.6 Root knot symptom in tobacco.

- Galls vary in size, coalesce, and produce multiple galls which give the seedlings a sickly appearance.
- Root knot affected seedlings do not establish well in the field resulting in heavy gaps.

18.2.8.2 CAUSAL ORGANISM

Meloidogyne incognita, Meloidogyne javanica,

18.2.8.3 MANAGEMENT

- Deep summer plowing during March and April months in the nursery area.
- Preparation of raised nursery bed soil solarization for 3 to 4 weeks during April and May.
- Rabbing the seed beds with paddy husk or slow burning farm waste material to destroy nematode population and egg masses.

- Avoiding the nursery site where or tomato was grown previously.
- Rotation of the nursery area with crops resistant to nematodes like groundnut, redgram, marigold, cotton, gingelly, and chillies is recommended. If such rotation crops are grown for more than three years the nematodes can be controlled.
- Nursery site should be changed every year. If one year the nursery is infected with nematodes, nursery should not be grown in that site next year.
- Application of organic amendments like vermicompost, neem cake, acacia compost at 1.0 ton/ha.
- Application of carbofuran 3G at 0.3% a. i/m^2 or *Trichoderma harzianum* or *Paecilomyces lilacinus* or *Pochonia chlamydosporia* or *Pseudomonas fluorescens* 20 g/m^2 to the nursery bed before sowing.
- Application of carbofuran 3G at 15 kg/ha or *Trichoderma harzianum* or *Paecilomyces lilacinus* or *Pochonia chlamydosporia* or *Pseudomonas fluorescens* 2.5 kg/ha along with 2.5 tons of FYM to the soil before transplanting.

18.2.9 BROOMRAPE (OROBANCHE)

18.2.9.1 SYMPTOMS

- Affected plants become stunted, leaves turn pale green and wilt.
- Initially leaf tips droop and as the attack intensifies, all the leaves wilt with characteristic ribbing of midribs.
- Orbanche shoots emerge from the ground in clusters near the base of the plant.

18.2.9.2 CAUSAL ORGANISM

Orobanche cernua complete (Holo) root parasite.

Orobanche cernua is a affecting the yield and quality of tobacco. The shoots emerge in clusters and their basal portion is attached to tobacco roots through which it draws nourishment and depletes the host resulting in 25–50% yield loss.

FIGURE 18.7 Orbanche in tobacco.

18.2.9.3 EPIDEMIOLOGY

High soil moisture due to irrigation or rain after planting, low soil temperature during winter months encourage heavy incidence of orobanche.

18.2.9.4 MANAGEMENT

- Deep plowing twice or thrice in summer for killing buries/seed orobanche is beneficial.
- In orobanche infested sick fields, tobacco crops do not growing for one or two seasons.

- Avoiding growing of brinjal, tomato, and bhendi crops in orobanche infested sick fields.
- Periodical removal of orobanche shoots before flowering and setting of seeds reduces the menace.
- The orobanche shoots should be destroyed by burning.
- Growing trap crops such as jowar, gingelly, blackgram, and greengram in kharif facilitates orobanche germination but will not allow it to grow. This reduces the orobanche seed load in the soil.
- CTRI developed a spear with a 2 m stick at the end of which 18 cm length, 8 cm breadth and 0.5 cm thick sharp iron blade is arranged. Cutting the orobanche shoots within 3–4 days after emergence out of the soil, before flowering of orobanche either up to the soil level or 2–3 cm below the soil level with spear effectively controls orobanche.

KEYWORDS

- ***Orobanche cernua***
- **powdery mildew**
- **root-knot nematode**
- **tobacco leaf curl**
- **tobacco mosaic**
- **wild fire disease**

REFERENCES

Amrinder, K. K. S., Vermaand, S. K., & Thind, (2009). Evaluation of fungicides against citrus foot rot (*Phytophthora nicotianae* var. *parasitica*). *Pl. Dis. Res., 24*(1), 19–22.

Cole, D. L., (1997). *The Efficacy of a Plant Activator CGA 245704 Against Field Diseases of Tobacco. Presented: CORESTA Agronomy & Phytopathology Group Meeting*. Montreux. Tobacco Research Board of Zimbabwe. Handbook of recommendations.

CTRI, (1994). *Management of Pests and Diseases of Tobacco Book Let*. By Central Tobacco Research Institute, Rajahmundry.

Dam, S. K., (2015). *Research Achievements, Proceedings of Institute Research Committee Meetings* (pp. 18, 19). ICAR-Central Tobacco Research Institute, Rajahmundry.

Erwin, D. C., & Ribeiro, O. K., (1996). *Phytophthora Diseases Worldwide* (Vol.1, pp. 225–300). American Phytopathological Society Press, St. Paul, MN.

Ghazanfar, M. U., Sahi, S. T., Wakil, W., & Iqbal, Z., (2010). Evaluation of various fungicides for the management of late blight of potato (*Phytophthora infestans*). *Pak. J. Phytopathol.*, *22*(2), 83–88.
Jones, K. J., & Shew, H. D., (1995). Early season root production and zoospore infection of cultivars of flue-cured tobacco that differ in level of partial resistance to *Phytophthora parasitica* var. *nicotianae*. *Plant Soil, 172*, 55–61.
Karegowda, C., (2014). *Research Achievements, Proceedings of Institute Research Committee Meetings* (pp. 20–21). ICAR-Central Tobacco Research Institute, Rajahmundry.
Karegowda, C., (2015). Integrated management of black shank disease of FCV tobacco in field and its effect on leaf yield and quality parameters. In: Narayanaswamy, H., Raju, J., Sharabasappa, H. G., Ganesh, N. R., & Murali, R., (eds.), *3rd International Symposium on Phytophthora: Taxonomy, Genomics, Pathogenicty, Resistance and Disease Management* (p. 51).
Karegowda, C., Venkatesh, & Gurumurthy, B. R., (2008). Integrated management of damping-off disease of FCV tobacco in nursery. *Karnataka J. Agric. Sci., 21*(2), 299–300.
Lucas, G. B., (1975). *Diseases of Tobacco* (3rd edn., Vol.3, pp. 125–129). Biological Consulting Associates, Raleigh, NC.
Melton, T. A., (1998). *Disease Management* (pp. 92–112). Flue-Cured Tobacco Information, North Carolina Cooperative Extension Service, North Carolina State University, College of Agriculture and Life Science.
Prinsloo, G. C., (1994). *Black Shank of Tobacco in the Republic of South Africa* (p. 89). Coresta Congress 1994, Harare, Zimbabwe.
Raju, J., Ravindra, H., Mohan, K., S., & Jayalakshmi, K., (2016). Efficacy of new molecules of fungicides against black shank of tobacco under field condition. In: *6th International Conference on Plant Pathogens and People* (p. 405). By, IPS, ICAR, New Delhi.
Reddy, P. R. S., (2006). *Central Tobacco Research Institute* (p. 71). Rajahmundry.
Shamarao, J., & Hundekar, A. R., (2008). In: *Proc First International Conference on Agrochemicals Protecting Crop, Health and Environment Held at IARI* (p. 327). New Delhi.
Shamarao, J., & Hundekar, A. R., (2009). Major diseases of tobacco and their management in Karnataka: A review. *Agric. Rev.*, *30*(3), 206–212.
Shenoi, M. M., Moses, J. S. L., Rao, S. V., Abdul, W. S. M., & Subrahmanya, K. N., (1992). Screening *Nicotiana* germplasm against tobacco mosaic virus and studies on TMV tolerant FCV special mutant line FCH. *Tobacco Res., 18*(1&2), 93–96.
Shew, H. D., & Lucas, G. B., (1991). *Compendium of Tobacco Diseases* (Vol.1, pp. 92–115). American Phytopathological Society Press, St. Paul, MN.
Shew, H. D., & Lucas, G. B., (1991). Wildfire and angular leaf spot. In: *Compendium of Tobacco Diseases* (pp. 30–32). APS Press. ISBN: 0-89054-117-5.
Shew, H. D., (1987). Effect of host resistance on spread of *Phytophthora parasitica* var. *nicotianae* and the subsequent development of tobacco black shank under field conditions. *Phytopathology, 77*, 1090–1093.
Suryanarayana, V. C., & Rajarao, D., (1988). *Indian Tobacco Literature* (Vol.1, pp. 26–30). Central Tobacco Research Institute.

Index

D

E

H

I

K

L

Q

R

S

X

Y

Z